basic mathematics:

revision and practice

basic mathematics:
revision and practice

R. Elvin

A. Ledsham

C. Oliver

Oxford University Press

Oxford University Press, Walton Street, Oxford OX2 6DP

Oxford New York Toronto
Delhi Bombay Calcutta Madras Karachi
Petaling Jaya Singapore Hong Kong Tokyo
Nairobi Dar es Salaam Cape Town
Melbourne Auckland

and associated companies in
Berlin Ibadan

Oxford is a trade mark of Oxford University Press

ISBN 0 19 914256 4

© Oxford University Press, 1979

First published 1979
Reprinted 1980 (twice), 1981 (twice), 1982, 1983, 1984 (twice), 1985

Second edition 1986
Reprinted 1986, 1987, 1990

Set by Quadra Graphics, Oxford
Printed and bound in Great Britain by
William Clowes Limited,
Beccles and London

This book covers the fundamental mathematical requirements of pupils following GCSE courses or their equivalent at age 16 plus. It includes a large number of questions designed to give practice in the basic skills needed for success in both traditional and modern mathematics examinations.

The book is divided into nine parts which allow the pupil (and the teacher) quickly to find the appropriate practice. The order of treatment does not presuppose any particular approach. All numerical answers are included at the end of the book and pupils are encouraged to check their own solutions, so enabling the teacher to spend more time in discussing particular problems with individuals.

Each section of the book contains groups of exercises which are preceded by brief notes and worked examples to explain the principles and methods required. In some cases the method shown is not the most elegant but the one which the authors have found is most easily understood by pupils. The careful grading of each of the many exercises allows the widest range of abilities to be catered for. Many of the questions are topical or sufficiently realistic to provide motivation in a subject often regarded as being divorced from life in the world outside the classroom.

The aim of the authors has been to provide ample practice in the mathematical techniques normally required by middle ability pupils. Only constant work on the fundamental processes emphasised in the book will help such pupils to gain confidence in mathematics and to face the demands they will meet on leaving school.

R. E.
A. H. C. L.
C. O.

Note to second edition

In this edition, section 2.4 has now been rewritten to cover work with a calculator and material on standard form has been transferred to an amended section 2.3. Certain other improvements have also been incorporated, but the pagination of the text remains unaltered.

CONTENTS

Part 1 Basic arithmetic

1.1 Addition and subtraction *1*
1.2 Multiplication and division *3*
1.3 Numbers and factors *6*
1.4 Decimal fractions *8*
1.5 The metric system *11*
1.6 Common fractions *12*
1.7 Fractions and decimals *16*
1.8 Ratio and proportion *18*
1.9 Percentage *21*

Part 2 Everyday arithmetic

2.1 Applications of percentage *24*
2.2 Averages *31*
2.3 Powers, roots and standard form *34*
2.4 The calculator *36*
2.5 Rates and taxes *40*
2.6 Wages and salaries *42*
2.7 Gas and electricity bills *44*
2.8 24-hour clock and timetables *45*
2.9 Foreign exchange *48*

Part 3 Mensuration

3.1 Simple area I *50*
3.2 Simple area II *52*
3.3 The circle *57*
3.4 Volume I *60*
3.5 Volume II *63*
3.6 Arc length and longitude *65*

Part 4 Algebra

4.1 Generalised arithmetic *68*
4.2 Directed numbers *69*
4.3 Substitution *70*
4.4 Like terms and brackets *74*
4.5 Simple equations *77*
4.6 Simultaneous equations *80*
4.7 Changing the subject *82*
4.8 Factorisation *84*
4.9 Algebraic fractions *89*
4.10 Indices *90*
4.11 Quadratic equations *93*
4.12 Simple problems *95*

Part 5 Graphs

5.1 Plotting graphs *98*
5.2 Data display *102*
5.3 Linear graphs *106*
5.4 Plotting coordinates *110*
5.5 Graphs and algebra *114*
5.6 Graphical solutions *122*
5.7 Travel graphs *124*

Part 6 Geometry

6.1 Constructions *130*
6.2 Angles and bearings *132*
6.3 Parallel lines *136*
6.4 Triangles *141*
6.5 Polygons and patterns *145*
6.6 Congruency and similarity *147*
6.7 Symmetry *158*
6.8 Quadrilaterals *161*
6.9 Pythagoras' Theorem *164*
6.10 The circle *167*
6.11 Tangents *172*
6.12 Loci *176*

Part 7 Trigonometry

7.1 Use of tables *178*
7.2 The tangent ratio *181*
7.3 The sine ratio *184*
7.4 The cosine ratio *186*
7.5 Right-angled triangles *188*
7.6 3-D problems *192*

Part 8 Modern mathematics

8.1 Number bases *196*
8.2 Sets *198*
8.3 Venn diagrams *200*
8.4 Inequalities *205*
8.5 Relations and mappings *208*
8.6 Transformations *212*
8.7 Matrices *215*
8.8 Vectors *218*
8.9 Flow charts *223*
8.10 Solids *229*
8.11 Networks *233*

Part 9 Statistics

9.1 Data collection *238*
9.2 Mode, median, mean *244*
9.3 Scatter diagrams *249*
9.4 Probability *253*

Answers *258*

Example 1

Find the 'odd answer out' of:
(a) 26 + 36 + 72
(b) 34 + 62 + 38
(c) 32 + 56 + 44

(a)   26   (b)   34   (c)   32
         36   62   56
       + 72   + 38   + 44
        ————   ————   ————
        134   134   132

So (c) is the 'odd answer out' because its answer is 132.

Exercise 1.1a

Find the 'odd answer out' for the following:
1. (a)   6   (b)   7   (c)   4
            8   5   15
         + 10   + 12   + 6

2. (a)   9   (b)   8   (c)   7
           11   9   12
         + 12   + 14   + 13

3. (a)   5   (b)   18   (c)   9
           16   14   10
         + 15   + 3   + 16

4. (a)   13   (b)   2   (c)   6
            8   15   13
         + 12   + 17   + 10

5. (a)   19   (b)   17   (c)   15
           11   9   5
         + 8   + 12   + 17

6. (a) 25 + 7 + 11   7. (a) 24 + 13 + 15
   (b) 22 + 11 + 9   (b) 31 + 5 + 17
   (c) 18 + 17 + 8   (c) 29 + 12 + 11
8. (a) 23 + 27 + 12   9. (a) 28 + 19 + 25
   (b) 19 + 22 + 21   (b) 17 + 32 + 24
   (c) 9 + 32 + 20   (c) 42 + 9 + 22

10. (a) 31 + 27 + 36   11. (a) 53 + 17 + 63
    (b) 42 + 13 + 40   (b) 48 + 29 + 56
    (c) 11 + 56 + 27   (c) 58 + 9 + 65

12. (a) 11 + 59 + 49   13. (a) 73 + 45 + 27
    (b) 7 + 71 + 40   (b) 69 + 55 + 22
    (c) 19 + 39 + 60   (c) 79 + 32 + 34

14. (a) 87 + 36 + 29   15. (a) 36 + 94 + 50
    (b) 59 + 44 + 49   (b) 29 + 97 + 53
    (c) 63 + 29 + 61   (c) 45 + 82 + 52

Example 2

Find the 'odd answer out' of:
(a) 317 + 232 + 387
(b) 255 + 318 + 362
(c) 297 + 421 + 218

(a)   317   (b)   255   (c)   297
        232   318   421
      + 387   + 362   + 218
       ————   ————   ————
        936   935   936

So (b) is the 'odd answer out' because its answer is 935.

Exercise 1.1b

Find the 'odd answer out' for the following:
1. (a)   287   (b)   309   (c)   187
           326   237   412
         + 233   + 299   + 246

2. (a)   336   (b)   256   (c)   307
           422   309   236
         + 148   + 340   + 362

3. (a)   311   (b)   294   (c)   325
           307   355   402
         + 257   + 227   + 148

4. (a)   435   (b)   396   (c)   287
           269   319   376
         + 277   + 266   + 319

5. (a)   401   (b)   339   (c)   375
           355   286   319
         + 138   + 270   + 201

6. (a) 324 + 331 + 131   7. (a) 439 + 203 + 173
   (b) 281 + 329 + 175   (b) 356 + 195 + 263
   (c) 199 + 321 + 265   (c) 503 + 138 + 174

8. (a) 396 + 237 + 272   9. (a) 517 + 218 + 215
   (b) 435 + 187 + 284   (b) 184 + 311 + 456
   (c) 511 + 209 + 185   (c) 396 + 305 + 249

10. (a) 455 + 302 + 187   11. (a) 376 + 319 + 179
    (b) 512 + 331 + 101   (b) 295 + 318 + 262
    (c) 432 + 217 + 296   (c) 337 + 291 + 247

12. (a) 1329 + 132 + 19
    (b) 1401 + 35 + 44
    (c) 1016 + 229 + 236

13. (a) 5276 + 81 + 97
    (b) 5085 + 88 + 281
    (c) 3540 + 1850 + 65

14. (a) 4327 + 424 + 155
    (b) 3215 + 1344 + 346
    (c) 3752 + 1060 + 94

15. (a) 3127 + 1346 + 362
    (b) 2955 + 1121 + 760
    (c) 2578 + 1592 + 666

Example 3

Find the 'odd answer out' of:
(a) $36 - 23$ (b) $45 - 32$ (c) $72 - 58$

(a)   36   (b)   45   (c)   72
    − 23   − 32   − 58
    ─────   ─────   ─────
      13   13   14
    ─────   ─────   ─────

So (c) is the 'odd answer out' because its answer is 14.

Exercise 1.1c

Find the 'odd answer out' for the following:

1. (a)   14   (b)   27   (c)   30
       − 9   − 21   − 25

2. (a)   27   (b)   38   (c)   42
       − 13   − 25   − 28

3. (a)   57   (b)   61   (c)   49
       − 34   − 38   − 25

4. (a)   82   (b)   68   (c)   53
       − 37   − 23   − 7

5. (a)   50   (b)   78   (c)   48
       − 19   − 46   − 17

6. (a) 93 − 27   7. (a) 95 − 78
   (b) 85 − 19   (b) 77 − 59
   (c) 74 − 9   (c) 53 − 36

8. (a) 80 − 42   9. (a) 90 − 58
   (b) 53 − 15   (b) 72 − 39
   (c) 61 − 24   (c) 51 − 18

10. (a) 93 − 84   11. (a) 70 − 56
    (b) 87 − 79   (b) 51 − 38
    (c) 46 − 38   (c) 63 − 49

12. (a) 45 − 9   13. (a) 92 − 59
    (b) 83 − 48   (b) 69 − 35
    (c) 62 − 27   (c) 81 − 48

14. (a) 90 − 23   15. (a) 71 − 48
    (b) 84 − 16   (b) 98 − 75
    (c) 75 − 8   (c) 80 − 56

Example 4

Find the 'odd answer out' of:
(a) $716 - 353$
(b) $551 - 187$
(c) $1175 - 811$

(a)   716   (b)   551   (c)   1175
      − 353   − 187   − 811
      ─────   ─────   ──────
        363   364   364
      ─────   ─────   ──────

So (a) is the 'odd answer out' because its answer is 363.

Exercise 1.1d

Find the 'odd answer out' for the following:

1. (a)   848   (b)   763   (c)   927
       − 613   − 527   − 691

2. (a)   464   (b)   675   (c)   733
       −319   − 529   −588

3. (a)   754   (b)   636   (c)   984
       − 538   − 419   − 768

4. (a)   823   (b)   653   (c)   742
       − 468   − 297   − 386

5. (a)   712   (b)   835   (c)   764
       − 234   −357   − 285

6. (a) 687 − 336   7. (a) 582 − 254
   (b) 538 − 186   (b) 843 − 516
   (c) 771 − 419   (c) 519 − 192

8. (a) 728 − 183   9. (a) 790 − 253
   (b) 813 − 269   (b) 802 − 264
   (c) 841 − 296   (c) 830 − 292

10. (a) 980 − 345   11. (a) 628 − 91
    (b) 908 − 273   (b) 619 − 83
    (c) 829 − 193   (c) 634 − 98

12. (a) $757 - 73$
    (b) $752 - 67$
    (c) $778 - 93$

13. (a) $3875 - 2532$
    (b) $3216 - 1872$
    (c) $2921 - 1577$

14. (a) $2346 - 1523$
    (b) $1756 - 932$
    (c) $1328 - 505$

15. (a) $1342 - 1267$
    (b) $1838 - 1762$
    (c) $1594 - 1518$

Exercise 1.1e

1. Jack has 26 marbles, Jim has 42 and Fred has 37. How many have they got altogether?
   If Henry has a large collection of 150, how many more has he got than the others put together?
2. Robert is 138 centimetres tall. If he is 39 centimetres shorter than his father, how tall is his father?
   If the father is 68 centimetres taller than Robert's younger brother Paul, how tall is Paul?
3. Halesowen is 126 metres above sea level and Adam's Hill is 304 metres above sea level. By how many metres does Adam's Hill stand above Halesowen?
   If Walton Hill stands 189 metres above Halesowen, how high is Walton Hill above sea level?
4. The population of Storrington is 2750. If this is 850 more than that of West Chiltington, what is the population of West Chiltington?
   If the population of West Chiltington is 3550 less than that of Steyning, what is the population of Steyning?
5. Ludlow is 87 metres above sea level and Titterstone Clee Hill stands 446 metres above Ludlow. How high is Titterstone Clee Hill above sea level?
   If Brown Clee Hill is 7 metres higher than Titterstone Clee Hill, by how many metres does Brown Clee Hill stand above Ludlow?

1.2 MULTIPLICATION AND DIVISION

To multiply by 10, move the figures one place to the left.

Example 1

$$36 \times 10 = 360$$

Hundreds	Tens	Units
	3	6

Hundreds	Tens	Units
3	6	0

To multiply by 100, move the figures two places to the left.

Example 2

$$36 \times 100 = 3600$$

Th	H	T	U	Th	H	T	U
		3	6	3	6	0	0

Exercise 1.2a

Multiply each of the following
(a) by 10;   (b) by 100:

1. 7
2. 14
3. 25
4. 37
5. 10
6. 70
7. 220
8. 400
9. 100
10. 786

Copy the following and fill in the empty spaces.
11. $8 \times 10 =$
12. $\quad \times 10 = 160$
13. $32 \times 100 =$
14. $46 \times \quad = 4600$
15. $\quad \times 100 = 4900$
16. $30 \times \quad = 3000$
17. $\quad \times 100 = 12\,000$
18. $500 \times 10 =$
19. $\quad \times 10 = 9000$
20. $\quad \times 100 = 30\,000$

To divide by 10, move the figures one place to the right.

Example 3

$$250 \div 10 = 25$$

H	T	U	H	T	U
2	5	0		2	5

To divide by 100, move the figures two places to the right.

Example 4

$$4600 \div 100 = 46$$

Th	H	T	U		Th	H	T	U
4	6	0	0				4	6

Exercise 1.2b

Divide each of the following
(a) by 10;   (b) by 100:

1. 6600	2. 9800	3. 10 500
4. 12 000	5. 20 000	6. 125 000
7. 301 000	8. 1000	9. 200 000
10. 10 000		

Copy the following and fill in the empty spaces.
11. $720 \div 10 =$	16. $7000 \div \quad = 70$
12. $\quad \div 10 = 36$	17. $\quad \div 100 = 180$
13. $2800 \div 100 =$	18. $4000 \div 10 =$
14. $5300 \div \quad = 53$	19. $\quad \div 10 = 300$
15. $\quad \div 100 = 65$	20. $\quad \div 100 = 900$

Example 5

Find the 'odd answer out' of:
(a) 232×4 (b) 34×27 (c) 51×18

(a)   232
    $\times \quad 4$

    $\overline{928}$

(b)   34
    $\times 27$

    $\overline{680}$ (34×20)
    238 (34×7)

    $\overline{918}$

(c)   51
    $\times 18$

    $\overline{408}$ (51×8)
    510 (51×10)

    $\overline{918}$

So (a) is the 'odd answer out' because its answer is 928.

Exercise 1.2c

Find the 'odd answer out' for the following:

1. (a)   6   (b)   12   (c)   7
      $\times 8$ $\times \; 4$ $\times 7$

2. (a)   36   (b)   29   (c)   16
      $\times \; 4$ $\times \; 5$ $\times \; 9$

3. (a)   75   (b)   124   (c)   93
      $\times \; 5$ $\times \; 3$ $\times \; 4$

4. (a)   117   (b)   146   (c)   195
      $\times \quad 5$ $\times \quad 4$ $\times \quad 3$

5. (a)   394   (b)   709   (c)   591
      $\times \quad 9$ $\times \quad 5$ $\times \quad 6$

6. (a)   459   (b)   328   (c)   287
      $\times \quad 5$ $\times \quad 7$ $\times \quad 8$

7. (a) 81×11 8. (a) 63×12
   (b) 223×4 (b) 69×11
   (c) 99×9 (c) 84×9

9. (a) 21×18 10. (a) 32×15
   (b) 22×17 (b) 30×16
   (c) 27×14 (c) 27×18

11. (a) 56×24 12. (a) 49×36
    (b) 64×21 (b) 104×17
    (c) 52×26 (c) 126×14

Example 6

Find the 'odd answer out' of:
(a) $161 \div 7$ (b) $207 \div 9$ (c) $192 \div 8$

(a) $7\overline{)161}$ = 23   (b) $9\overline{)207}$ = 23   (c) $8\overline{)192}$ = 24

So (c) is the 'odd answer out' because its answer is 24.

Exercise 1.2d

Find the 'odd answer out' for the following:

1. (a) $330 \div 6$ 2. (a) $220 \div 5$
   (b) $270 \div 5$ (b) $352 \div 8$
   (c) $495 \div 9$ (c) $270 \div 6$

3. (a) $2630 \div 5$ 4. (a) $2275 \div 7$
   (b) $3162 \div 6$ (b) $2916 \div 9$
   (c) $2108 \div 4$ (c) $2600 \div 8$

5. (a) $944 \div 4$
    (b) $1180 \div 5$
    (c) $711 \div 3$

6. (a) $1309 \div 7$
    (b) $1488 \div 8$
    (c) $930 \div 5$

7. (a) $3708 \div 3$
    (b) $4940 \div 4$
    (c) $6180 \div 5$

8. (a) $7242 \div 6$
    (b) $6030 \div 5$
    (c) $8442 \div 7$

9. (a) $6510 \div 5$
    (b) $5212 \div 4$
    (c) $9121 \div 7$

10. (a) $2928 \div 4$
    (b) $2196 \div 3$
    (c) $4386 \div 6$

11. (a) $2808 \div 12$
    (b) $2585 \div 11$
    (c) $2106 \div 9$

12. (a) $1023 \div 11$
    (b) $1104 \div 12$
    (c) $736 \div 8$

Example 7

Find the 'odd answer out' of:
(a) $1357 \div 23$
(b) $1020 \div 17$
(c) $4602 \div 78$

$$
\begin{array}{r}
59 \\
23\overline{)1357} \\
115 \quad (5 \times 23) \\
\hline
207 \\
207 \quad (9 \times 23) \\
\hline
\end{array}
$$
(a)

$$
\begin{array}{r}
60 \\
17\overline{)1020} \\
102 \quad (17 \times 6) \\
\hline
0 \\
0 \quad (17 \times 0) \\
\hline
\end{array}
$$
(b)

$$
\begin{array}{r}
59 \\
78\overline{)4602} \\
390 \quad (78 \times 5) \\
\hline
702 \\
702 \quad (78 \times 9) \\
\hline
\end{array}
$$
(c)

So (b) is the 'odd answer out' because its answer is 60.

Exercise 1.2e

Find the 'odd answer out' for the following:

1. (a) $540 \div 15$
    (b) $630 \div 18$
    (c) $468 \div 13$

2. (a) $602 \div 14$
    (b) $688 \div 16$
    (c) $756 \div 18$

3. (a) $715 \div 13$
    (b) $1026 \div 19$
    (c) $935 \div 17$

4. (a) $736 \div 23$
    (b) $672 \div 21$
    (c) $858 \div 26$

5. (a) $1056 \div 24$
    (b) $990 \div 22$
    (c) $1100 \div 25$

6. (a) $925 \div 25$
    (b) $1008 \div 28$
    (c) $936 \div 26$

7. (a) $1921 \div 17$
    (b) $2147 \div 19$
    (c) $1792 \div 16$

8. (a) $2546 \div 19$
    (b) $2144 \div 16$
    (c) $2430 \div 18$

9. (a) $3749 \div 23$
    (b) $4727 \div 29$
    (c) $3444 \div 21$

10. (a) $1664 \div 32$
    (b) $1908 \div 36$
    (c) $1855 \div 35$

11. (a) $1107 \div 41$
    (b) $980 \div 35$
    (c) $1512 \div 54$

12. (a) $2013 \div 61$
    (b) $1760 \div 55$
    (c) $1485 \div 45$

Exercise 1.2f

1. John weighs 21 kilograms. If his father is four times heavier, how much does his father weigh? If the father is three times heavier than John's older brother David, how heavy is David?

2. Mary is four times heavier than her baby sister Julie. If Mary weighs 60 kilograms, how much does Julie weigh? If their mother is five times heavier than Julie, how much does their mother weigh?

3. The distance from Birmingham to Evesham is 48 kilometres. If Bristol is three times further away from Birmingham than Evesham, how far is it from Birmingham to Bristol? If Bristol is eight times further away from Birmingham than Alvechurch, how far is it from Birmingham to Alvechurch?

4. Ann, Janet and Christine have picked some flowers. Ann has picked 54, three times as many as Janet has. How many has Janet picked? How many has Christine picked if she has picked five times as many as Janet? How many less than 200 have they picked together?

The *multiples* of 6 are:
6, (1 × 6); 12, (2 × 6); 18; 24; etc.

Example 1

List the first five multiples of 12.
The multiples are:
12, (1 × 12); 24, (2 × 12); 36; 48; 60.

Exercise 1.3a

List the first five multiples of:
1. 2	**2.** 5	**3.** 8	**4.** 7
5. 11	**6.** 20	**7.** 30	**8.** 60
9. 15	**10.** 25	**11.** 16	**12.** 18
13. 14	**14.** 13	**15.** 24	**16.** 21
17. 45	**18.** 36	**19.** 51	**20.** 72

6 can be divided by 1, 2, 3, and 6.
These numbers are the *factors* of 6.

Example 2

Find the factors of 72.
72 = 1 × 72 or 2 × 36 or 3 × 24
     or 4 × 18 or 6 × 12 or 8 × 9.
So the factors of 72 are
1, 2, 3, 4, 6, 8, 9, 12, 18, 24, 36, and 72.

Exercise 1.3b

Find the factors of:
1. 3	**2.** 8	**3.** 10	**4.** 12
5. 15	**6.** 18	**7.** 30	**8.** 27
9. 24	**10.** 32	**11.** 45	**12.** 40
13. 54	**14.** 42	**15.** 60	**16.** 48
17. 63	**18.** 84	**19.** 66	**20.** 72

A *square* number has an odd number of
different factors.

Example 3

16 = 1 × 16 or 2 × 8 or 4 × 4.
So the factors of 16 are, 1, 2, 4, 8, 16
(that is, five different factors).
So 16 is a square number (4 × 4).
Note that 1 is also a square number.

Exercise 1.3c

List the factors of the following numbers and
state which are square numbers.
1. 2	**2.** 5	**3.** 6	**4.** 7

5. 11	**6.** 20	**7.** 35	**8.** 4
9. 50	**10.** 25	**11.** 16	**12.** 22
13. 14	**14.** 13	**15.** 28	**16.** 21
17. 49	**18.** 36	**19.** 51	**20.** 70

A *prime* number has only two different
factors.

Example 4

(a) 2 = 1 × 2, so 2 is a prime number.
(b) 13 = 1 × 13, so 13 is a prime number.
(c) 51 = 1 × 51 or 3 × 17.
    So 51 is not a prime number.
(d) 1 has only one factor, so 1 is *not* a
    prime number.

Exercise 1.3d

List the factors of the following numbers and
state which are prime numbers.
1. 17	**2.** 33	**3.** 26	**4.** 29
5. 39	**6.** 57	**7.** 37	**8.** 91
9. 87	**10.** 53	**11.** 111	**12.** 97
13. 61	**14.** 67	**15.** 119	**16.** 103
17. 73	**18.** 123	**19.** 101	**20.** 117

The *prime factors* of 6 are 2 and 3;
these are the prime numbers which are
also factors of 6.

Example 5

Find the prime factors of 72.
72 = 2 × 36
   = 2 × 2 × 18
   = 2 × 2 × 2 × 9
   = 2 × 2 × 2 × 3 × 3
So the prime factors of 72 are
2 × 2 × 2 × 3 × 3.

Exercise 1.3e

Find the prime factors of:
1. 10	**2.** 15	**3.** 28	**4.** 24
5. 36	**6.** 40	**7.** 45	**8.** 54
9. 70	**10.** 63	**11.** 78	**12.** 60
13. 66	**14.** 112	**15.** 96	**16.** 162
17. 154	**18.** 180	**19.** 168	**20.** 216

The largest factor of both 16 and 28 is 4;
4 is called the *highest common factor*
(H.C.F.) of 16 and 28.

Example 6

Find the H.C.F. of 48 and 60.
48 = 2 × 24
   = 2 × 2 × 12
   = 2 × 2 × 2 × 6
   = 2 × 2 × 2 × 2 × **3**

60 = 2 × 30
   = 2 × 2 × 15
   = 2 × 2 × 3 × 5

The common factors (heavy type) are
2, 2, and 3.
So the H.C.F. of 48 and 60 is
2 × 2 × 3 = 12.

Exercise 1.3f

Find the H.C.F. of:

1. 6 and 8	2. 12 and 18
3. 30 and 24	4. 36 and 27
5. 56 and 42	6. 72 and 48
7. 54 and 36	8. 63 and 84
9. 75 and 105	10. 64 and 80
11. 54, 42, and 78	12. 36, 44, and 68
13. 70, 84, and 98	14. 64, 96, and 112
15. 56, 72, and 104	16. 66, 84, and 90
17. 108, 132, and 156	18. 52, 65, and 91
19. 96, 128, and 144	20. 24, 36, and 52

The smallest multiple of both 16 and 28
is 112; 112 is called the *lowest common
multiple* (L.C.M.) of 16 and 28.

Example 7

Find the L.C.M. of 48 and 60.

1 × 48 = 48	1 × 60 = 60
2 × 48 = 96	2 × 60 = 120
3 × 48 = 144	3 × 60 = 180
4 × 48 = 192	4 × 60 = **240**
5 × 48 = **240**	

So 240 is the L.C.M. of 48 and 60.

Exercise 1.3g

Find the L.C.M. of:

1. 6 and 8	2. 10 and 15
3. 8 and 12	4. 6 and 9
5. 9 and 12	6. 10 and 12
7. 6 and 15	8. 8 and 20
9. 9 and 15	10. 12 and 16

11. 3, 4, and 8	12. 3, 9, and 12
13. 5, 8, and 10	14. 4, 5, and 6
15. 6, 9, and 12	16. 10, 12, and 15
17. 9, 12, and 16	18. 15, 20, and 24
19. 15, 18, and 30	20. 24, 36, and 48

Exercise 1.3h

Find the answer to each part of the following and
name the 'odd answer out'.

1. (a) Sum of the first six prime numbers.
   (b) L.C.M. of 5 and 8.
   (c) Sum of both square numbers between 10
      and 30.
2. (a) Sum of all prime numbers less than 10.
   (b) H.C.F. of 56 and 70.
   (c) Sum of the first three square numbers.
3. (a) Sum of both prime numbers between 15 and
      20.
   (b) L.C.M. of 12 and 9.
   (c) Sum of the 3rd and 5th square numbers.
4. (a) Difference between both prime numbers
      between 30 and 40.
   (b) H.C.F. of 36, 42, and 54.
   (c) Difference between the 4th and the 3rd
      square numbers.
5. (a) Sum of the nearest two prime numbers to 30.
   (b) L.C.M. of 6, 9, and 27.
   (c) Sum of all square numbers which are greater
      than 1 but less than 30.
6. (a) Sum of all prime numbers between 20 and 40.
   (b) L.C.M. of 8, 15, and 20.
   (c) Sum of all square numbers between 15 and 50.
7. (a) Product of the 1st and the 4th prime numbers.
   (b) H.C.F. of 60, 90, and 105.
   (c) Difference between the 8th and the 7th
      square numbers.
8. (a) Sum of the first three prime numbers that end
      with a 7.
   (b) L.C.M. of 15, 20, and 30.
   (c) Difference between the first two square num-
      bers that end with a 4.
9. (a) Product of the highest prime number which
      is less than 30 and the only even prime
      number.
   (b) L.C.M. of 7, 8, and 14.
   (c) Sum of the first two square numbers that
      end with a 9.
10. (a) Difference between the highest prime num-
      ber which is less than 40 and the lowest one
      which is greater than 20.
   (b) H.C.F. of 80, 96, and 144.
   (c) Difference between the 5th and the 3rd
      square numbers.

The decimal point separates whole numbers from fractions.

Example 1

3·5 is 3 units and 5 tenths.
3·55 is 3 units, 5 tenths, and 5 hundredths,
(or 3 units and 55 hundredths).
3·555 is 3 units, 5 tenths, 5 hundredths
and 5 thousandths,
(or 3 units and 555 thousandths).

Exercise 1.4a

Give the value of the underlined figures:
1. 2·75   2. 10·64
3. 523·92   4. 5·385
5. 9·236   6. 31·567
7. 25·328   8. 286·4
9. 325·272   10. 200·003

When adding or subtracting decimal fractions, the decimal points are put underneath each other to make sure that each figure is in its correct column.

Example 2

Find the 'odd answer out' of:
(a) 13·46 + 1·2 + 75·36
(b) 7·34 + 24 + 58·78
(c) 32·14 + 25·08 + 32·9

(a)   13·46   (b)   7·34   (c)   32·14
          1·2   24   25·08
      + 75·36   + 58·78   + 32·9
      _____   _____   _____
        90·02   90·12   90·12
      ========   ========   ========

So (a) is the 'odd answer out' because its answer is 90·02.

Example 3

Find the 'odd answer out' of:
(a) 24·68 − 13·24
(b) 39 − 27·46   (c) 59·8 − 48·36

(a)   24·68   (b)   39·00   (c)   59·80
      − 13·24   − 27·46   − 48·36
      _____   _____   _____
        11·44   11·54   11·44
      ========   ========   ========

So (b) is the 'odd answer out' because its answer is 11·54.

Exercise 1.4b

1.   36·2   2.   27·8
     + 1·4   + 3·4

3.   32·1   4.   53·25
     + 3·75   + 18·86

5. 81·25 + 6·584
6. 39·4 + 3·785
7. 21·6 + 9·3 + 7·26
8. 23·4 + 5·87 + 32·115
9. 12·06 + 4·5 + 0·375
10. 1·26 + 4·735 + 19
11.   3·8   12.   4·3
      − 2·4   − 2·5

13.   6·75   14.   8·37
      − 2·9   − 4·65

15. 9·2 − 5·45   16. 20·32 − 13·17
17. 13·76 − 9·38   18. 38·2 − 19·35
19. 135·35 − 28·7   20. 101·26 − 59·37

Find the 'odd answer out' for the following:
21. (a) 38·7 + 1·9 + 3·25
    (b) 25 + 1·65 + 17·1
    (c) 32·35 + 1·95 + 9·55
22. (a) 21·25 + 12 + 3·85
    (b) 24·325 + 9·2 + 3·575
    (c) 19·3 + 4·35 + 13·55
23. (a) 4·375 + 0·585 + 12·32
    (b) 8·36 + 5·325 + 3·495
    (c) 9·8 + 5 + 2·38
24. (a) 3·27 + 5·19 + 11·38
    (b) 8 + 1·345 + 10·395
    (c) 12·3 + 0·975 + 6·465
25. (a) 10·2 + 1·065 + 15·005
    (b) 11·32 + 5 + 9·85
    (c) 6·005 + 0·845 + 19·42
26. (a) 25·29 − 11·56
    (b) 37·13 − 23·5
    (c) 45·2 − 31·47
27. (a) 8·325 − 3·65
    (b) 9·34 − 4·665
    (c) 8 − 3·425
28. (a) 13·285 − 9·29
    (b) 7·6 − 3·615
    (c) 10·36 − 6·375
29. (a) 14·27 − 9·145   30. (a) 153·23 − 28·78
    (b) 13·065 − 7·84   (b) 159·4 − 35·05
    (c) 5·6 − 0·475   (c) 132 − 7·65

To multiply by 10, move the figures one place to the left.

Example 4

$$3 \cdot 6 \times 10 = 36$$

Units	tenths
3	6

T	U	t
3	6	

$$0 \cdot 36 \times 10 = 3 \cdot 6$$

U	t	h
0	3	6

U	t
3	6

To multiply by 100, move the figures two places to the left.

Example 5

$$3 \cdot 6 \times 100 = 360$$
$$0 \cdot 36 \times 100 = 36$$

Exercise 1.4c

Multiply each of the following:
(a) by 10; (b) by 100:

1. $2 \cdot 4$ 2. $12 \cdot 2$ 3. $3 \cdot 75$
4. $15 \cdot 36$ 5. $2 \cdot 135$ 6. $18 \cdot 576$
7. $0 \cdot 85$ 8. $0 \cdot 7$ 9. $0 \cdot 2368$
10. $0 \cdot 0139$

Copy the following and fill in the empty spaces.
11. $3 \cdot 6 \times 10 =$
12. $\quad \times 10 = 45$
13. $2 \cdot 9 \times 100 =$
14. $\quad \times 100 = 320$
15. $0 \cdot 53 \quad = 53$
16. $0 \cdot 9 \times 100 =$
17. $\quad \times 100 = 80$
18. $0 \cdot 02 \times \quad = 2$
19. $0 \cdot 004 \times 10 =$
20. $\quad \times 10 = 0 \cdot 015$

To divide by 10, move the figures one place to the right.

Example 6

$$36 \div 10 = 3 \cdot 6$$
$$3 \cdot 6 \div 10 = 0 \cdot 36$$

To divide by 100, move the figures two places to the right.

Example 7

$$360 \div 100 = 3 \cdot 6$$
$$3 \cdot 6 \div 100 = 0 \cdot 036$$

Exercise 1.4d

Divide each of the following
(a) by 10; (b) by 100:

1. $25 \cdot 3$ 2. $38 \cdot 16$ 3. $6 \cdot 25$
4. $7 \cdot 35$ 5. 36 6. 60
7. $20 \cdot 4$ 8. $100 \cdot 3$ 9. $0 \cdot 85$
10. $0 \cdot 032$

Copy the following and fill in the empty spaces.
11. $4 \cdot 2 \div 10 =$
12. $\quad \div 10 = 0 \cdot 51$
13. $36 \div 100 =$
14. $\quad \div 100 = 4 \cdot 8$
15. $65 \div \quad = 0 \cdot 65$
16. $60 \div 100 =$
17. $\quad \div 100 = 0 \cdot 3$
18. $5 \div \quad = 0 \cdot 05$
19. $0 \cdot 09 \div 10 =$
20. $\quad \div 10 = 0 \cdot 0012$

Multiply a decimal by a whole number as follows.

Example 8

$$3 \cdot 6 \times 3 = 10 \cdot 8$$

(a) Multiply 36 by 3 to give 108.
(b) Count the number of decimal places, i.e. one.
(c) Put this number of places (one) in the answer to give $10 \cdot 8$.

Exercise 1.4e

1. $1 \cdot 2 \times 3$ 2. $0 \cdot 8 \times 2$
3. $3 \cdot 8 \times 4$ 4. $4 \cdot 3 \times 8$
5. $7 \cdot 5 \times 5$ 6. $2 \cdot 16 \times 7$
7. $3 \cdot 12 \times 9$ 8. $5 \cdot 15 \times 12$
9. $0 \cdot 24 \times 8$ 10. $1 \cdot 36 \times 16$
11. $2 \cdot 34 \times 15$ 12. $0 \cdot 92 \times 16$
13. $0 \cdot 015 \times 9$ 14. $0 \cdot 018 \times 7$
15. $0 \cdot 028 \times 5$ 16. $0 \cdot 0036 \times 8$
17. $0 \cdot 0045 \times 6$ 18. $0 \cdot 24 \times 30$
19. $0 \cdot 018 \times 50$ 20. $0 \cdot 026 \times 15$

Multiply a decimal by a decimal as follows.

Example 9

$$0 \cdot 136 \times 0 \cdot 23 = 0 \cdot 031\ 28$$

(a) Multiply 136 by 23 to give 3128.
(b) Count the number of decimal places, i.e. five.
(c) Put this number of places (five) in the answer to give $0 \cdot 031\ 28$.

Exercise 1.4f

1. $3 \cdot 7 \times 0 \cdot 6$	**2.** $9 \cdot 2 \times 0 \cdot 3$
3. $8 \cdot 6 \times 0 \cdot 7$	**4.** $3 \cdot 12 \times 0 \cdot 8$
5. $8 \cdot 25 \times 0 \cdot 4$	**6.** $6 \cdot 31 \times 0 \cdot 12$
7. $51 \cdot 5 \times 0 \cdot 9$	**8.** $32 \cdot 4 \times 0 \cdot 6$
9. $86 \cdot 5 \times 0 \cdot 12$	**10.** $4 \cdot 25 \times 0 \cdot 13$
11. $5 \cdot 12 \times 0 \cdot 16$	**12.** $4 \cdot 32 \times 0 \cdot 15$
13. $23 \cdot 2 \times 0 \cdot 14$	**14.** $52 \cdot 1 \times 0 \cdot 15$
15. $0 \cdot 246 \times 0 \cdot 4$	**16.** $0 \cdot 113 \times 0 \cdot 8$
17. $1 \cdot 24 \times 0 \cdot 05$	**18.** $5 \cdot 25 \times 0 \cdot 012$
19. $6 \cdot 34 \times 0 \cdot 015$	**20.** $4 \cdot 82 \times 0 \cdot 016$

Divide a decimal by a whole number as follows.

Example 10

$$6 \cdot 25 \div 25 = 0 \cdot 25$$

$$
\begin{array}{r}
0 \cdot 25 \\
25\overline{)6 \cdot 25} \\
5\ 0 \\
\hline
1\ 25 \\
\underline{1\ 25}
\end{array}
$$

Exercise 1.4g

1. $0 \cdot 6 \div 2$	**2.** $0 \cdot 8 \div 4$
3. $0 \cdot 72 \div 3$	**4.** $0 \cdot 98 \div 7$
5. $0 \cdot 858 \div 6$	**6.** $1 \cdot 274 \div 7$
7. $2 \cdot 448 \div 9$	**8.** $3 \cdot 675 \div 15$
9. $6 \cdot 048 \div 24$	**10.** $7 \cdot 072 \div 32$
11. $1 \cdot 032 \div 12$	**12.** $1 \cdot 245 \div 15$
13. $0 \cdot 095 \div 5$	**14.** $0 \cdot 0868 \div 7$
15. $0 \cdot 336 \div 8$	**16.** $0 \cdot 837 \div 9$
17. $0 \cdot 585 \div 13$	**18.** $0 \cdot 756 \div 18$
19. $0 \cdot 325 \div 25$	**20.** $0 \cdot 451 \div 41$

Divide a decimal by a decimal as follows.

Example 11

$$3 \cdot 128 \div 2 \cdot 3$$

(a) Turn the divisor into a whole number by multiplying it by 10, 100, etc:
$$2 \cdot 3 \times 10 = 23$$
(b) Multiply the dividend by the same number: $\quad 3 \cdot 128 \times 10 = 31 \cdot 28$
(c) Then divide as shown in Example 10.
$$3 \cdot 128 \div 2 \cdot 3 = 31 \cdot 28 \div 23 = 1 \cdot 36.$$

Exercise 1.4h

1. $8 \cdot 48 \div 0 \cdot 4$	**2.** $8 \cdot 37 \div 0 \cdot 9$
3. $7 \cdot 02 \div 0 \cdot 6$	**4.** $1 \cdot 05 \div 0 \cdot 7$
5. $5 \cdot 45 \div 0 \cdot 5$	**6.** $14 \cdot 4 \div 0 \cdot 8$
7. $11 \cdot 7 \div 0 \cdot 9$	**8.** $0 \cdot 42 \div 0 \cdot 03$
9. $1 \cdot 26 \div 0 \cdot 09$	**10.** $3 \cdot 25 \div 0 \cdot 05$
11. $252 \div 0 \cdot 4$	**12.** $365 \div 0 \cdot 5$
13. $828 \div 0 \cdot 9$	**14.** $325 \div 0 \cdot 05$
15. $0 \cdot 285 \div 0 \cdot 15$	**16.** $0 \cdot 196 \div 0 \cdot 14$
17. $2 \cdot 16 \div 0 \cdot 18$	**18.** $6 \cdot 75 \div 0 \cdot 25$
19. $22 \cdot 4 \div 0 \cdot 16$	**20.** $19 \cdot 5 \div 0 \cdot 15$

Decimal places

To write a number to a given number of decimal places:
(a) work out the answer to one more place than needed;
(b) if this last figure is smaller than 5, remove it; if it is 5 or more, add 1 to the figure in front of it.

Example 12

$3 \cdot 724 = 3 \cdot 72$ (to two decimal places)
$4 \cdot 6867 = 4 \cdot 687$ (3 D.P.)

Exercise 1.4i

Correct each of the following to the number of decimal places indicated.

1. $2 \cdot 643$ (2 D.P.)	**2.** $1 \cdot 338$ (2 D.P.)
3. $17 \cdot 64$ (1 D.P.)	**4.** $42 \cdot 79$ (1 D.P.)
5. $1 \cdot 7342$ (3 D.P.)	**6.** $1 \cdot 5628$ (3 D.P.)
7. $13 \cdot 65$ (1 D.P.)	**8.** $25 \cdot 375$ (2 D.P.)
9. $8 \cdot 054$ (2 D.P.)	**10.** $24 \cdot 03$ (1 D.P.)
11. $5 \cdot 104$ (2 D.P.)	**12.** $4 \cdot 507$ (2 D.P.)
13. $27 \cdot 08$ (1 D.P.)	**14.** $9 \cdot 509$ (2 D.P.)

15. 6·305 (2 D.P.)   **16.** 3·899 (2 D.P.)
17. 5·099 (2 D.P.)   **18.** 14·799 (2 D.P.)
19. 13·99 (1 D.P.)   **20.** 1·999 (2 D.P.)

Significant figures

In any number, the first non-zero figure is the first significant figure.

Example 13

1·006 has 4 significant figures
0·006 has only 1 significant figure.
1·006 written correct to 3 significant
  figures is 1·01.
5246 written correct to 2 significant
  figures (2 s.f.) is 5200.

Exercise 1.4j

State the number of significant figures in each of the following.

1. 1·325   **2.** 3·26   **3.** 0·853
4. 0·057   **5.** 5·01   **6.** 4·250
7. 320   **8.** 200   **9.** 5·0
10. 8·00   **11.** 5·24   **12.** 8·357

Correct each of the following to the number of significant figures indicated.

13. 3·223 (2 S.F.)   **14.** 7·574 (2 S.F.)
15. 13·36 (3 S.F.)   **16.** 17·21 (3 S.F.)
17. 31·4 (2 S.F.)   **18.** 36·9 (2 S.F.)
19. 15·19 (2 S.F.)   **20.** 19·87 (2 S.F.)
21. 3574 (3 S.F.)   **22.** 4285 (3 S.F.)
23. 5486 (2 S.F.)   **24.** 20·04 (3 S.F.)
25. 3·99 (2 S.F.)   **26.** 9·999 (3 S.F.)
27. 0·0637 (2 S.F.)   **28.** 0·00724 (2 S.F.)
29. 0·0088 (1 S.F.)   **30.** 0·00769 (1 S.F.)

1.5 THE METRIC SYSTEM

The unit of length is the metre.
1000 millimetres (mm) = 1 metre (m)
100 centimetres (cm)   = 1 metre (m)
1000 metres (m)   = 1 kilometre (km)

The unit of mass is the kilogram
1000 milligrams (mg)   = 1 gram (g)
1000 grams (g)   = 1 kilogram (kg)
1000 kilograms (kg)   = 1 tonne (t)

Example 1

Change (a) 426 cm to m
       (b) 2·64 km to m
(a) 426 cm = 426 ÷ 100 = 4·26 m
(b) 2·64 km = 2·64 × 1000 = 2640 m

Example 2

Change (a) 276 mg to g
       (b) 52·63 g to mg
(a) 276 mg = 276 ÷ 1000 = 0·276 g
(b) 52·63 g = 52·63 × 1000 = 52 630 mg

Exercise 1.5a

Change the following as indicated.

1. 357 cm to m   **2.** 5329 cm to m
3. 3760 cm to m   **4.** 49 cm to m
5. 60 cm to m   **6.** 9 cm to m
7. 5276 mm to m   **8.** 752 mm to m
9. 80 mm to m   **10.** 7 mm to m
11. 9137 m to km   **12.** 830 m to km
13. 3·372 km to m   **14.** 2·49 km to m
15. 19·6 km to m   **16.** 3·45 m to cm
17. 9·2 m to cm   **18.** 5·936 m to mm
19. 8·21 m to mm   **20.** 7·9 m to mm

Exercise 1.5b

Change the following as indicated.

1. 1328 mg to g   **2.** 536 mg to g
3. 780 mg to g   **4.** 90 mg to g
5. 8 mg to g   **6.** 1500 g to kg
7. 590 g to kg   **8.** 30 g to kg
9. 2 g to kg   **10.** 1320 kg to t
11. 800 kg to t   **12.** 4·536 g to mg
13. 8·98 g to mg   **14.** 3·4 g to mg
15. 5·26 kg to g   **16.** 8·5 kg to g
17. 0·7 kg to g   **18.** 3·71 t to kg
19. 5·6 t to kg   **20.** 0·3 t to kg

The unit for the measurement of capacity is the litre.

1000 millilitres (ml) = 1 litre (l)

Example 3

Change 2646 ml to litres.
2646 ml = 2646 ÷ 1000
= 2·646 litres

Exercise 1.5c

Change the following as indicated.
1. 3278 ml to litres
2. 8250 ml to litres
3. 9300 ml to litres
4. 6035 ml to litres
5. 5020 ml to litres
6. 1·332 litres to ml
7. 7·6 litres to ml
8. 0·755 litres to ml
9. 0·32 litres to ml
10. 0·1 litres to ml

Example 4

Find the cost of 25 cm of tubing at £5 per m.

$$25 \text{ cm} = \frac{25}{100} \text{ m at £5}$$

$$= \frac{1}{4} \text{ m at £5}$$

$$= £1·25$$

Example 5

Find the cost of 650 g of bacon at £2 per kg.

$$650 \text{ g} = \frac{650}{1000} \text{ kg at £2}$$

$$= \frac{13}{20} \text{ kg at £2}$$

$$= £1·30$$

Exercise 1.5d

Find the cost of the following.
1. 400 g of soap powder at 50p per kg.
2. 600 g of flour at 25p per kg.
3. 450 g of sugar at 40p per kg.
4. 350 g of jam at 60p per kg.
5. 800 g of cheese at £1·50 per kg.
6. 600 g of butter at £1.25 per kg.
7. 45 cm of dress cloth at 80p per m.
8. 50 m of thread at 60p per km.
9. 600 ml of milk at 25p per l.
10. A 400 ml can of beer at 65p per l.
11. A 350 ml can of lemonade at 40p per l.
12. 150 ml of white spirit at 20p per l.
13. 400 kg of cement at £50 per t.
14. 350 kg of potatoes at £120 per t.
15. 750 kg of garden soil at £5 per t.

1.6 COMMON FRACTIONS

A fraction keeps the same value when its numerator (top line) and denominator (bottom line) are both either:

(i) multiplied by the same number,
or (ii) divided by the same number.

Example 1

(a) $\frac{1}{2} = \frac{1 \times 3}{2 \times 3} = \frac{3}{6}$

(b) $\frac{10}{12} = \frac{10 \div 2}{12 \div 2} = \frac{5}{6}$

Exercise 1.6a

Copy and complete the following.
1. $\frac{1}{6} = \frac{}{12}$
2. $\frac{5}{8} = \frac{10}{}$
3. $\frac{}{3} = \frac{4}{6}$
4. $\frac{9}{} = \frac{18}{26}$
5. $\frac{17}{51} = \frac{1}{}$
6. $\frac{24}{36} = \frac{}{3}$

Write each of the following in its simplest form.
7. $\frac{3}{6}$
8. $\frac{4}{12}$
9. $\frac{5}{20}$
10. $\frac{4}{6}$
11. $\frac{9}{12}$
12. $\frac{18}{24}$
13. $\frac{27}{45}$
14. $\frac{54}{63}$
15. $\frac{26}{39}$

An *improper* fraction has its numerator bigger than its denominator.

Example 2

(a) $\frac{7}{5}$
(b) $\frac{14}{3}$
(c) $\frac{15}{6}$

Improper fractions can be written as *mixed numbers*.

Example 3

(a) $\frac{7}{5} = 1\frac{2}{5}$
(b) $\frac{14}{3} = 4\frac{2}{3}$
(c) $\frac{15}{6} = 2\frac{3}{6} = 2\frac{1}{2}$

Mixed numbers can be changed into improper fractions.

Example 4

(a) $1\frac{1}{4} = \frac{5}{4}$ (b) $3\frac{2}{7} = \frac{23}{7}$

(c) $8\frac{1}{10} = \frac{81}{10}$

Exercise 1.6b

Write as mixed numbers:

1. $\frac{3}{2}$ 2. $\frac{4}{3}$ 3. $\frac{7}{4}$ 4. $\frac{5}{2}$

5. $\frac{7}{3}$ 6. $\frac{18}{7}$ 7. $\frac{9}{6}$ 8. $\frac{32}{24}$

Write as improper fractions:

9. $1\frac{1}{4}$ 10. $1\frac{1}{5}$ 11. $1\frac{3}{7}$ 12. $1\frac{2}{5}$

13. $2\frac{1}{4}$ 14. $3\frac{1}{4}$ 15. $6\frac{1}{4}$ 16. $3\frac{3}{6}$

17. $3\frac{2}{3}$ 18. $7\frac{3}{5}$ 19. $3\frac{5}{7}$ 20. $7\frac{6}{9}$

Fractions can only be added or subtracted when their denominators are the same: only their numerators are then added or subtracted.

Example 5

(a) $\frac{1}{3} + \frac{1}{3} = \frac{2}{3}$ (b) $\frac{3}{5} + \frac{4}{5} = \frac{7}{5} = 1\frac{2}{5}$

(c) $\frac{3}{7} - \frac{2}{7} = \frac{1}{7}$ (d) $\frac{5}{6} - \frac{1}{6} = \frac{4}{6} = \frac{2}{3}$

Exercise 1.6c

1. $\frac{1}{5} + \frac{1}{5}$ 2. $\frac{2}{5} + \frac{1}{5}$ 3. $\frac{3}{10} + \frac{4}{10}$

4. $\frac{3}{8} + \frac{1}{8}$ 5. $\frac{2}{3} + \frac{2}{3}$ 6. $\frac{3}{10} + \frac{7}{10}$

7. $\frac{5}{7} + \frac{6}{7}$ 8. $\frac{3}{8} + \frac{7}{8}$ 9. $\frac{2}{3} - \frac{1}{3}$

10. $\frac{4}{5} - \frac{2}{5}$ 11. $\frac{5}{7} - \frac{2}{7}$ 12. $\frac{7}{8} - \frac{1}{8}$

13. $\frac{9}{10} - \frac{3}{10}$ 14. $\frac{7}{15} - \frac{2}{15}$ 15. $\frac{15}{16} - \frac{9}{16}$

To add or subtract fractions with different denominators, a common denominator is found.
A common denominator of $\frac{1}{2}$, $\frac{1}{3}$ and $\frac{1}{4}$ is 12, this is the smallest number which can be divided by 2, 3 and 4.

Example 6

(a) $\frac{1}{2} + \frac{2}{5} = \frac{5}{10} + \frac{4}{10} = \frac{9}{10}$

(b) $\frac{5}{6} - \frac{1}{2} = \frac{5}{6} - \frac{3}{6} = \frac{2}{6} = \frac{1}{3}$

Exercise 1.6d

1. $\frac{1}{3} + \frac{1}{5}$ 2. $\frac{1}{3} + \frac{1}{4}$ 3. $\frac{1}{5} + \frac{1}{10}$

4. $\frac{2}{3} + \frac{1}{4}$ 5. $\frac{3}{4} + \frac{1}{8}$ 6. $\frac{1}{3} + \frac{1}{6}$

7. $\frac{1}{4} + \frac{1}{12}$ 8. $\frac{2}{3} + \frac{1}{12}$ 9. $\frac{2}{9} + \frac{5}{18}$

10. $\frac{1}{8} + \frac{1}{4} + \frac{3}{16}$ 11. $\frac{1}{2} - \frac{1}{4}$ 12. $\frac{1}{5} - \frac{1}{10}$

13. $\frac{1}{2} - \frac{1}{3}$ 14. $\frac{1}{4} - \frac{1}{5}$ 15. $\frac{2}{3} - \frac{1}{4}$

16. $\frac{7}{8} - \frac{3}{4}$ 17. $\frac{5}{6} - \frac{1}{2}$ 18. $\frac{11}{12} - \frac{3}{4}$

19. $\frac{5}{6} - \frac{1}{4}$ 20. $\frac{9}{10} - \frac{11}{15}$

To add or subtract mixed numbers, first change them to improper fractions.

Example 7

(a) $1\frac{3}{4} + 1\frac{1}{2} = \frac{7}{4} + \frac{3}{2}$

$\qquad = \frac{7}{4} + \frac{6}{4} = \frac{13}{4} = 3\frac{1}{4}$

(b) $2\frac{2}{3} + 2\frac{5}{6} = \frac{8}{3} + \frac{17}{6}$

$\qquad = \frac{16}{6} + \frac{17}{6} = \frac{33}{6} = 5\frac{3}{6} = 5\frac{1}{2}$

(c) $4\frac{2}{3} - 2\frac{1}{2} = \frac{14}{3} - \frac{5}{2}$

$\qquad = \frac{28}{6} - \frac{15}{6} = \frac{13}{6} = 2\frac{1}{6}$

Exercise 1.6e

1. $2\frac{1}{4} + 1\frac{1}{4}$ 2. $3\frac{1}{3} + 1\frac{1}{5}$

3. $2\frac{1}{3} + 3\frac{1}{4}$ 4. $4\frac{1}{5} + 1\frac{1}{10}$

5. $6\frac{3}{4} + 1\frac{1}{8}$ 6. $1\frac{7}{10} + 1\frac{3}{5}$

7. $3\frac{2}{3} + 1\frac{3}{4}$ 8. $2\frac{2}{5} + 3\frac{7}{10}$

9. $1\frac{2}{3} + 1\frac{5}{6}$ 10. $1\frac{5}{12} + 1\frac{1}{3}$

11. $1\frac{3}{4} + 1\frac{7}{12}$ 12. $2\frac{9}{10} + 1\frac{1}{2}$

13. $4\frac{3}{4} - 3\frac{1}{2}$ 14. $2\frac{5}{6} - 1\frac{2}{3}$

15. $3\frac{7}{8} - 2\frac{1}{2}$ 16. $8\frac{4}{5} - 3\frac{7}{10}$

17. $8 - 2\frac{3}{4}$ 18. $2\frac{1}{2} - 1\frac{2}{3}$

19. $4\frac{1}{5} - 2\frac{2}{3}$ 20. $5\frac{1}{6} - 2\frac{2}{3}$

21. $4\frac{1}{2} - 3\frac{5}{6}$ 22. $4\frac{3}{4} - 1\frac{11}{12}$

23. $1\frac{9}{10} - \frac{14}{15}$ 24. $3\frac{1}{4} - 1\frac{5}{6}$

25. $7\frac{2}{3} - 4\frac{7}{9}$

To multiply fractions, multiply the numerators together and the denominators together.

Example 8

$$\frac{2}{3} \times \frac{5}{9} = \frac{2 \times 5}{3 \times 9} = \frac{10}{27}$$

Cancelling can make this easier.

Example 9

$$\frac{7}{9} \times \frac{12}{35} = \frac{\cancel{7}^1 \times \cancel{12}^4}{\cancel{9}_3 \times \cancel{35}_5} = \frac{1 \times 4}{3 \times 5} = \frac{4}{15}$$

Exercise 1.6f

1. $\frac{1}{2} \times \frac{1}{3}$ 2. $\frac{1}{4} \times \frac{2}{5}$ 3. $\frac{3}{4} \times \frac{1}{2}$

4. $\frac{3}{7} \times \frac{1}{2}$ 5. $\frac{2}{3} \times \frac{4}{5}$ 6. $\frac{1}{3} \times \frac{3}{5}$

7. $\frac{1}{3} \times \frac{6}{7}$ 8. $\frac{3}{5} \times \frac{10}{21}$ 9. $\frac{4}{7} \times \frac{21}{32}$

10. $\frac{5}{6} \times \frac{9}{11}$ 11. $\frac{1}{9} \times \frac{12}{13}$ 12. $\frac{3}{8} \times \frac{4}{21}$

13. $\frac{5}{16} \times \frac{6}{25}$ 14. $\frac{5}{7} \times \frac{14}{15}$ 15. $\frac{12}{13} \times \frac{39}{48}$

To multiply mixed numbers, first change them to improper fractions.

Example 10

(a) $1\frac{2}{5} \times 1\frac{1}{2} = \frac{7}{5} \times \frac{3}{2}$

$$= \frac{7 \times 3}{5 \times 2} = \frac{21}{10} = 2\frac{1}{10}$$

(b) $3\frac{1}{3} \times 1\frac{1}{5} = \frac{10}{3} \times \frac{6}{5}$

$$= \frac{\cancel{10}^2 \times \cancel{6}^2}{\cancel{3}_1 \times \cancel{5}_1} = \frac{2 \times 2}{1 \times 1} = 4$$

Exercise 1.6g

1. $1\frac{1}{4} \times 2\frac{1}{3}$ 2. $1\frac{2}{3} \times 1\frac{1}{4}$

3. $2\frac{1}{2} \times 2\frac{1}{2}$ 4. $1\frac{3}{4} \times 1\frac{2}{3}$

5. $3\frac{1}{4} \times 1\frac{1}{5}$ 6. $1\frac{1}{4} \times 2\frac{2}{3}$

7. $1\frac{1}{15} \times 2\frac{1}{2}$ 8. $3\frac{3}{4} \times 1\frac{1}{5}$

9. $2\frac{1}{2} \times 5$ 10. $7\frac{1}{2} \times 4$

11. $2\frac{1}{7} \times 1\frac{1}{3}$ 12. $2\frac{5}{8} \times 3\frac{2}{7}$

13. $4\frac{4}{7} \times 2\frac{5}{8}$ 14. $3\frac{3}{5} \times 3\frac{1}{3}$

15. $1\frac{1}{4} \times 1\frac{1}{2} \times 1\frac{1}{3}$

To divide by a fraction, multiply by its inverse; e.g. the inverse of $\frac{1}{2}$ is $\frac{2}{1}$ and the inverse of $4\frac{1}{2}$ (i.e. $\frac{9}{2}$) is $\frac{2}{9}$.

Example 11

(a) $\frac{2}{3} \div \frac{3}{4} = \frac{2}{3} \times \frac{4}{3}$

$$= \frac{2 \times 4}{3 \times 3} = \frac{8}{9}$$

(b) $2\frac{1}{2} \div 1\frac{1}{4} = \frac{5}{2} \div \frac{5}{4}$

$$= \frac{5}{2} \times \frac{4}{5}$$

$$= \frac{\cancel{5}^1 \times \cancel{4}^2}{\cancel{2}_1 \times \cancel{5}_1} = \frac{1 \times 2}{1 \times 1} = 2$$

Exercise 1.6h

1. $\frac{1}{4} \div \frac{1}{3}$ 2. $\frac{2}{5} \div \frac{2}{7}$

3. $\frac{4}{5} \div \frac{3}{4}$ 4. $\frac{3}{7} \div \frac{2}{5}$

5. $\frac{5}{12} \div \frac{3}{5}$ 6. $\frac{1}{3} \div \frac{5}{9}$

7. $\frac{2}{5} \div \frac{9}{10}$ 8. $\frac{3}{7} \div \frac{11}{14}$

9. $\frac{4}{9} \div \frac{2}{3}$ 10. $\frac{2}{5} \div \frac{4}{5}$

11. $5 \div 1\frac{1}{4}$ 12. $6 \div 1\frac{1}{2}$

13. $7\frac{1}{2} \div 2\frac{1}{2}$ 14. $3\frac{1}{2} \div 1\frac{3}{4}$

15. $1\frac{1}{10} \div 1\frac{1}{5}$ 16. $1\frac{3}{8} \div 2\frac{1}{4}$

17. $2\frac{6}{7} \div 1\frac{1}{14}$ 18. $2\frac{2}{3} \div 1\frac{7}{9}$

19. $1\frac{5}{12} \div 3\frac{3}{16}$ 20. $3\frac{3}{5} \div 2\frac{1}{4}$

Exercise 1.6i

In questions **1** to **10** find the 'odd answer out'

1. (a) $\frac{4}{5} + \frac{1}{30}$ 2. (a) $\frac{9}{10} - \frac{11}{40}$

(b) $\frac{9}{20} + \frac{7}{15}$ (b) $\frac{1}{3} + \frac{7}{24}$

(c) $\frac{4}{9} + \frac{7}{18}$ (c) $\frac{3}{5} + \frac{3}{20}$

3. (a) $\frac{5}{12} - \frac{4}{15}$ 4. (a) $\frac{4}{15} - \frac{1}{6}$

(b) $\frac{1}{15} + \frac{1}{10}$ (b) $\frac{1}{5} - \frac{3}{40}$

(c) $\frac{2}{5} - \frac{1}{4}$ (c) $\frac{5}{6} - \frac{17}{24}$

5. (a) $1\frac{1}{2} + \frac{2}{3}$ 6. (a) $1\frac{9}{20} + 1\frac{3}{10}$

(b) $2\frac{5}{8} - \frac{11}{24}$ (b) $2\frac{1}{12} + \frac{2}{3}$

(c) $3\frac{3}{4} - 1\frac{5}{12}$ (c) $3\frac{1}{6} - \frac{2}{3}$

7. (a) $2\frac{7}{10} - \frac{3}{40}$

   (b) $1\frac{3}{5} + \frac{9}{10}$

   (c) $4\frac{1}{4} - 1\frac{5}{8}$

8. (a) $2\frac{3}{4} - 1\frac{11}{20}$

   (b) $\frac{1}{2} + \frac{1}{4} + \frac{5}{12}$

   (c) $\frac{1}{5} + \frac{2}{3} + \frac{3}{10}$

9. (a) $3\frac{1}{3} - \frac{8}{15}$

   (b) $1\frac{3}{20} + \frac{7}{10} + \frac{3}{4}$

   (c) $1\frac{1}{2} + 1\frac{1}{10} + \frac{1}{5}$

10. (a) $1\frac{1}{6} + \frac{1}{5} + \frac{1}{3}$

    (b) $1\frac{4}{5} + \frac{3}{4} - \frac{17}{20}$

    (c) $1\frac{1}{4} + \frac{11}{15} - \frac{1}{12}$

11. The distance from Halesowen to Quinton is $2\frac{1}{2}$ miles. I am given a lift for part of the way and the car's mileometer records $1\frac{3}{10}$ miles. If I walk the rest of the way, how far do I walk?

12. The distance from Halesowen to Stourbridge is $4\frac{1}{2}$ miles. I am given a lift for part of the way and the car's mileometer records $2\frac{9}{10}$ miles. If I walk the rest of the way, how far do I walk?

13. A one-litre flask is filled with milk, and it is used to fill two glasses, one of capacity $\frac{1}{2}$ of a litre and the other of capacity $\frac{1}{6}$ of a litre. What fraction of a litre will remain in the flask?

14. In a certain class $\frac{3}{4}$ of the children have dark hair, $\frac{1}{12}$ have ginger hair and the remainder are blonde. What fraction of the class have blonde hair?

15. At a polling-station $\frac{1}{2}$ of the people voted Labour, $\frac{1}{3}$ voted Conservative, $\frac{1}{15}$ voted Liberal and the remainder voted for an independent candidate. What fraction voted for the independent candidate?

16. A group of people travelled from Stourbridge to Kidderminster, $\frac{1}{20}$ of them decided to walk, $\frac{1}{12}$ went by car, $\frac{2}{5}$ went by train and all the rest travelled by bus. What fraction went by bus?

17. Two tonnes of coal are shared between three men. The first receives $\frac{3}{4}$ of a tonne, the second $\frac{4}{5}$ of a tonne. What fraction of a tonne does the third receive?

18. Three tonnes of sand are shared between three builders. The first receives $1\frac{1}{2}$ tonnes, the second $\frac{5}{6}$ of a tonne. What fraction of a tonne does the third builder receive?

In questions **19** to **28** find the 'odd answer out'.

19. (a) $\frac{20}{33} \times \frac{44}{45}$

    (b) $\frac{27}{40} \times \frac{32}{45}$

    (c) $\frac{14}{15} \times \frac{18}{35}$

20. (a) $\frac{6}{7} \times \frac{21}{32}$

    (b) $\frac{7}{9} \times \frac{24}{35}$

    (c) $\frac{6}{25} \div \frac{9}{20}$

21. (a) $\frac{24}{35} \div \frac{20}{21}$

    (b) $\frac{5}{6} \times \frac{44}{45}$

    (c) $\frac{33}{40} \times \frac{48}{55}$

22. (a) $\frac{8}{21} \div \frac{9}{14}$

    (b) $\frac{10}{21} \div \frac{8}{9}$

    (c) $\frac{14}{15} \times \frac{40}{63}$

23. (a) $1\frac{7}{20} \times \frac{35}{36}$

    (b) $1\frac{11}{24} \div \frac{14}{15}$

    (c) $3\frac{3}{4} \times \frac{5}{12}$

24. (a) $1\frac{1}{20} \times 2\frac{1}{12}$

    (b) $\frac{11}{18} \times 3\frac{6}{7}$

    (c) $1\frac{9}{35} \div \frac{8}{15}$

25. (a) $3\frac{11}{15} \div 1\frac{7}{25}$

    (b) $2\frac{2}{9} \times 1\frac{5}{16}$

    (c) $2\frac{7}{24} \times 1\frac{7}{33}$

26. (a) $1\frac{7}{20} \div 4\frac{4}{5}$

    (b) $\frac{25}{27} \times \frac{9}{22} \times \frac{4}{5}$

    (c) $\frac{18}{25} \times \frac{15}{16} \times \frac{5}{12}$

27. (a) $4\frac{4}{9} \div 2\frac{1}{12}$

    (b) $2\frac{2}{9} \times 2\frac{1}{10} \times \frac{16}{35}$

    (c) $1\frac{1}{24} \times 2\frac{1}{4} \times \frac{14}{15}$

28. (a) $1\frac{13}{27} \times 2\frac{11}{12} \div 3\frac{1}{9}$

    (b) $3\frac{3}{14} \times \frac{15}{36} \div 1\frac{1}{8}$

    (c) $1\frac{1}{4} \times 1\frac{8}{27} \div 1\frac{1}{6}$

29. The distance from Halesowen to Dudley is $5\frac{1}{4}$ miles. If I walk $\frac{2}{3}$ of the way, how far do I walk?

30. The distance from Stourbridge to Birmingham is $11\frac{1}{4}$ miles. If I walk $\frac{2}{5}$ of the way, how far do I walk?

31. $8\frac{1}{4}$ tonnes of coal are to be shared between a number of men. If one man receives $\frac{6}{11}$ of the load, how many tonnes does he get?

32. $10\frac{1}{2}$ tonnes of sand are to be shared between a number of builders. If one of them receives $\frac{4}{7}$ of the load, how many tonnes does he get?

33. An urn contains $12\frac{1}{2}$ litres of lemonade. How many glasses, each of capacity $\frac{5}{16}$ litre, will it be able to fill?

34. A grain merchant has only $13\frac{1}{2}$ tonnes in his stock. If he has several customers who are all ordering $\frac{3}{4}$ of a tonne, how many can he supply?

Change a decimal fraction into a common fraction or a mixed number as follows.
(i) Write each separate figure after the decimal point as a common fraction.
(ii) Add these fractions together.
(iii) Make sure this sum is given in its simplest form.

Example 1

(a) $0{\cdot}27 = \frac{2}{10} + \frac{7}{100} = \frac{20}{100} + \frac{7}{100} = \frac{27}{100}$

(b) $0{\cdot}35 = \frac{3}{10} + \frac{5}{100} = \frac{30}{100} + \frac{5}{100}$

$$= \frac{35}{100} = \frac{7}{20}$$

(c) $1{\cdot}025 = 1 + \frac{0}{10} + \frac{2}{100} + \frac{5}{1000}$

$$= 1 + \frac{20}{1000} + \frac{5}{1000}$$

$$= 1\frac{25}{1000} = 1\frac{1}{40}$$

Exercise 1.7a

Change the following into common fractions or mixed numbers.

1. 0·7	2. 0·5	3. 0·4
4. 0·9	5. 0·45	6. 0·26
7. 0·85	8. 0·32	9. 0·58
10. 0·72	11. 0·875	12. 0·325
13. 0·625	14. 0·475	15. 0·03
16. 0·08	17. 0·05	18. 0·07
19. 0·055	20. 0·024	21. 0·062
22. 0·0275	23. 0·0625	24. 0·003
25. 0·006	26. 0·002	27. 0·0035
28. 0·0048	29. 3·2	30. 8·25
31. 12·16	32. 41·02	33. 5·018
34. 13·012	35. 2·0125	36. 8·0175
37. 14·004	38. 11·005	39. 16·0025
40. 21·0015		

Change a common fraction into a decimal fraction by dividing the numerator (top line) by the denominator (bottom line). The answer will be either
(i) an exact decimal (see example 2),
or (ii) a recurring decimal (see example 3),
or (iii) a non-terminating decimal when the answer is usually written to a given number of decimal places (see example 4).

Example 2

Change (a) $\frac{3}{4}$ into a decimal fraction

(b) $\frac{7}{40}$ into a decimal fraction

(a) $\frac{3}{4} = 3 \div 4$

$$= 0{\cdot}75$$

```
      0·75
  4) 3·00
     2 8
      20
      20
      . .
```

(b) $\frac{7}{40} = 7 \div 40$

$$= 0{\cdot}175$$

```
       0·175
  40) 7·000
      40
      300
      280
      200
      200
      . . .
```

Exercise 1.7b

Change the following to decimal fractions.

1. $\frac{3}{8}$	2. $\frac{1}{8}$	3. $\frac{1}{4}$	4. $\frac{3}{20}$
5. $\frac{11}{20}$	6. $\frac{7}{20}$	7. $\frac{13}{20}$	8. $\frac{19}{20}$
9. $\frac{3}{5}$	10. $\frac{1}{5}$	11. $\frac{4}{5}$	12. $\frac{11}{40}$
13. $\frac{9}{40}$	14. $\frac{17}{40}$	15. $\frac{21}{40}$	16. $\frac{3}{16}$
17. $\frac{11}{16}$	18. $\frac{7}{50}$	19. $\frac{9}{50}$	20. $\frac{31}{50}$

Example 3

Change (a) $\frac{1}{3}$ into a decimal fraction

(b) $\frac{2}{11}$ into a decimal fraction

(a) $\frac{1}{3} = 1 \div 3$

$$= 0{\cdot}333$$

$$= 0{\cdot}\dot{3}$$

```
      0·333
  3) 1·000
```

(b) $\frac{2}{11} = 2 \div 11$

$$= 0{\cdot}1818$$

$$= 0{\cdot}\dot{1}\dot{8}$$

```
       0·1818
  11) 2·0000
```

Exercise 1.7c

Change the following to decimal fractions.

1. $\frac{2}{3}$ 2. $\frac{5}{6}$ 3. $\frac{1}{6}$ 4. $\frac{5}{12}$

5. $\frac{7}{12}$ 6. $\frac{11}{12}$ 7. $\frac{5}{11}$ 8. $\frac{9}{11}$

9. $\frac{3}{11}$ 10. $\frac{7}{11}$ 11. $\frac{4}{15}$ 12. $\frac{11}{15}$

13. $\frac{7}{15}$ 14. $\frac{13}{15}$ 15. $\frac{8}{15}$ 16. $\frac{7}{30}$

17. $\frac{11}{30}$ 18. $\frac{13}{30}$ 19. $\frac{5}{18}$ 20. $\frac{7}{18}$

Example 4

Change $\frac{5}{17}$ into a decimal fraction correct to 3 decimal places

$\frac{5}{17} = 5 \div 17$

$= 0.2941\ldots$

$= 0.294$ (to 3 D.P.)

$$
\begin{array}{r}
0.2941 \\
17\overline{)5.0000} \\
3\,4 \\
\hline
1\,60 \\
1\,53 \\
\hline
70 \\
68 \\
\hline
20 \\
13 \\
\hline
7
\end{array}
$$

Exercise 1.7d

Change the following to decimal fractions correct to 3 decimal places.

1. $\frac{5}{7}$ 2. $\frac{6}{7}$ 3. $\frac{4}{7}$ 4. $\frac{2}{7}$

5. $\frac{1}{7}$ 6. $\frac{3}{7}$ 7. $\frac{7}{13}$ 8. $\frac{10}{13}$

9. $\frac{9}{13}$ 10. $\frac{11}{13}$ 11. $\frac{10}{21}$ 12. $\frac{2}{21}$

13. $\frac{16}{21}$ 14. $\frac{9}{17}$ 15. $\frac{4}{17}$ 16. $\frac{3}{17}$

17. $\frac{6}{17}$ 18. $\frac{3}{19}$ 19. $\frac{5}{19}$ 20. $\frac{7}{19}$

Example 5

Find the difference between the fractions $\frac{1}{3}$ and $\frac{19}{300}$ giving the answer (a) as a fraction (b) as a decimal.

(a) $\frac{1}{3} - \frac{19}{300} = \frac{100}{300} - \frac{19}{300} = \frac{81}{300} = \frac{27}{100}$

(b) $\frac{27}{100} = 0.27$

Exercise 1.7e

1. A club has 56 members, of whom 35 are men. What fraction of the membership are men? Express this fraction also as a decimal.
2. One day 120 trains arrive at a station and 42 of them are late. What fraction of the trains arrive late? Express this fraction also as a decimal.
3. A village school has 150 pupils and 135 are present on a certain day. What fraction of the pupils are present? Express this fraction also as a decimal.
4. Find the sum of the fractions $\frac{1}{3}$ and $\frac{17}{300}$, giving the answer (a) as a fraction, (b) as a decimal.
5. Find the difference between the fractions $\frac{1}{7}$ and $\frac{3}{70}$, giving the answer (a) as a fraction, (b) as a decimal.
6. Find the sum of the fractions $\frac{2}{3}$ and $\frac{7}{30}$, giving the answer (a) as a fraction, (b) as a decimal.
7. Find the difference between the fractions $\frac{2}{3}$ and $\frac{53}{300}$, giving the answer (a) as a fraction, (b) as a decimal.
8. A piece of string is $\frac{1}{3}$ of a metre in length, if a piece equal to $\frac{7}{30}$ of a metre in length is cut off, what length remains? Express the answer (a) as a fraction, (b) as a decimal.
9. I have $\frac{2}{3}$ of a kilogram of sand. If $\frac{1}{30}$ of a kilogram is then added, how much will I then have? Express the answer (a) as a fraction, (b) as a decimal.
10. A flask holds $\frac{2}{3}$ of a litre of coffee and some is poured out so as to fill a cup of capacity $\frac{11}{30}$ of a litre. How much remains in the flask? Express the answer (a) as a fraction, (b) as a decimal.

In questions 11 to 15, find which is the greater and by how much. Express the answer (i) as a fraction, (ii) as a decimal.

11. (a) the sum of $\frac{3}{7}$ and $\frac{19}{70}$

    (b) the sum of $\frac{2}{3}$ and $\frac{7}{300}$

12. (a) the sum of $\frac{2}{7}$ and $\frac{3}{700}$

    (b) the sum of $\frac{2}{9}$ and $\frac{7}{90}$

13. (a) the difference of $\frac{8}{9}$ and $\frac{17}{90}$

    (b) the difference of $\frac{5}{7}$ and $\frac{3}{700}$

14. (a) the sum of $\frac{5}{9}$ and $\frac{13}{900}$

    (b) the difference of $\frac{2}{3}$ and $\frac{11}{300}$

15. (a) the sum of $\frac{2}{3}$ and $\frac{61}{300}$

    (b) the difference of $\frac{6}{7}$ and $\frac{19}{700}$

Ratio compares quantities of the *same* kind and is found by writing one as a fraction of the other in its simplest form.

Example 1

Give each of the following ratios in their simplest form.

(a) 15 to 75

$$\frac{15}{75} = \frac{1}{5}$$

∴ ratio is 1:5

(b) 50 cm to 2 m

$$\frac{50 \text{ cm}}{2 \text{ m}} = \frac{50 \text{ cm}}{200 \text{ cm}}$$

$$= \frac{50}{200} = \frac{1}{4}$$

∴ ratio is 1:4

(c) £2·75 to £1·25

$$\frac{£2·75}{£1·25} = \frac{275p}{125p}$$

$$= \frac{275}{125} = \frac{11}{5}$$

∴ ratio is 11:5

Exercise 1.8a

Give each of the following ratios in their simplest form.

1. 24 to 96	**2.** 18 to 108
3. 25 to 75	**4.** 16 to 40
5. 36 to 54	**6.** 45 to 60
7. 60 to 72	**8.** 32 to 56
9. 60 to 24	**10.** 32 to 12
11. 28 to 16	**12.** 45 to 25
13. 36 to 21	**14.** 60 cm to 4 m
15. 80 cm to 6 m	**16.** 120 m to 2 km
17. 450 kg to 1 t	**18.** 250 g to 2 kg
19. 450 ml to 6 litres	**20.** 24 mm to 30 cm
21. 75p to £2	**22.** 3 m to 45 cm
23. 3 km to 800 m	**24.** 4 kg to 720 g
25. 2 l to 150 ml	**26.** £5 to 40p
27. £1·35 to £1·80	**28.** £1·26 to £1·44
29. £1·20 to £2·80	**30.** 19p to £1·14
31. 48p to £1·32	**32.** £4·32 to £1·80
33. £2·52 to £1·96	**34.** £2·70 to 36p

35. £2·25 to £1·00	**36.** 80 cm to 1·28 m
37. 2·4 mm to 1·6 cm	**38.** 900 g to 1·26 kg
39. 1·08 m to 24 cm	**40.** 1·32 litres to 550 ml

Example 2

In a class there are 27 boys and 15 girls. Find the ratio of the number of boys to the number of girls.

$$\frac{\text{number of boys}}{\text{number of girls}} = \frac{27}{15} = \frac{9}{5}$$

∴ the ratio of the number of boys to the number of girls is 9:5.

Exercise 1.8b

1. In a certain month there were 12 wet days and 18 fine days. Find the ratio of wet days to fine days.
2. Amongst a group of boys there were 20 Aston Villa supporters and 24 West Bromwich Albion supporters. Find the ratio of Aston Villa supporters to West Bromwich Albion supporters.
3. On an overcrowded bus there were 9 standing passengers and 30 seated passengers. Find the ratio of standing passengers to seated passengers.
4. In a certain street there were 20 bungalows and 35 houses. Find the ratio of bungalows to houses.
5. In a cricket match a batsman scored 54 runs in his first innings and 96 runs in his second innings. Find the ratio of his first score relative to the second.
6. There are 20 people present in a class, but 12 are absent because they have influenza. Find the ratio of those present to those absent.
7. Amongst a group of girls, 24 chose to play hockey and 21 chose to play netball. Find the ratio of those who played hockey to those who played netball.
8. A group of people travelled from Birmingham to Shrewsbury; 54 went by train and 12 went by bus. Find the ratio of those who used the train to those who used the bus.
9. In a group of men there were 56 smokers and 40 non-smokers. Find the ratio of smokers to non-smokers.
10. A local bus from Birmingham to Kidderminster takes 64 minutes whereas an express bus takes only 48 minutes. Find the ratio of the slow time to the fast time.

Example 3

Divide £50 in the ratio 2:3:5

Number of shares = 2 + 3 + 5 = 10

So 1 share = $\dfrac{£50}{10}$ = £5

The amounts are
    2 × £5 = £10
    3 × £5 = £15
    5 × £5 = £25
Check: £10 + £15 + £25 = £50.

Exercise 1.8c

1. Divide £90 in the ratio 3:7.
2. Divide £84 in the ratio 2:5.
3. Divide £45 in the ratio 4:5.
4. Share out 60 sweets between Michael and Margaret in the ratio 5:7.
5. Share out 48 kg of sand between Mr. Johnson and Mr. Walker in the ratio 3:5.
6. Mary and Julie receive £66 from a rich uncle. They share the money between them in the ratio 5:6. How much does each receive?
7. Divide £120 in the ratio 3:4:5.
8. Divide £72 in the ratio 2:3:4.
9. Divide £132 in the ratio 2:3:6.
10. Divide £56 in the ratio 1:2:5.
11. Share out 60 marbles between Tom, Jack and Bill in the ratio 2:3:7.
12. Share out 45 beads between Ann, Jane and Susan in the ratio 1:3:5.
13. Share out 96 kg coal between Mr. Smith, Mr. Jones and Mr. Thompson in the ratio 1:5:6.
14. David, Peter and John organize a fête which raises £110. The money is shared between charities for the old, for the blind, and for animals in the ratio 5:4:2. How much does each charity receive?
15. At a local election 120 people voted. They voted Liberal, Conservative and Labour in the respective ratio 1:3:6. How many people voted for each party?

Example 4

£185 is to be shared between John, Mary and Jane (whose ages are 10, 12, and 15 years respectively) in the same ratio as their ages. How much does each receive?

Ratio of ages = 10:12:15
Number of shares = 10 + 12 + 15 = 37

∴ 1 share = £185 ÷ 37 = £5
    John gets £5 × 10 = £50
    Mary gets £5 × 12 = £60
    Jane gets £5 × 15 = £75
Check: £50 + £60 + £75 = £185.

Exercise 1.8d

1. £160 is to be shared between Bill, Jane and Wendy (whose ages are 9, 10, and 13 years respectively) in the same ratio as their ages. How much does each receive?
2. Mr. Andrews, Mr. Bailey and Mr. Carter own 4, 5, and 6 parts respectively of the same business. If the business makes a profit of £120 in a certain week, how much does each receive?
3. In a television quiz contest the prize money is £108. It is shared out between the top three contestants in the same ratio as that of the marks they obtain. If their marks are 15, 13, and 8, how much does each receive?
4. A football club pays out £140 in bonus money to its three top-scoring players in the same ratio as that of the goals they have scored in a certain tournament. If they score 16, 11, and 8 goals, how much does each receive?
5. A cricket club pays £120 in bonus money to its three most successful batsmen in the same ratio as the number of centuries they have scored during the season. If they have scored 11, 7, and 6 centuries, how much does each receive?
6. 72 kg of compost is to be spread over three small garden plots in a ratio equal to that of their areas. If their areas are 7 m², 8 m², and 9 m², how much is used on each plot?
7. 540 kg of dry concrete mixture is to be spread over three surfaces in a ratio equal to that of their areas. If their areas are 9 m², 11 m², and 16 m², how much is used on each surface?
8. £10 is to be shared between three fruit pickers in a ratio equal to that of the number of bags they are able to fill. If they fill 5, 6, and 9 bags, how much does each receive?
9. Three boys sell some magazines for a total profit of £2. The profit is shared between them in a ratio equal to that of the number of magazines they sell. If they sell 7, 8, and 10 magazines, how much does each receive?
10. 15 t of surface soil is to be spread over three small lawns in a ratio equal to that of their areas. If their areas are 7 m², 11 m² and 12 m², how much is used on each surface?

Two quantities are in *direct proportion* when an increase in one is in the same ratio as an increase in the other. e.g. if 5 oranges cost 20p, 10 oranges cost 40p: the cost is in direct proportion to the quantity. The cost varies directly as the quantity.

Example 5

If 4 m of cloth cost £10, what will 10 m cost?
The number of metres of cloth is increased in the ratio $\frac{10}{4}$.
So the cost is increased in the ratio $\frac{10}{4}$.
∴ 10 m costs $\frac{10}{4} \times £10 = £25$.

Example 6

If 12 eggs cost 54p, how many eggs could be bought for 18p?
The cost of the eggs decreases in the ratio $\frac{18}{54}$
So the number of eggs bought decreases in the ratio $\frac{18}{54}$.
∴ number of eggs bought $= \frac{18}{54} \times 12$

$$= 4 \text{ eggs.}$$

Exercise 1.8e

1. Curtain hooks cost 30p for 12. Find the cost of
   (a) 40 curtain hooks,
   (b) 8 curtain hooks.
2. If 12 oranges cost 88p, find the cost of
   (a) 15 oranges,
   (b) 9 oranges.
3. If 20 m² of carpet costs £36, find the cost of
   (a) 35 m²,
   (b) 15 m².
4. Ceiling tiles cost £1·56 for 12. Find the cost of
   (a) 7, (b) 30, (c) 9, (d) 20.
5. It takes me 44 minutes to cycle from Birmingham to Halesowen, a distance of 8 miles. How long will it take to cycle.
   (a) from Birmingham to Droitwich (20 miles)?
   (b) from Birmingham to West Bromwich (6 miles)?
6. If 12 brass screws cost 16p, how many can be bought for (a) 40p? (b) 12p?
7. 12 grapefruits cost 90p. How many can be bought for (a) £2·10? (b) 75p?
8. If 20 m² of flooring cost £15, what area of the same flooring can be bought for (a) £33? (b) £9?

9. Ceramic tiles cost £3·24 for 12. How many can be bought for
   (a) £8·64?   (b) £5·94?
   (c) £2·70?   (d) £2·43?
10. It takes me 80 minutes to cycle from Oldham to Walkden, a distance of 15 miles. Which of the distances below can I cycle in
    (a) 96 minutes?
    (b) 64 minutes?
    Oldham to Manchester (7 miles)
    Oldham to Bury (10 miles)
    Oldham to Glossop (12 miles)
    Oldham to Huddersfield (18 miles)

Two quantities are in *inverse proportion* when an increase in one is in the same ratio as a *decrease* in the other. e.g. if 5 combine harvesters reap a field in 2 days, then 10 combine harvesters would reap the same field in 1 day.
The time for harvesting varies inversely as the number of harvesters used.

Example 7

A car travelling at an average speed of 60 km/h completes a journey in 38 minutes. How long would the same journey take at an average speed of 40 km/h?
The speed decreases in the ratio $\frac{40}{60}$,
so the time increases in the ratio $\frac{60}{40}$.
∴ time taken $= \frac{60}{40} \times 38 = 57$ minutes.

Example 8

If 5 men can unload a lorry in 18 minutes, how long would it take 9 men to unload the same lorry?
The number of men increases in the ratio $\frac{9}{5}$,
so the time taken decreases in the ratio $\frac{5}{9}$.
∴ time taken $= \frac{5}{9} \times 18 = 10$ minutes.

Exercise 1.8f

1. If 12 dockers can load a barge in 75 minutes, how long will it take (a) 10 dockers, (b) 25 dockers to load the same barge?
2. If 15 men can unload a ship's cargo in 100 minutes, how long will it take (a) 6 men, (b) 20 men to unload the same cargo?

3. If 15 bricklayers can build a wall in 30 hours of working time, how long will it take (a) 9 brick-layers, (b) 25 bricklayers to build the same wall?

4. An electric heater raises the temperature of 8 kilograms of water by 18 degrees in a certain amount of time. By how many degrees would it raise the temperature of
(a) 6 kilograms, (b) 9 kilograms of water in the same time?

5. An electric heater raises the temperature of 900 grams of water by 20 degrees in a certain amount of time. By how many degrees would it raise the temperature of (a) 750 grams, (b) 1·2 kilograms of water in the same time?

6. A 500 watt electric heater boils a given amount of water in 12 minutes. How long would it take (a) a less powerful heater of 300 watts, (b) a more powerful heater of 750 watts to boil the same amount of water?

7. A 1·2 kilowatt electric heater boils a given amount of water in 10 minutes. How long would it take (a) a less powerful heater of 800 watts, (b) a more powerful heater of 2 kilowatts to boil the same amount of water?

8. Water flows constantly into a cylindrical beaker of base area 24 cm^2. The water level rises by 10 cm every second. How quickly would the water level rise in a beaker of base area (a) 20 cm^2, (b) 30 cm^2 if the flow rate remained the same?

9. A car travelling at 90 km/h completes a certain journey in 40 minutes. How long would it take a car travelling at (a) 100 km/h, (b) 75 km/h to complete the same journey?

10. A car travelling at 96 km/h takes 35 minutes to travel between two service areas on a motorway. How long would it take a car travelling at (a) 112 km/h, (b) 80 km/h to travel between the same two service areas?

1.9 PERCENTAGE

A percentage is a fraction with a denominator of 100.

Example 1

Write each percentage as a fraction in its simplest form.

(a) $7\% \quad = \frac{7}{100}$

(b) $25\% \quad = \frac{25}{100} = \frac{25 \div 25}{100 \div 25} = \frac{1}{4}$

(c) $84\% \quad = \frac{84}{100} = \frac{84 \div 4}{100 \div 4} = \frac{21}{25}$

(d) $160\% = \frac{160}{100} = \frac{8}{5} = 1\frac{3}{5}$

Example 2

Write each percentage as a decimal fraction.

(a) $17\% \quad = \frac{17}{100} = 0\cdot17$

(b) $75\% \quad = \frac{75}{100} = 0\cdot75$

(c) $3\% \quad = \frac{3}{100} = 0\cdot03$

(d) $283\% = \frac{283}{100} = 2\cdot83$

Exercise 1.9a

Write each percentage as a fraction in its simplest form.

1. 9%	2. 15%	3. 45%	4. 20%
5. 70%	6. 32%	7. 68%	8. 56%
9. 42%	10. 6%	11. 140%	12. 180%
13. 130%	14. 210%	15. 275%	16. 335%
17. 236%	18. 308%	19. 405%	20. 465%

Write each percentage as a decimal fraction.

21. 15%	22. 35%	23. 90%	24. 80%
25. 60%	26. 5%	27. 16%	28. 72%
29. 44%	30. 58%	31. 125%	32. 156%
33. 170%	34. 230%	35. 225%	36. 340%
37. 212%	38. 304%	39. 408%	40. 401%

Example 3

Write each percentage as a fraction in its simplest form.

(a) $12\frac{1}{2}\% \quad = \frac{12\frac{1}{2}}{100} = \frac{12\frac{1}{2} \times 2}{100 \times 2} = \frac{25}{200} = \frac{1}{8}$

(b) $32\frac{1}{4}\% \quad = \frac{32\frac{1}{4}}{100} = \frac{32\frac{1}{4} \times 4}{100 \times 4} = \frac{129}{400}$

(c) $6\frac{2}{3}\% \quad = \frac{6\frac{2}{3}}{100} = \frac{20}{300} = \frac{1}{15}$

(d) $128\frac{4}{7}\% = \dfrac{128\frac{4}{7}}{100} = \dfrac{900}{700} = \dfrac{9}{7} = 1\frac{2}{7}$

Example 4

(a) $37\frac{1}{2}\% = \dfrac{37\cdot5}{100}$ (as $\frac{1}{2} = 0\cdot5$) $= 0\cdot375$

(b) $6\frac{1}{4}\% = \dfrac{6\cdot25}{100}$ (as $\frac{1}{4} = 0\cdot25$) $= 0\cdot0625$

(c) $133\frac{1}{3}\% = \dfrac{133\cdot\dot{3}}{100}$ (as $\frac{1}{3} = 0\cdot\dot{3}$) $= 1\cdot33\dot{3}$

Exercise 1.9b

Write each percentage as a fraction in its simplest form.

1. $17\frac{1}{2}\%$ 2. $18\frac{3}{4}\%$ 3. $63\frac{1}{3}\%$ 4. $53\frac{1}{3}\%$

5. $20\frac{5}{6}\%$ 6. $41\frac{2}{3}\%$ 7. $42\frac{1}{2}\%$ 8. $5\frac{5}{9}\%$

9. $61\frac{1}{9}\%$ 10. $21\frac{7}{8}\%$ 11. $102\frac{1}{2}\%$ 12. $147\frac{1}{2}\%$

13. $123\frac{1}{3}\%$ 14. $204\frac{1}{6}\%$ 15. $427\frac{1}{2}\%$ 16. $308\frac{1}{3}\%$

17. $313\frac{1}{3}\%$ 18. $206\frac{2}{3}\%$ 19. $129\frac{1}{6}\%$ 20. $115\frac{5}{8}\%$

Write each percentage as a decimal fraction.

21. $62\frac{1}{2}\%$ 22. $87\frac{1}{2}\%$ 23. $32\frac{1}{2}\%$ 24. $47\frac{1}{2}\%$

25. $56\frac{1}{4}\%$ 26. $81\frac{1}{4}\%$ 27. $9\frac{3}{8}\%$ 28. $66\frac{2}{3}\%$

29. $11\frac{1}{9}\%$ 30. $83\frac{1}{3}\%$ 31. $112\frac{1}{2}\%$ 32. $137\frac{1}{2}\%$

33. $107\frac{1}{2}\%$ 34. $222\frac{1}{2}\%$ 35. $131\frac{1}{4}\%$ 36. $306\frac{1}{4}\%$

37. $203\frac{1}{8}\%$ 38. $333\frac{1}{3}\%$ 39. $403\frac{1}{3}\%$ 40. $216\frac{2}{3}\%$

To find the value of a percentage of a quantity, change the percentage to a common fraction; then find that fraction of the quantity.

Example 5

(a) 28% of 50 $= \dfrac{\overset{14}{\cancel{28}}}{\underset{\underset{1}{2}}{\cancel{100}}} \times \dfrac{\overset{1}{\cancel{50}}}{1} = \dfrac{14}{1} = 14$

(b) 66% of £1·25 $= \dfrac{\overset{33}{\cancel{66}}}{\underset{\underset{2}{4}}{\cancel{100}}} \times \dfrac{\overset{5}{\cancel{125}}}{1}$ pence

$= \dfrac{165}{2} = 82\frac{1}{2}$ pence

(c) $6\frac{1}{4}\%$ of 1 m 44 cm $= \dfrac{6\frac{1}{4}}{100} \times 144$ cm

$= \dfrac{\overset{1}{\cancel{25}}}{\underset{\underset{1}{16}}{\cancel{400}}} \times \dfrac{\overset{9}{\cancel{144}}}{1} = 9$ cm

Exercise 1.9c

Find the 'odd answer out'.

1. (a) 20% of £400
   (b) 25% of £300
   (c) 30% of £250

2. (a) 24% of £150
   (b) 25% of £140
   (c) 15% of £240

3. (a) 16% of £250
   (b) 36% of £125
   (c) 25% of £160

4. (a) 32% of £7·50
   (b) 15% of £16·00
   (c) 20% of £12·50

5. (a) 64% of £5·00
   (b) 44% of £7·50
   (c) 60% of £5·50

6. (a) 25% of 2 m 16 cm
   (b) 40% of 1 m 30 cm
   (c) 30% of 1 m 80 cm

7. (a) 15% of 11 m
   (b) 35% of 4 m 80 cm
   (c) 12% of 14 m

8. (a) 15% of 4 m
   (b) 12% of 5 m 50 cm
   (c) 8% of 7 m 50 cm

9. (a) 36% of 2 km 850 m
   (b) 30% of 3 km 420 m
   (c) 35% of 2 km 920 m

10. (a) $12\frac{1}{2}\%$ of £272
    (b) $16\frac{2}{3}\%$ of £198
    (c) $18\frac{3}{4}\%$ of £176

11. (a) $62\frac{1}{2}\%$ of £192
    (b) $58\frac{1}{3}\%$ of £204
    (c) $66\frac{2}{3}\%$ of £180

12. (a) $6\frac{2}{3}\%$ of £1350
    (b) $5\frac{5}{9}\%$ of £1710
    (c) $7\frac{1}{2}\%$ of £1200

13. (a) 120% of £60
    (b) 140% of £50
    (c) 125% of £56

14. (a) 160% of £90
    (b) 175% of £84
    (c) 180% of £80

15. (a) 150% of 76 cm
    (b) 135% of 80 cm
    (c) 144% of 75 cm

To write one quantity as a percentage of another quantity, first put them as a fraction. Each quantity must be in the same units. Then change the fraction into a percentage by multiplying by 100.

Example 6

Find (a) 15 as a percentage of 60.

      (b) £1·20 as a percentage of £6·00.

      (c) 225 mm as a percentage of 20 cm.

(a) Fraction is $\dfrac{15}{60}$.

$$\text{Percentage is } \frac{\cancel{15}^{1}}{\cancel{60}_{4}} \times \frac{\cancel{100}^{25}}{1}\% = \frac{25}{1} = 25\%$$

(b) Fraction is $\dfrac{£1\cdot20}{£6\cdot00} = \dfrac{120p}{600p}$

$$\text{Percentage is } \frac{\cancel{120}^{1}}{\cancel{600}_{5}\,_{1}} \times \frac{\cancel{100}^{20}}{1}\% = \frac{20}{1} = 20\%$$

(c) Fraction is $\dfrac{225\text{ mm}}{20\text{ cm}} = \dfrac{225\text{ mm}}{200\text{ mm}}$

$$\text{Percentage is } \frac{225}{\cancel{200}_{2}} \times \frac{\cancel{100}^{1}}{1}\% = \frac{225}{2}$$

$$= 112\tfrac{1}{2}\%$$

Exercise 1.9d

Find the 'odd answer out' by changing each to a percentage.

1. (a) £42 of £168
   (b) £45 of £225
   (c) £48 of £192
2. (a) £112 of £160
   (b) £120 of £150
   (c) £144 of £180
3. (a) £4·20 of £5·60
   (b) £4·50 of £6·00
   (c) £3·60 of £5·00
4. (a) £1·89 of £4·20
   (b) £2·10 of £5·25
   (c) £1·96 of £4·90
5. (a) 2 m 16 cm of 3 m 60 cm
   (b) 2 m 24 cm of 4 m
   (c) 2 m 22 cm of 3 m 70 cm
6. (a) 3 m 40 cm of 4 m
   (b) 3 m 24 cm of 3 m 60 cm
   (c) 3 m 42 cm of 3 m 80 cm
7. (a) £35 of £315
   (b) £36 of £288
   (c) £33 of £264
8. (a) £54 of £144
   (b) £49 of £147
   (c) £48 of £128
9. (a) 32 cm of 1 m 92 cm
   (b) 30 cm of 2 m 25 cm
   (c) 33 cm of 1 m 98 cm
10. (a) 33 cm of 3 m 96 cm
    (b) 28 cm of 3 m 36 cm
    (c) 25 cm of 4 m
11. (a) 1 km 680 m of 2 km 520 m
    (b) 1 km 750 m of 2 km 800 m
    (c) 1 km 800 m of 2 km 880 m
12. (a) £216 of £180
    (b) £242 of £220
    (c) £198 of £165
13. (a) £4·41 of £4·20
    (b) £4·50 of £3·60
    (c) £4·40 of £3·52
14. (a) 1 kg 800 g of 1 kg 250 g
    (b) 1 kg 820 g of 1 kg 300 g
    (c) 1 kg 680 kg of 1 kg 200 g
15. (a) 6720 ml of 3840 ml
    (b) 6930 ml of 3960 ml
    (c) 6750 ml of 3750 ml

Exercise 1.9e

1. In a box of 150 eggs, 20% were broken. How many whole eggs were there?
2. A factory has 1600 workers, but during a bus strike 15% were absent. Find the number of workers absent.
3. A school has 1850 pupils, and one day 4% of them arrived late. Find the number of those arriving late.
4. 2400 people live in the village of Bilton, and the results of a local election are given below.
   25% voted Conservative, 15% voted Liberal, 40% voted Labour, 20% did not vote.
   Convert these percentages to actual numbers.
5. A school has 1200 pupils and the percentage attendance records for a certain week are given below.
   Monday 90%   Thursday 98%
   Tuesday 85%   Friday 95%
   Wednesday 92%
   Find the number of pupils present on each day.
6. 720 buses operate from a depot, but 144 are being repaired. What percentage of the total are in service?
7. One day 132 trains arrive at a railway station and 99 of them are on time. What percentage of them are late?
8. An examination is marked out of 144 and one pupil gets 108. Find his mark as a percentage.
9. 48 000 live in Scarborough and 7200 are council house dwellers. What percentage of the population live in council houses?
10. 3200 people live in Swanland and the results of a local election are given below.
    1280 voted Conservative, 800 voted Liberal 960 voted Labour, 160 did not vote.
    Convert the figures to percentages.

A *discount* is a sum of money taken off an account. It is usually written as a percentage.

Example 1

Find (a) the discount of 5% on a bill of £6·50,
     (b) the amount actually paid.

(a)   Discount = 5% of £6·50

$$= \frac{5}{100} \times \frac{650}{1} \text{ pence}$$

$$= \frac{65}{2} = 32\frac{1}{2} \text{ pence}$$

(b)   Amount to be paid = £6·50 − £0·32$\frac{1}{2}$
$$= £6·17\frac{1}{2}$$

Exercise 2.1a

Find (a) the discount for the following cases and (b) the amount actually paid.

1. A car is priced in a catalogue at £2400, but the dealer offers a 20% discount to the purchaser.
2. A three-piece suite which has a catalogue price of £600, but is sold with a 15% discount during a sale.
3. A colour television, with a marked price of £400, which is sold with a 9% discount to a man who pays immediate cash.
4. A contractor makes an estimate of £250 for a cavity-wall insulation, but then offers an 8% discount to the house owner for immediate payment.
5. A man is to have his windows double-glazed and the firm's representative offers him a 12% discount on an estimate of £350.
6. A black-and-white television, with a marked price of £150 which is offered for sale at a 16% discount rate because the woodwork is slightly defaced.
7. An electric cooker of catalogue price £125, which is sold with a 12% discount to a man who is willing to collect it.

8. A record player of marked price £80, which is sold to a purchaser with a 15% discount in exchange for handing in his old record player to the dealer.
9. A washing machine of catalogue price £180 which is sold at a discount rate of 25% to a man who is willing to do his own delivering and installing.
10. A refrigerator of marked price £85, which is offered for sale with a 20% discount because the dealer is over-stocked.
11. A spin drier of marked price £60, which is sold with an 8% discount for immediate cash payment.
12. A man is to have his loft insulated. A contractor makes an estimate of £54, and offers a 5% discount for prompt payment on completion of the contract.
13. A bicycle of catalogue price £63 which is sold with a 6% discount because the paintwork is slightly scratched.
14. A radio of catalogue price £22 which is sold to a man with a 5% discount in exchange for his being able to do his own repairs and therefore forfeit the guarantee.
15. A steam iron is priced at £15, but is sold at 4% discount because it was part of a bulk purchase.

A *commission* is a sum of money received by an agent on the value of goods he sells. It is usually written as a percentage.

Example 2

The table shows the commission charged by an estate agent for selling a house.

Sales price	Commission
first £5000	2%
all over £5000	$1\frac{1}{4}$%

What is the total commission on a house sold for £12 500?

Commission on £5000   = 2% of £5000

$$= \frac{2}{100} \times \frac{5000}{1}$$

$$= £100$$

Remainder $= £12\,500 - £5000 = £7500$

Commission on remainder $= 1\frac{1}{4}\%$ of £7500

$$= \frac{1\frac{1}{4}}{100} \times \frac{7500}{1}$$

$$= \frac{5}{\underset{4}{\cancel{400}}} \times \frac{\cancel{7500}^{75}}{1}$$

$$= £\frac{375}{4}$$

$$= £93\frac{3}{4} = £93.75$$

So total commission $\quad = £100 + £93.75$

$$= £193.75$$

Exercise 2.1b

1. A man who sells newspapers receives for his wages 15% of the money he takes. If he sells 800 newspapers at 5p a copy, how much does he receive?
2. A man who sells refreshments at a kiosk receives a 5% commission on the value of his sales. If his takings are £180 on a certain day, how much commission does he receive?
3. A stock auctioneer receives a 4% commission on his sales. His takings for five days in a certain week are given below. Find the commission he receives for each day.
   Monday £1250   Wednesday £925   Friday £1025.
   Tuesday £1100   Thursday £775
4. An ice cream seller receives for his wages 16% of the money he takes. Over the three days of the Summer Bank Holiday his takings were:
   Saturday £150; Sunday £105; Monday £120.
   Find how much he earns on each day.
5. A door-to-door sales agent is paid commission on his sales at a rate of 2%. Find how much commission he receives for the sale of each of the following:
   (a)   Carpets costing £450
   (b)   A lounge suite costing £720
   (c)   A dining-room suite costing £635.
6. The commission that an estate agent charges on house sales is $2\frac{1}{2}\%$ up to £10 000 and 2% above that figure. Find the commission he receives for the sale of each property below.
   (a)   Selling price $= £7200$
   (b)   Selling price $= £9400$
   (c)   Selling price $= £12\,750$
   (d)   Selling price $= £15\,650$

Interest is the amount earned on money which is invested or the sum charged on money which is borrowed. It is usually written as a percentage of the money lent or borrowed for each year.

Example 3

(a)   £200 is invested at 4% interest. How much interest is earned in each year?

$$4\% \text{ of } £200 = \frac{4}{\cancel{100}} \times \frac{\cancel{200}}{1}$$

$$= £8 \text{ yearly.}$$

(b)   £480 is borrowed for 3 years at $7\frac{1}{2}\%$ interest. What is the total interest paid?

The interest is $7\frac{1}{2}\%$ of £480 each year.
The total interest is $3 \times (7\frac{1}{2}\%$ of £480)

$$= \frac{3}{1} \times \frac{7\frac{1}{2}}{100} \times \frac{480}{1}$$

$$= \frac{3}{1} \times \frac{\cancel{15}^{3}}{\underset{1}{\cancel{200}}} \times \frac{\cancel{480}^{12}}{1}$$

$$= £108$$

Exercise 2.1c

For the following find
(a)   the yearly interest,
(b)   the total interest chargeable after the time stated,
(c)   the debt (total interest + loan).

	Loan	Interest rate	Time
1.	£150	10%	4 years
2.	£120	15%	4 years
3.	£250	12%	3 years
4.	£350	16%	2 years
5.	£120	$12\frac{1}{2}\%$	4 years
6.	£240	$18\frac{3}{4}\%$	3 years
7.	£160	$21\frac{1}{4}\%$	3 years
8.	£210	15%	4 years
9.	£315	12%	2 years
10.	£125	9%	2 years

For the following find
(a)   the yearly interest,
(b)   the total interest payable after the time stated,
(c)   the value of the investor's savings (total interest + investment).

	Investment	Interest rate	Time
11.	£350	6%	4 years
12.	£280	5%	4 years
13.	£425	8%	3 years
14.	£360	$7\frac{1}{2}$%	3 years
15.	£112	$6\frac{1}{4}$%	5 years
16.	£256	$3\frac{1}{8}$%	4 years
17.	£160	$9\frac{3}{8}$%	5 years
18.	£450	9%	4 years
19.	£135	8%	2 years
20.	£360	4%	2 years

Simple interest is the interest calculated at a fixed yearly rate on the sum of money first borrowed or invested.

Example 4

Find the simple interest on £420 at 12% for 2 years 9 months.
Interest for 1 year = 12% of £420
Interest for 2 years 9 months (i.e. $2\frac{3}{4}$ years)

$$= 2\frac{3}{4} \times (12\% \text{ of } £420)$$

$$= \frac{11}{4} \times \frac{12}{100} \times \frac{420^{21}}{1}$$

$$= £\frac{693}{5} = £138\frac{3}{5} = £138 \cdot 60$$

Exercise 2.1d

For the following find
(a) the yearly simple interest,
(b) the total interest chargeable after the time stated,
(c) the debt (total interest + loan).

	Loan	Interest rate	Time
1.	£250	12%	1 year 6 months
2.	£320	15%	1 year 3 months
3.	£450	16%	2 years 3 months
4.	£150	12%	1 year 4 months
5.	£350	18%	2 years 4 months
6.	£180	15%	1 year 8 months
7.	£225	16%	1 year 2 months
8.	£160	15%	2 years 2 months

For the following find
(a) the yearly simple interest,
(b) the total interest payable after the time stated,
(c) the value of the investor's savings (total interest + investment).

	Investment	Interest rate	Time
9.	£150	8%	2 years 6 months
10.	£320	10%	1 year 9 months
11.	£120	5%	3 years 6 months
12.	£750	8%	3 years 4 months
13.	£250	6%	2 years 8 months
14.	£720	5%	1 year 10 months
15.	£450	8%	3 years 2 months

Compound interest is the interest calculated on the amount of money at the beginning of each year, where the amount is
original sum + interest earned.

Example 5

Find the compound interest on £200 for 3 years at 10%.

original sum   = £200
1st year interest = £ 20 (10% of £200)

amount = £220 (£200 + £20)
2nd year interest = £ 22 (10% of £220)

amount = £242 (£220 + £22)
3rd year interest = £ 24·20 (10% of £242)

amount = £266·20
original sum   = £200

∴ interest   = £ 66·20

Exercise 2.1e

For the following find (for the time stated):
(a) the total debt,
(b) the compound interest chargeable.

For the following find (for the time stated):
(a) the total value of the investor's savings,
(b) the compound interest payable.

	Loan	Compound interest rate	Time		Loan	Compound interest rate	Time
1.	£300	20%	4 years	6.	£500	10%	4 years
2.	£400	15%	3 years	7.	£400	5%	3 years
3.	£600	12%	2 years	8.	£1000	6%	2 years
4.	£500	16%	2 years	9.	£200	9%	2 years
5.	£200	18%	2 years	10.	£500	4%	2 years

A deposit is required when goods are bought on hire purchase. Simple interest is then charged on the remaining sum owing. The deposit is often stated as a percentage of the cash price.

Example 6

The cash price of a bicycle is £64. It can be bought on hire purchase with a deposit of 25% and 18 monthly payments of £3. What is the difference between the hire purchase price and the cash price?

$$\text{Deposit} = 25\% \text{ of } £64 = \frac{\cancel{25}^{1}}{\cancel{100}_{1}} \times \frac{\cancel{64}^{16}}{1} = £16$$

$$\text{Total monthly payments} = 18 \times 3 = £54$$

$$\therefore \text{ total hire purchase price} = £70 \qquad \text{i.e. } £54 + £16$$

$$\text{and difference} = £70 - £64 = £6$$

Exercise 2.1f

For the following find:
(a) the hire purchase price,
(b) the difference between the hire purchase price and the cash price.

	Article	Cash price	Deposit	Time to pay	Monthly amount
1.	cooker	£90	20%	1 year	£7
2.	spin dryer	£60	25%	2 years	£2
3.	record player	£75	12%	1 year	£6
4.	television	£135	20%	2 years	£5
5.	washing machine	£240	15%	2 years	£9
6.	kitchen suite	£450	16%	1 year 6 months	£24
7.	colour television	£350	12%	2 years	£15
8.	car	£2500	15%	2 years	£95
9.	motor cycle	£900	25%	1 year 6 months	£40
10.	van	£1800	16%	2 years	£70

Example 7

A second-hand car is priced at £1800. It may be bought on the following hire purchase terms:

deposit $33\frac{1}{3}$%; interest 12% per annum; repayment period 2 years in equal monthly instalments.

Find (a) the cost of each monthly instalment,
       (b) the total hire purchase price.

$$\text{Deposit} = 33\frac{1}{3}\% \text{ of } £1800 = \frac{33\frac{1}{3}}{100} \times \frac{1800}{1}$$

$$= \frac{100^1}{300} \times \frac{1800^{600}}{1} = £600$$

Amount remaining = £1800 − £600 = £1200

Interest on amount remaining = 2 × (12% of £1200)

$$= \frac{2}{1} \times \frac{12}{100} \times \frac{1200^{12}}{1}$$

$$= £288$$

Amount (excluding deposit) to be paid = £1200 + £288
$$= £1488$$
Cost of each monthly instalment = £1488 ÷ 24
$$= £62 \text{ per month.}$$
Total hire purchase price = deposit + balance + interest
$$= £600 + £1200 + £288$$
$$= £2088$$

Exercise 2.1g

For the following find:
(a) the cost of each monthly instalment,
(b) the total hire purchase price.

Article	Cash price	Deposit	Interest rate	Time to pay
1. car	£2400	20%	10%	2 years
2. car	£1800	15%	20%	1 year
3. van	£1200	30%	10%	2 years
4. motor cycle	£960	25%	15%	2 years
5. three-piece suite	£600	40%	10%	2 years
6. colour television	£300	20%	15%	2 years
7. cooker	£120	25%	20%	1 year
8. refrigerator	£75	20%	10%	2 years
9. television	£100	25%	12%	1 year
10. motor mower	£300	30%	20%	1 year

Cost price (wholesale price) is the amount a shopkeeper *pays* for his goods.

Selling price (retail price) is the amount a shopkeeper *receives* for his goods.

profit = (selling price) − (cost price)

A loss is made when the selling price is *less* than the cost price.

loss = (cost price) − (selling price)

Example 8

Find the profit or loss when
(a) eggs are bought at 5p each and sold for 65p a dozen.
(b) a tonne of potatatoes is bought for £60 and sold for 5p a kg.

(a)   Cost price of eggs = 12 × 5p = 60p a dozen
Selling price of eggs = 65p a dozen
∴ profit = (65 − 60)p = 5p a dozen

(b)   Cost price of potatoes = £60 a tonne
Selling price of potatoes = 1000 × 5p = £50 a tonne
∴ loss = £60 − £50 = £10 a tonne

Exercise 2.1h

For each question find the profit or loss on each single item.

Commodity	Retail price	Wholesale price
1. bars of chocolate	12p each	£9 per 100 bars
2. pencils	10p each	£40 per 500
3. post cards	9p each	£32 per 400
4. oranges	5p each	£12 per 200
5. grapefruits	8p each	£9 per 150
6. cigarettes	52p for twenty	£125 for 5000
7. sheets of paper	40p for eighty	£9 for 2000
8. tomatoes	60p per kg	£540 per tonne
9. apples	50p per kg	£400 per tonne
10. bananas	30p per kg	£360 per tonne
11. sugar	28p per kg	£15 for 50 kg
12. cheese	£1·30 per kg	£30 for 20 kg
13. butter	£1·20 per kg	£54 for 50 kg
14. wire	20p per m	£120 for $\frac{1}{2}$ km
15. carpet	£5·40 per m²	£240 for 50 m²
16. linoleum	£1·20 per m²	£75 for 50 m²
17. paraffin	12p per litre	£4·00 for 50 l
18. beer	60p per litre	£21·60 for 40 l
19. lemonade	18p per can	£3·60 for 24 cans
20. milk	10p per $\frac{1}{4}$ l cup	£2·40 for 15 l

Profit (or loss) is usually written as a percentage of the cost price (wholesale price).

Example 9

Find the percentage profit or loss when
(a) pens are bought at 6p each and sold at 75p a dozen
(b) cheese is bought at £1·25 per kg and sold at 10p per 100g

(a)   Cost price = 72p per dozen
     Selling price = 75p per dozen
     ∴ profit = 3p per dozen

$$\text{Profit as a fraction of cost price} = \frac{3}{72}$$

$$\therefore \text{percentage profit} = \frac{3^1}{72} \times \frac{100^{25}}{1} = \frac{25}{6} = 4\tfrac{1}{6}\%$$

(b)   Cost price = £1·25 per kg
     Selling price = £1·00 per kg
     ∴ loss = 25p per kg

$$\text{Loss as a fraction of cost price} = \frac{25}{125}$$

$$\therefore \text{percentage loss} = \frac{25^1}{125} \times \frac{100^{20}}{1} = 20\%$$

Exercise 2.1i

Find the percentage profit (or loss) for each question

Commodity	Retail price	Wholesale price
1. bars of chocolate	12p each	£9 per 100 bars
2. pencils	10p each	£40 per 500
3. post cards	9p each	£32 per 400
4. oranges	5p each	£12 per 200
5. grapefruits	8p each	£9 per 150
6. cigarettes	52p for twenty	£125 for 5000
7. sheets of paper	40p for eighty	£9 for 2000
8. tomatoes	60p per kg	£540 per tonne
9. apples	50p per kg	£400 per tonne
10. bananas	30p per kg	£360 per tonne
11. sugar	28p per kg	£15 for 50 kg
12. cheese	£1·30 per kg	£30 for 20 kg
13. butter	£1·20 per kg	£54 for 50 kg
14. wire	20p per m	£120 for $\tfrac{1}{2}$ km
15. carpet	£5·40 per m²	£240 for 50 m²
16. linoleum	£1·20 per m²	£75 for 50 m²
17. paraffin	12p per litre	£4·00 for 50 l
18. beer	60p per litre	£21·60 for 40 l
19. lemonade	18p per can	£3·60 for 24 cans
20. milk	10p per $\tfrac{1}{4}$ l cup	£2·40 for 15 l

The *average* (or *mean*) of a set of numbers is the sum of all the numbers divided by how many numbers there are in the set.

Example 1

Find the average of 2, 4, 6, 8, 10, 12.

Sum = 2 + 4 + 6 + 8 + 10 + 12 = 42
number = 6
∴ average = 42 ÷ 6 = 7

Exercise 2.2a

1. The weights of four boys are 53, 50, 55, and 58 kilograms. Find their average weight.
2. The weights of four girls are 36, 31, 34, and 35 kilograms. Find their average weight.
3. The heights of six men are 180, 175, 161, 185, 190, and 183 centimetres. Find their average height.
4. The heights of six women are 162, 156, 165, 153, 150 and 156 cm. Find their average height.
5. The numbers of pupils present in class on the five days of a certain week were: Monday 35; Tuesday 32; Wednesday 34; Thursday 36; Friday 33. Find the average daily attendance figure for the week.
6. The number of pupils who had a school lunch on the five days of a certain week were: Monday 331; Tuesday 327; Wednesday 331; Thursday 329; Friday 332. Find the average daily number for the week.
7. Over two cricket matches a batsman scores the following number of runs: 87, 33, 14, and 18. Find his average score.
8. Over three cricket matches a batsman makes the following scores: 15, 12, 16, 0, 23, and 18 runs. Find the average score.
9. On five consecutive days a certain train arrived at the same station and was late by the following numbers of minutes: 5, 0, 11, 1, and 58. Find the average lateness.
10. Six bus journeys from Manchester to Bury took the following times: 34, 38, 39, 36, 34, and 35 minutes. Find the average journey time.
11. Four train journeys from London to Brighton took the following times: 1 h 7 min, 1 h 4 min, 1 h 2 min, and 1 h 3 min. Find the average journey time.
12. Over a four-week period a man earned the following wages: £37·02; £39·53, £41·21, and £40·76. Find his average weekly wage.
13. The chest dimensions of five boys are 83, 81, 85, 88 and 83 cm. Find their average chest dimension.
14. The waist dimensions of five girls are 68, 61, 59, 58 and 64 cm. Find their average waist dimension.
15. Over a six month period the number of wet days in each of the months were: 5, 15, 3, 7, 12, and 6. Find the average number of wet days per month.
16. Over a six week period the number of hours of sunshine in each of the weeks were: 34, 21, 11, 4, 0, and 32. Find the weekly average number of hours of sunshine.
17. I visit a greengrocer's shop on five occasions and buy 1 kg of apples every time. The prices I have to pay are 52p, 50p, 42p, 46p, and 50p. What is the average price I have to pay?
18. My bills for electricity in each quarter of last year were as follows: £15·10, £24·16, £49·95, and £28·35. What was the average quarterly charge?
19. A shop employs eight staff. Their wages are as follows:
    2 are paid £34 per week.
    2 are paid £24 per week.
    4 are paid £15 per week.
    Find the total wages bill of the shop and the average weekly wage of the staff.
20. A restaurant employs twelve staff. Their wages are as follows:
    2 are paid £42 per week.
    7 are paid £36 per week.
    3 are paid £24 per week.
    Find the total wages bill of the restaurant and the average weekly wage of the staff.

If the average of a set of numbers is known, then the sum of all these numbers can be calculated.

Example 2

The average of a set of 15 numbers is 12·4. What is the sum of the numbers?

average = 12·4
number of numbers = 15
∴ sum = 12·4 × 15 = 186

Exercise 2.2b

1. The average score per match of a rugby club over a whole season was 19 points. If 32 matches were played during that season, find the total number of points scored.

2. A shop employs 12 staff and the average weekly wage is £28 per person. Find the total weekly wage bill.

3. A factory employs 150 men and the average weekly wage is £39 per man. Find the total weekly wage bill.

4. Over a six week period a man's average weekly wage is £41·32. Find the total amount that he earns over this period.

5. I make six journeys from London to Dover by train and the average time of the journeys is 1 h 25 min. What is my total journey time?

6. The average attendance at a football club for one season is 12 560. What is the total attendance figure if the club plays 30 home matches in that season?

7. I find that I have to replace a light bulb in my house on average every 73 days. If I buy 15 bulbs at once, how long approximately should I expect them to last?

8. After being planted, a tree is found to grow at an average rate of 5 cm per month. How tall would you expect the tree to be after 50 years?

9. Over one season the average scoring rate of a footballer is 1·2 goals per match. What is his seasonal total if he makes 45 appearances.

10. I make 12 bus journeys to and from work each week. In one week I find that the average journey time is 32·5 minutes. Find the total time that I spend travelling to and from work during that particular week.

Example 3

The average height of a football team is 176 cm. The average height without the goalkeeper is 175 cm. What is the height of the goalkeeper?

Total height of team
$$= 176 \times 11 = 1936 \text{ cm}$$
Total height of 10 players
$$= 175 \times 10 = 1750 \text{ cm}$$
∴ height of goalkeeper
$$= 1936 - 1750 = 186 \text{ cm}$$

Exercise 2.2c

1. The average height of four boys is 164 cm. One of the boys is taller than the other three: if he is left out, the average height of the other three is 161 cm. Find the height of the tall boy.

2. The average weight of six girls is 50 kg. If the lightest girl is left out, the average weight of the other five is 53 kg. Find the weight of the lightest girl.

3. For the five months May to September there were on average 9 wet days per month. If September is left out the average number of wet days per month was 5. How many wet days were there in September?

4. In one week from Monday to Saturday it took me on average 23 minutes to drive to work. If Saturday is left out, the average time for the other five days was 24 minutes. Find how long my journey took on Saturday.

5. From Monday to Saturday the average number of cars using a ferry was 66. On Saturday there were many day trippers, so the average for the other five days was only 52. How many cars used the ferry on Saturday?

6. Over five days the average midday temperature was 21 °C. However on one of these days the thermometer was placed in direct sunlight so the reading was faulty. If the average temperature for the other four days was 19 °C, what was the faulty reading?

7. An evening newspaper appears Monday to Saturday inclusive. During a certain week the average daily sales were 39 000. If the Saturday figure was left out, the average was only 32 000. How many copies were sold on the Saturday?

8. Over a four week period the average weekly wage of a bricklayer was £51·64. One of these weeks was very wet and his average weekly wage for the other three was £56·40. Find his wage for the wet week.

9. Over ten matches a footballer scored an average of 1·4 goals. In one of these matches the opposing goalkeeper received an injury. If his average for the other nine matches was 1·0 goals, find how many goals he scored when the goalkeeper was injured.

10. During a certain week the average number of hours of sunshine each day was 6·0. If Saturday is left out, the average for the other six days was 6·8. Find the number of hours of sunshine for the Saturday.

Average speed is the ratio of the distance travelled to the time taken. It is usually calculated in kilometres per hour (km/h).

Example 4

A car travels 144 km in $2\frac{1}{4}$ h. What is the average speed?

$$\text{Average speed} = \frac{\text{distance}}{\text{time}}$$

$$= \frac{144}{2\frac{1}{4}}$$

$$= \frac{\overset{16}{144} \times 4}{\underset{1}{9}} = 64 \text{ km/h.}$$

Exercise 2.2d

Find the average speed for the following air flights:

1. London to Gibraltar, 1896 km in 3 h.
2. London to Rome, 1464 km in 3 h.
3. London to Montreal, 5040 km in 7 h.
4. London to New York, 5184 km in 6 h.
5. London to Havana, 6952 km in 11 h.

Find the average speed for the following sea voyages:

6. Hull to Hamburg, 600 km in 15 h.
7. Hull to Rotterdam, 336 km in 14 h.
8. Liverpool to New York, 4800 km in 5 days.
9. Liverpool to Monte Video, 10 080 km in 10 days.
10. Southampton to Lisbon, 2112 km in 2 days.

Find the average speed for the following rail journeys:

11. London to Bristol, 192 km in $1\frac{1}{2}$ h.
12. London to Peterborough, 120 km in $1\frac{1}{4}$ h.
13. London to Bournemouth, 175 km in $1\frac{3}{4}$.
14. London to Manchester, 300 km in $2\frac{1}{2}$ h.
15. London to Plymouth, 364 km in $3\frac{1}{2}$ h.
16. Liverpool to Scunthorpe, 190 km in 2 h 30 min.
17. Lincoln to Northampton, 153 km in 2 h 15 min.
18. Leeds to Hull, 84 km in 1 h 20 min.
19. Leeds to Penzance, 646 km in 9 h 30 min.
20. Doncaster to Inverness, 625 km in 8 h 20 min.

Example 5

A lorry leaves Scunthorpe at noon and arrives at Grimsby Docks at 1.18 p.m. If the distance is 52 km, what is the average speed of the lorry?

Distance = 52 km
Time = 1 h 18 min $= 1\frac{18}{60}$ h

$$\text{Average speed} = \frac{\text{distance}}{\text{time}} = \frac{52}{1\frac{18}{60}}$$

$$= 52 \div 1\frac{18}{60}$$

$$= 52 \div \frac{78}{60}$$

$$= \frac{\overset{4}{52}}{1} \times \frac{\overset{10}{60}}{\underset{1}{78}} = 40 \text{ km/h}$$

Exercise 2.2e

1. A train leaves London at 9 a.m. and arrives in Southampton at 10.20 a.m. If the distance is 120 km, find the average speed.
2. A bus leaves Leeds at 10 a.m. and arrives in York at 11.10 a.m. If the distance is 42 km, find the average speed.
3. A lorry leaves a London warehouse at 8.30 a.m. and arrives at a factory in Reading at 9.36 a.m. If the distance is 55 km, find the average speed.
4. A car leaves Manchester at 2.00 p.m. and arrives at Lancaster, 80 km away, at 3.40 p.m. Find the average speed.
5. A train leaves London at 8.00 a.m. and travels 231 km to Cardiff by 10.20 a.m. Find the average speed.
6. A train leaves London at 5.00 p.m. and arrives in Liverpool at 7.40 p.m. If the distance is 312 km, find the average speed.
7. A boat leaves Stranraer at 6.30 a.m. and reaches Larne at 8.35 a.m. having sailed a distance of 50 km. Find the average speed.
8. A bus leaves Newcastle at 7.00 a.m. and travels 27 km to Durham, arriving there at 7.54 a.m. Find the average speed.
9. A car leaves London at 6.00 p.m. and arrives in Leicester at 7.48 p.m. If the car travels 153 km between the two cities, find its average speed.
10. A bus leaves Birmingham at 9.05 a.m. and arrives in Ludlow at 10.55 a.m. If the distance is 66 km, find the average speed.

The square of $3 = 3^2 = 3 \times 3 = 9$
The square of $40 = 40^2 = 40 \times 40 = 1600$
The square of $9000 = 9000^2 = 9000 \times 9000$
$$= 81\,000\,000$$
The square of $0 \cdot 05 = 0 \cdot 05^2 = 0 \cdot 05 \times 0 \cdot 05$
$$= 0 \cdot 0025$$
The square of $2\frac{1}{4} = (2\frac{1}{4})^2 = 2\frac{1}{4} \times 2\frac{1}{4}$
$$= \frac{9}{4} \times \frac{9}{4} = \frac{81}{16} = 5\frac{1}{16}$$

Exercise 2.3a

Find the squares of the following.

1. 2	2. 5	3. 9	4. 20
5. 60	6. 70	7. 400	8. 800
9. 100	10. 2000	11. 5000	12. 6000
13. $0 \cdot 5$	14. $0 \cdot 8$	15. $0 \cdot 1$	16. $0 \cdot 03$
17. $0 \cdot 02$	18. $0 \cdot 06$	19. $\frac{1}{4}$	20. $\frac{1}{7}$
21. $\frac{1}{12}$	22. $\frac{3}{4}$	23. $\frac{2}{5}$	24. $\frac{4}{11}$
25. $1\frac{1}{5}$	26. $1\frac{1}{3}$	27. $1\frac{1}{4}$	28. $1\frac{2}{3}$
29. $2\frac{1}{5}$	30. $3\frac{1}{3}$	31. $5\frac{1}{2}$	32. $3\frac{1}{2}$
33. $4\frac{1}{2}$	34. $1\frac{1}{10}$	35. $3\frac{2}{3}$	36. $1\frac{4}{5}$
37. $2\frac{2}{5}$	38. $1\frac{3}{5}$	39. $1\frac{2}{9}$	40. $1\frac{1}{11}$
41. 13	42. 16	43. 23	44. 29
45. 31	46. $1 \cdot 7$	47. $1 \cdot 9$	48. $2 \cdot 6$
49. $3 \cdot 5$	50. $4 \cdot 1$		

The square root of $9 = \sqrt{9} = 3$
$$\text{because } 3 \times 3 = 9$$
The square root of $400 = \sqrt{400} = 20$
$$\text{because } 20 \times 20 = 400$$
The square root of $0 \cdot 01 = \sqrt{0 \cdot 01} = 0 \cdot 1$
$$\text{because } 0 \cdot 1 \times 0 \cdot 1 = 0 \cdot 01$$
The square root of $2\frac{1}{4} = \sqrt{2\frac{1}{4}}$
$$= \sqrt{\frac{9}{4}}$$
$$= \frac{\sqrt{9}}{\sqrt{4}} = \frac{3}{2} = 1\frac{1}{2}$$
$$\text{because } 1\frac{1}{2} \times 1\frac{1}{2} = 2\frac{1}{4}$$
The square root of $256 = \sqrt{256} = \sqrt{(4 \times 64)}$
$$= 2 \times 8$$
$$= 16$$
$$\text{because } 16 \times 16 = 256$$

The square root of $\sqrt{12 \cdot 25} = \sqrt{(0 \cdot 25 \times 49)}$
$$= 0 \cdot 5 \times 7$$
$$= 3 \cdot 5$$
$$\text{because } 3 \cdot 5 \times 3 \cdot 5 = 12 \cdot 25$$

Exercise 2.3b

Find the square roots of the following.

1. 16	2. 36	3. 49	4. 144
5. 900	6. 100	7. 2500	8. 12 100
9. 360 000	10. 810 000	11. 490 000	12. 250 00(
13. $0 \cdot 36$	14. $0 \cdot 49$	15. $0 \cdot 04$	16. $0 \cdot 0081$
17. $0 \cdot 0064$	18. $0 \cdot 0016$	19. $\frac{1}{9}$	20. $\frac{1}{64}$
21. $\frac{1}{121}$	22. $\frac{9}{25}$	23. $\frac{49}{64}$	24. $\frac{25}{144}$
25. $6\frac{1}{4}$	26. $7\frac{1}{9}$	27. $3\frac{1}{16}$	28. $1\frac{24}{25}$
29. $5\frac{4}{9}$	30. $3\frac{13}{36}$	31. $1\frac{13}{36}$	32. $1\frac{19}{81}$
33. $2\frac{2}{49}$	34. $1\frac{17}{64}$	35. $1\frac{15}{49}$	36. $7\frac{9}{16}$
37. $44\frac{4}{9}$	38. 324	39. 196	40. 576
41. 441	42. 729	43. 1089	44. 225
45. 625	46. 2025	47. $1 \cdot 44$	48. $1 \cdot 21$
49. $2 \cdot 56$	50. $4 \cdot 84$		

Exercise 2.3c

In each of the following, state which is the greater, and by how much.

1. $\sqrt{81}$ or 2^2	2. $\sqrt{144}$ or 3^2
3. $\sqrt{196}$ or 5^2	4. $\sqrt{625}$ or 4^2
5. $\sqrt{1 \cdot 44}$ or $1 \cdot 1^2$	6. $\sqrt{0 \cdot 81}$ or $0 \cdot 8^2$
7. $\sqrt{12 \cdot 96}$ or $1 \cdot 9^2$	8. $\sqrt{98 \cdot 01}$ or $3 \cdot 1^2$
9. $\sqrt{39 \cdot 69}$ or $2 \cdot 5^2$	10. $\sqrt{7 \cdot 29}$ or $1 \cdot 6^2$
11. $\sqrt{23 \cdot 04}$ or $2 \cdot 2^2$	12. $\sqrt{29 \cdot 16}$ or $2 \cdot 3^2$
13. $\sqrt{10 \cdot 24}$ or $1 \cdot 8^2$	14. $\sqrt{19 \cdot 36}$ or $2 \cdot 1^2$
15. $\sqrt{31 \cdot 36}$ or $2 \cdot 4^2$	16. $\sqrt{17 \cdot 64}$ or $2 \cdot 05^2$
17. $\sqrt{12 \cdot 25}$ or $1 \cdot 9^2$	18. $\sqrt{100}$ or $3 \cdot 2^2$
19. $\sqrt{10\,000}$ or $9 \cdot 9^2$	20. $\sqrt{2500}$ or $7 \cdot 1^2$
21. $\sqrt{1600}$ or $6 \cdot 3^2$	22. $\sqrt{400}$ or $4 \cdot 4^2$
23. $\sqrt{12\,100}$ or $10 \cdot 5^2$	24. $\sqrt{19\,600}$ or 12^2
25. $\sqrt{6400}$ or 9^2	26. $\sqrt{14\,400}$ or 11^2
27. $\sqrt{48\,400}$ or 15^2	28. $\sqrt{90\,000}$ or 17^2
29. $\sqrt{40\,000}$ or 14^2	30. $\sqrt{4225}$ or 8^2

The cube of $7 = 7^3 = 7 \times 7 \times 7 = 343$

The cube of $0.8 = 0.8^3 = 0.8 \times 0.8 \times 0.8 = 0.512$

The cube of $\frac{1}{3} = \left(\frac{1}{3}\right)^3 = \frac{1}{3} \times \frac{1}{3} \times \frac{1}{3} = \frac{1}{27}$

Exercise 2.3d

Find the cubes of the following.

1. 3   2. 5   3. 9   4. 11
5. 12   6. 15   7. 30   8. 60
9. $\frac{1}{2}$ 10. $\frac{1}{4}$ 11. $\frac{1}{8}$ 12. $\frac{1}{7}$
13. $\frac{2}{3}$ 14. $\frac{4}{5}$ 15. $1\frac{1}{2}$ 16. $2\frac{1}{2}$
17. 0.1   18. 0.4   19. 0.9   20. 0.6

The cube root of $125 = \sqrt[3]{(125)} = 5$

because $5 \times 5 \times 5 = 125$

The cube root of $0.216 = \sqrt[3]{(0.216)} = 0.6$

because $0.6 \times 0.6 \times 0.6 = 0.216$

The cube root of $\frac{27}{125} = \sqrt[3]{\left(\frac{27}{125}\right)} = \frac{\sqrt[3]{27}}{\sqrt[3]{125}} = \frac{3}{5}$

because $\frac{3}{5} \times \frac{3}{5} \times \frac{3}{5} = \frac{27}{125}$

Exercise 2.3e

Find the cube roots of the following.

1. 8   2. 64   3. 512   4. 216
5. 1000   6. 8000   7. 64 000   8. 125 000
9. $\frac{1}{216}$ 10. $\frac{1}{343}$ 11. $\frac{1}{729}$ 12. $\frac{27}{64}$
13. $\frac{8}{729}$ 14. $\frac{125}{216}$ 15. $1\frac{61}{64}$ 16. $4\frac{12}{125}$
17. 0.008   18. 0.027   19. 0.125   20. 0.343

Exercise 2.3f

In each of the following, state which is the greater and by how much.

1. $\sqrt[3]{216\,000}$ or 4^3 2. $\sqrt[3]{27\,000}$ or 3^3
3. $\sqrt[3]{1\,000\,000}$ or 5^3 4. $\sqrt[3]{0.343}$ or 0.9^3
5. $\sqrt[3]{0.008}$ or 0.6^3 6. $\sqrt[3]{0.064}$ or 0.7^3

Numbers may be written in standard form thus: $a \times 10^n$ where a is a number between 1 and 10 and n is a whole number.

Example 1

Write in standard form:

(a) 186 000;   (b) 32.76

(a) $186\,000 = 1.86 \times 100\,000$
$= 1.86 \times 10 \times 10 \times 10$
$\times 10 \times 10$
$= 1.86 \times 10^5$

(b) $32.76 = 3.276 \times 10$
$= 3.276 \times 10^1$

Exercise 2.3g

Write the following numbers in standard form:

1. 800 000   2. 360 000   3. 548 000
4. 50 000   5. 35 000   6. 22 400
7. 9000   8. 7500   9. 2840
10. 1563   11. 700   12. 290
13. 342   14. 186.3   15. 80
16. 36   17. 29.8   18. 4
19. 8.2   20. 7.36   21. 1.23
22. 427.4   23. 1169.2   24. 4000.6
25. 1111 111   26. 2020 206   27. 10 000 001
28. 40 060 007   29. 186 000   30. 10.0001

Example 2

Write as ordinary numbers:

(a) 2.34×10^3;   (b) 4.00×10^6

(a) $2.34 \times 10^3 = 2.34 \times 10 \times 10 \times 10$
$= 2.34 \times 1000$
$= 2340$

(b) 4.00×10^6
$= 4.00 \times 10 \times 10 \times 10 \times 10 \times 10 \times 10$
$= 4.00 \times 1\,000\,000$
$= 4\,000\,000$

Exercise 2.3h

Write the following as ordinary numbers.

1. 7×10^5 2. 8.25×10^5
3. 9.663×10^5 4. 5.001×10^5
5. 8×10^4 6. 6.3×10^4
7. 7.506×10^4 8. 6×10^3
9. 3.91×10^3 10. 5.376×10^3
11. 4.5×10^2 12. 9.38×10^2
13. 8.837×10^2 14. 6.206×10^2
15. 5×10^1 16. 7.28×10^1
17. 1.532×10^1 18. 5.086×10^1
19. 9.51×10^0 20. 4.537×10^0
21. 1.6×10^0 22. 1.06×10^1
23. 4.82×10^2 24. 4.007×10^5
25. 2.7×10^6 26. 9.91×10^5
27. 1.02×10^7 28. 9.436×10^7
29. 8.42×10^8 30. 9.63×10^{10}

Always press the $\boxed{\text{AC}}$ key on your calculator before you start a new calculation.

If your calculator does not have an $\boxed{\text{AC}}$ key, press the $\boxed{\text{C}}$ key instead.

To add, subtract, multiply or divide numbers, enter the numbers and signs in the given order and finally press the $\boxed{\text{=}}$ key.

Example 1

Calculate (a) $827 + 286 + 45$
         (b) $203 + 191 - 99$

and check your answers with a rough estimate.

(a) $\boxed{\text{AC}}$ $\boxed{8}\boxed{2}\boxed{7}$ $\boxed{+}$ $\boxed{2}\boxed{8}\boxed{6}$ $\boxed{+}$ $\boxed{4}\boxed{5}$ $\boxed{=}$ 1158

(b) $\boxed{\text{AC}}$ $\boxed{2}\boxed{0}\boxed{3}$ $\boxed{+}$ $\boxed{1}\boxed{9}\boxed{1}$ $\boxed{-}$ $\boxed{9}\boxed{9}$ $\boxed{=}$ 295

For the rough estimate, write each number correct to one significant figure and repeat the calculation without using your calculator.

(a) $827 + 286 + 45$ $800 + 300 + 50 = 1150$ ✓
(b) $203 + 191 - 99$ $200 + 200 - 100 = 300$ ✓

Exercise 2.4a

Use your calculator to evaluate each of the following. Check your answers with a rough estimate.

1. $92 + 77 + 43$	**2.** $58 + 37 + 61$	**3.** $22 + 86 + 14$
4. $73 + 44 + 55$	**5.** $61 + 35 + 99$	**6.** $76 + 32 + 8$
7. $320 + 540 + 790$	**8.** $430 + 810 + 690$	**9.** $413 + 331 + 604$
10. $121 + 237 + 719$	**11.** $487 + 396 + 578$	**12.** $867 + 189 + 392$
13. $583 + 291 + 632$	**14.** $476 + 322 + 213$	**15.** $918 + 484 + 191$
16. $568 + 734 + 87$	**17.** $327 + 261 + 93$	**18.** $754 + 229 + 35$
19. $687 + 86 + 7$	**20.** $970 + 99 + 6$	**21.** $860 - 390$
22. $776 - 369$	**23.** $532 - 244$	**24.** $629 - 148$
25. $821 - 478$	**26.** $792 - 506$	**27.** $913 - 84$
28. $312 + 487 - 129$	**29.** $534 + 279 - 325$	**30.** $292 + 987 - 315$

Example 2

Calculate (a) $3 \cdot 7 + 28 \cdot 64 + 1 \cdot 23$
         (b) $49 \cdot 1 - 5 \cdot 276$

and check your answers with a rough estimate.

(a) $\boxed{\text{AC}}$ $\boxed{3}\boxed{\cdot}\boxed{7}$ $\boxed{+}$ $\boxed{2}\boxed{8}\boxed{\cdot}\boxed{6}\boxed{4}$ $\boxed{+}$ $\boxed{1}\boxed{\cdot}\boxed{2}\boxed{3}$ $\boxed{=}$ 33·57

(b) $\boxed{\text{AC}}$ $\boxed{4}\boxed{9}\boxed{\cdot}\boxed{1}$ $\boxed{-}$ $\boxed{5}\boxed{\cdot}\boxed{2}\boxed{7}\boxed{6}$ $\boxed{=}$ 43·824

For the rough estimate, write each number correct to one significant figure and repeat the calculation without using your calculator.

(a) $3 \cdot 7 + 28 \cdot 64 + 1 \cdot 23 \approx 4 + 30 + 1 = 35$ ✓
(b) $48 \cdot 1 - 5 \cdot 276 \approx 50 - 5 = 45$ ✓

Exercise 2.4b

Use your calculator to evaluate each of the following. Check your answers with a rough estimate.

1. $1 \cdot 8 + 3 \cdot 7 + 2 \cdot 4$
2. $5 \cdot 3 + 2 \cdot 9 + 4 \cdot 1$
3. $8 \cdot 1 + 6 \cdot 5 + 3 \cdot 9$
4. $4 \cdot 5 + 7 \cdot 2 + 9 \cdot 7$
5. $13 \cdot 2 + 17 \cdot 8 + 31 \cdot 4$
6. $58 \cdot 9 + 36 \cdot 7 + 21 \cdot 3$
7. $24 \cdot 7 + 41 \cdot 9 + 67 \cdot 2$
8. $48 \cdot 6 + 32 \cdot 4 + 60 \cdot 8$
9. $62 \cdot 9 + 89 \cdot 1 + 45 \cdot 3$
10. $77 \cdot 5 + 52 + 49$
11. $32 \cdot 8 + 78 + 45$
12. $1 \cdot 73 + 5 \cdot 24 + 6 \cdot 41$
13. $4 \cdot 26 + 6 \cdot 93 + 2 \cdot 67$
14. $5 \cdot 62 + 3 \cdot 16 + 2 \cdot 58$
15. $6 \cdot 18 + 7 \cdot 9 + 3 \cdot 8$
16. $5 \cdot 19 + 6 \cdot 7 + 8 \cdot 3$
17. $23 \cdot 24 + 31 \cdot 42 + 29 \cdot 73$
18. $18 \cdot 47 + 43 \cdot 14 + 24 \cdot 36$
19. $52 \cdot 76 + 37 \cdot 25 + 45 \cdot 19$
20. $56 \cdot 35 + 31 \cdot 2 + 43 \cdot 7$
21. $87 \cdot 22 + 44 \cdot 1 + 36$
22. $66 \cdot 42 + 8 \cdot 83 + 4 \cdot 27$
23. $51 \cdot 6 + 4 \cdot 73 + 9 \cdot 24$
24. $48 \cdot 79 + 6 \cdot 9 + 8 \cdot 3$
25. $63 + 13 \cdot 27 + 8 \cdot 44$
26. $38 \cdot 89 - 11 \cdot 27$
27. $92 \cdot 82 - 49 \cdot 27$
28. $52 \cdot 86 - 19 \cdot 4$
29. $78 \cdot 6 - 27 \cdot 35$
30. $92 - 54 \cdot 63$
31. $8 \cdot 456 - 2 \cdot 723$
32. $6 \cdot 589 - 1 \cdot 047$
33. $57 \cdot 81 - 8 \cdot 632$
34. $91 \cdot 7 - 4 \cdot 916$
35. $60 \cdot 2 - 6 \cdot 86$
36. $82 - 8 \cdot 814$
37. $5 \cdot 32 - 1 \cdot 156$
38. $6 \cdot 7 - 2 \cdot 612$
39. $9 \cdot 8 - 2 \cdot 053$
40. $8 - 3 \cdot 598$

Answers to calculations often include a large number of figures. However, answers are usually required to a suitable number of significant figures or decimal places.

Example 3

(a) Calculate 517×36 and give the answer correct to three significant figures.
(b) Calculate $3 \cdot 25 \div 0 \cdot 32$ and give the answer correct to three significant figures.

(a) [AC] [5] [1] [7] [X] [3] [6] [=] $18\ 612 = 18\ 600$ to three S.F.

(b) [AC] [3] [.] [2] [5] [÷] [0] [.] [3] [2] [=] $10 \cdot 156\ 25 = 10 \cdot 2$ to three S.F.

Exercise 2.4c

Calculate each of the following and give the answers correct to three significant figures.

1. 896×712
2. 742×652
3. 516×492
4. 867×125
5. 532×224
6. 992×196
7. 584×368
8. 672×294
9. 864×169
10. 2204×29
11. 3004×32
12. 1036×94
13. 1325×47
14. 4216×19
15. 1382×52
16. 844×63
17. $971\ 540 \div 155$
18. $857\ 152 \div 472$
19. $702\ 168 \div 136$
20. $993\ 882 \div 302$
21. $536\ 400 \div 225$
22. $861\ 102 \div 198$
23. $81\ 465 \div 250$
24. $38\ 773 \div 350$
25. $303\ 615 \div 540$
26. $117\ 366 \div 150$
27. $6322 \cdot 32 \div 18$
28. $9907 \cdot 48 \div 46$
29. $8068 \cdot 84 \div 59$
30. $9177 \cdot 12 \div 36$
31. $9719 \cdot 34 \div 23$
32. $10\ 970 \cdot 3 \div 55$
33. $69 \cdot 9616 \div 1 \cdot 6$
34. $97 \cdot 5117 \div 3 \cdot 3$
35. $95 \cdot 0508 \div 2 \cdot 7$
36. $73 \cdot 0422 \div 4 \cdot 2$
37. $25 \cdot 9792 \div 0 \cdot 8$
38. $85 \cdot 284 \div 1 \cdot 5$
39. $86 \cdot 502 \div 3 \cdot 9$
40. $99 \cdot 456 \div 2 \cdot 8$
41. $65 \cdot 756 \div 3 \cdot 4$
42. $52 \cdot 825 \div 1 \cdot 25$
43. $36 \cdot 06 \times 0 \cdot 12$
44. $18 \cdot 58 \times 0 \cdot 46$
45. $15 \cdot 26 \times 0 \cdot 51$
46. $59 \cdot 91 \times 0 \cdot 09$
47. $40 \cdot 06 \times 0 \cdot 22$
48. $17 \cdot 17 \times 0 \cdot 35$
49. $10 \cdot 05 \times 0 \cdot 54$
50. $6 \cdot 72 \times 0 \cdot 95$

A calculator can easily be used for changing a common fraction to a decimal. The numerator is simply divided by the denominator.

Example 4

Change into decimal form (a) $\frac{9}{25}$ (b) $2\frac{3}{32}$

(a) $\boxed{\text{AC}}$ $\boxed{9}$ $\boxed{\div}$ $\boxed{2}\boxed{5}$ $\boxed{=}$ 0·36

(b) $\boxed{\text{AC}}$ $\boxed{3}$ $\boxed{\div}$ $\boxed{3}\boxed{2}$ $\boxed{+}$ $\boxed{2}$ $\boxed{=}$ 2·09375

Exercise 2.4d

Change each of the following into decimal form.

1. $\frac{6}{25}$ 2. $\frac{8}{25}$ 3. $\frac{18}{25}$ 4. $\frac{2}{25}$ 5. $\frac{21}{50}$ 6. $\frac{33}{50}$ 7. $\frac{49}{50}$ 8. $\frac{3}{50}$

9. $\frac{36}{125}$ 10. $\frac{64}{125}$ 11. $\frac{112}{125}$ 12. $\frac{1}{125}$ 13. $\frac{81}{500}$ 14. $\frac{361}{500}$ 15. $\frac{5}{32}$ 16. $\frac{27}{32}$

17. $\frac{1}{32}$ 18. $1\frac{4}{25}$ 19. $1\frac{16}{25}$ 20. $1\frac{1}{25}$ 21. $1\frac{9}{50}$ 22. $1\frac{32}{125}$ 23. $2\frac{3}{25}$ 24. $2\frac{27}{50}$

Example 5

Change into decimal form (a) $\frac{4}{15}$ (b) $1\frac{25}{36}$

(a) $\boxed{\text{AC}}$ $\boxed{4}$ $\boxed{\div}$ $\boxed{1}\boxed{5}$ $\boxed{=}$ 0·266 666 6 = $0.2\dot{6}$

(b) $\boxed{\text{AC}}$ $\boxed{2}\boxed{5}$ $\boxed{\div}$ $\boxed{3}\boxed{6}$ $\boxed{+}$ $\boxed{1}$ $\boxed{=}$ 1·694 444 4 = $1.69\dot{4}$

Exercise 2.4e

Change each of the following into decimal form.

1. $\frac{17}{30}$ 2. $\frac{29}{30}$ 3. $\frac{14}{15}$ 4. $\frac{2}{15}$ 5. $\frac{1}{30}$ 6. $\frac{1}{15}$ 7. $\frac{13}{18}$ 8. $\frac{11}{18}$

9. $\frac{17}{18}$ 10. $\frac{32}{75}$ 11. $\frac{56}{75}$ 12. $\frac{13}{75}$ 13. $\frac{49}{150}$ 14. $\frac{121}{150}$ 15. $\frac{13}{24}$ 16. $\frac{7}{24}$

17. $\frac{5}{24}$ 18. $\frac{25}{72}$ 19. $\frac{1}{72}$ 20. $1\frac{11}{30}$ 21. $1\frac{7}{30}$ 22. $1\frac{19}{36}$ 23. $1\frac{64}{75}$ 24. $2\frac{8}{15}$

A square root can easily be found by using a calculator. Pressing the $\boxed{\sqrt{}}$ key gives the answer immediately.

Example 6

Find, correct to three decimal places, the square root of (a) 45 (b) 1·01

(a) $\boxed{\text{AC}}$ $\boxed{4}\boxed{5}$ $\boxed{\sqrt{}}$ $\boxed{=}$ 6·708 203 9 = 6·708 to three decimal places.

(b) $\boxed{\text{AC}}$ $\boxed{1}\boxed{\cdot}\boxed{0}\boxed{1}$ $\boxed{\sqrt{}}$ $\boxed{=}$ 1·004 987 5 = 1·005 to three decimal places.

Exercise 2.4f

Find the square root of each of the following.

In questions **1** to **15**, give the answers correct to three decimal places.

1. 15   2. 18   3. 7   4. 2   5. 10   6. 3   7. 5   8. 58
9. 90   10. 86   11. 4·9   12. 3·6   13. 2·5   14. 14·4   15. 12·1

In questions **16** to **30**, give the answers correct to two decimal places.

16. 150   17. 180   18. 320   19. 120   20. 960   21. 125   22. 315   23. 425
24. 384   25. 2000   26. 3000   27. 5000   28. 1200   29. 1800   30. 3250

In section 2.2, average speeds were calculated for journeys of known distance and time. The time for a journey can be calculated if the distance and average speed are known. The formula used is:

time = distance ÷ average speed .

A calculator can be conveniently used for this purpose.

Example 7

A car is driven for 180 km at an average speed of 48 km/h.
Find the time that the journey takes.

AC 1 8 0 ÷ 4 8 =   3·75 hours

The figures after the decimal point must now be changed to minutes.

AC 0 · 7 5 X 6 0 =   45 minutes

Therefore the time taken is 3 hours 45 minutes.

Exercise 2.4g

Find the time taken for each of the following car journeys.

	distance covered	average speed		distance covered	average speed		distance covered	average speed
1.	187 km	68 km/h	**5.**	180 km	75 km/h	**9.**	54 km	72 km/h
2.	266 km	56 km/h	**6.**	152 km	95 km/h	**10.**	34 km	40 km/h
3.	299 km	92 km/h	**7.**	201 km	60 km/h	**11.**	52 km	65 km/h
4.	135 km	108 km/h	**8.**	364 km	80 km/h	**12.**	51 km	85 km/h

Distances may be calculated from known average speeds and journey times, and a calculator is useful for this purpose also. The formula used is:

distance = average speed X time taken

Example 8

A train travels at an average speed of 110 km/h for 1 h 24 min.
Find the distance covered.

AC 2 4 ÷ 6 0 + 1 = X 1 1 0 =   154 km

Exercise 2.4h

Find the distance covered on each of the following train journeys.

	average speed	time taken		average speed	time taken		average speed	time taken
1.	148 km/h	1 h 45 min	**5.**	110 km/h	3 h 12 min	**9.**	160 km/h	1 h 9 min
2.	112 km/h	3 h 45 min	**6.**	105 km/h	2 h 36 min	**10.**	150 km/h	54 min
3.	128 km/h	2 h 15 min	**7.**	120 km/h	1 h 27 min	**11.**	180 km/h	39 min
4.	130 km/h	1 h 48 min	**8.**	140 km/h	2 h 57 min	**12.**	200 km/h	33 min

Rates are paid on the rateable value (R.V.) of a property.

Example 1

Find the rates paid on a shop with a rateable value of £725 if the rate levied is 72p in the £.

R.V. = £725; rate = 72p in the £

∴ rates paid = 725 × 72p

$$= £\overset{29}{\cancel{725}} \times \frac{\overset{18}{\cancel{72}}}{\underset{1}{\cancel{100}}}$$

$$= £522$$

Exercise 2.5a

Find the rates that are payable by the owners of the following properties.

Rateable value	Rate levied
1. £200	50p in the £
2. £300	60p in the £
3. £250	80p in the £
4. £180	70p in the £
5. £360	75p in the £
6. £400	84p in the £
7. £350	72p in the £
8. £450	66p in the £
9. £480	75p in the £
10. £320	55p in the £
11. £225	50p in the £
12. £366	80p in the £
13. £424	60p in the £
14. £516	70p in the £
15. £186	90p in the £
16. £242	75p in the £
17. £324	55p in the £
18. £508	85p in the £
19. £335	75p in the £
20. £495	65p in the £

Example 2

Find the rateable value of a property where the owner paid rates of £153 when the rate levied was 90p in the £.

Rates paid = £153; rate = 90p in the £

$$∴ R.V. = \frac{£153}{90p} = £\frac{\overset{17}{\cancel{153}} \times \overset{10}{\cancel{100}}}{\underset{1}{\underset{9}{\cancel{90}}}}$$

$$= £170$$

Exercise 2.5b

Find the rateable value of the following properties.

	Rates paid	Rate levied
1.	£180	50p in the £
2.	£360	80p in the £
3.	£144	60p in the £
4.	£288	90p in the £
5.	£210	75p in the £
6.	£320	64p in the £
7.	£140	56p in the £
8.	£484	88p in the £
9.	£315	75p in the £
10.	£221	65p in the £
11.	£187·50	50p in the £
12.	£171·60	60p in the £
13.	£409·60	80p in the £
14.	£302·40	90p in the £
15.	£177·80	70p in the £
16.	£406·50	75p in the £
17.	£124·80	65p in the £
18.	£216·70	55p in the £
19.	£213·75	75p in the £
20.	£276·25	85p in the £

Example 3

A small town has a total rateable value of £250 000. If the town council wishes to raise £180 000 in rates, what rate in the £ has it to levy?

Rateable value = £250 000
Amount to be raised in rates = £180 000

$$∴ \text{rate levied} = \frac{£180\,000}{£250\,000} = £0·72$$

so rate is 72p in the £.

Exercise 2.5c

Calculate the rate that each of the following towns or villages levy on the inhabitants from the details given below.

Town or village	Amount to be raised	Rateable value
1. Cleobury Mortimer	£75 000	£100 000
2. Tenbury Wells	£160 000	£200 000
3. Bewdley	£240 000	£400 000
4. Ludlow	£350 000	£500 000
5. Leominster	£270 000	£300 000
6. Grasmere	£128 000	£160 000

7.	Ambleside	£270 000	£360 000
8.	Sedbergh	£192 000	£320 000
9.	Kirkby Lonsdale	£297 000	£330 000
10.	Keswick	£315 000	£450 000
11.	Hawes	£108 000	£180 000
12.	Leyburn	£135 000	£150 000
13.	Reeth	£52 000	£80 000
14.	Richmond	£432 000	£540 000
15.	Aysgarth	£119 000	£140 000
16.	Albury	£162 000	£216 000
17.	Shere	£252 000	£315 000
18.	Gomshall	£196 000	£280 000
19.	Peaslake	£135 000	£225 000
20.	Cranleigh	£495 000	£550 000

Income tax is paid at a given rate on income after various allowances are deducted.

Allowances = (Tax Code Number) $\times$ 10
Taxable income
= (total income) $-$ (allowances)

Example 4

A man earns £5000 per year. If his Tax Code Number is 192, calculate his taxable income.

Total allowances = 192 $\times$ 10 = £1920
Taxable income = £5000 $-$ £1920 = £3080

Exercise 2.5d

In each case find the taxable income from the details given.

	Annual earnings	Tax Code Number
1.	£3000	150
2.	£4000	175
3.	£3800	125
4.	£3600	115
5.	£3200	95
6.	£4500	132
7.	£3750	114
8.	£4150	96
9.	£2850	112
10.	£3360	128

The income tax to be paid is calculated as a percentage of the taxable income.

Example 5

A man earns £75 per week and he has a

Tax Code Number of 190. Find the total amount of tax paid in a year when the tax rate is 34%.

Amount earned in year
$$= £75 \times 52 = £3900$$
Total allowances = 190 $\times$ 10 = £1900
$\therefore$ taxable income
$$= £3900 - £1900 = £2000$$
tax paid = 34% of £2000

$$= \frac{34}{\cancel{100}} \times \frac{\overset{20}{\cancel{2000}}}{1}$$

$$= £680$$

Exercise 2.5e

Find the yearly income tax paid if the tax rate is 35%.

	Wage or salary	Tax code no.
1.	£315 per month	108
2.	£280 per month	116
3.	£335 per month	92
4.	£405 per month	166
5.	£325 per month	148
6.	£70 per week	114
7.	£65 per week	128
8.	£85 per week	82
9.	£60 per week	164
10.	£80 per week	136

Income tax is deducted by the employer in equal weekly or monthly amounts.

Example 6

A man has a Tax Code Number of 173. If he earns £4330 a year, find the amount of tax he pays each week when the tax rate is 34%

Total amount earned = £4330
Total allowances = 173 $\times$ 10 = £1730
Taxable income = £4330 $-$ £1730 = £2600
Tax paid = 34% of £2600

$$= \frac{34}{\cancel{100}} \times \frac{\overset{26}{\cancel{2600}}}{1} = £884$$

Weekly tax = £884 $\div$ 52 = £17 per week.

Exercise 2.5f

Find (a) the yearly tax (b) the monthly tax if the tax rate is 35%

Salary	Tax code no.
1. £270 per month	180
2. £400 per month	120
3. £345 per month	174
4. £540 per month	192
5. £305 per month	102

Calculate (a) the yearly tax (b) the weekly tax paid by the following wage earners if the tax rate is 35%

Wage	Tax code no.
6. £55 per week	182
7. £90 per week	156
8. £95 per week	78
9. £65 per week	130
10. £120 per week	104

2.6 WAGES AND SALARIES

A wage is the amount earned at a fixed rate per hour for a given number of hours per week. Any extra hours worked are called *overtime*.

Example 1

Find the wage of a man who earns £1·17 per hour for a 40 hour week.

wage = £1·17 × 40 = £46·80

Exercise 2.6a

In each case find the weekly wage for the information given.

1. £1·20 per hour, 40 hours
2. £1·35 per hour, 40 hours
3. £1·05 per hour, 40 hours
4. £1·24 per hour, 40 hours
5. £1·48 per hour, 40 hours
6. £1·50 per hour, 42 hours
7. £1·10 per hour, 42 hours
8. £1·40 per hour, 42 hours
9. £1·15 per hour, 42 hours
10. £1·32 per hour, 42 hours
11. £1·25 per hour, 36 hours
12. £1·30 per hour, 36 hours
13. £1·60 per hour, 36 hours
14. £1·45 per hour, 36 hours
15. £1·04 per hour, 36 hours
16. £1·20 per hour, 44 hours
17. £1·40 per hour, 44 hours
18. £1·55 per hour, 44 hours
19. £1·02 per hour, 44 hours
20. £1·16 per hour, 44 hours

Overtime is paid at a higher rate per hour, e.g. 'Time and a quarter', 'time and a half', etc.

'Time and a half' means that for each hour worked, you are paid for $1\frac{1}{2}$ hours work.

Example 2

A man earns £1·20 per hour. What is his overtime pay per hour
(a) at 'time and a quarter'?
(b) 'at double time'?

(a) Time and a quarter = £1·20 × $1\frac{1}{4}$
$\qquad\qquad\qquad\quad$ = £1·50 per hour
(b) Double time $\qquad$ = £1·20 × 2
$\qquad\qquad\qquad\quad$ = £2·40 per hour.

Exercise 2.6b

In each case find the pay per hour for
(a) overtime rate of 'time and a half'
(b) rest-day rate of 'double time'.

1. £1·10	2. £1·24	3. £1·70
4. £1·32	5. £1·40	6. £1·36

In each case find the pay per hour for
(a) overtime rate of 'time and a quarter'
(b) rest-day rate of 'double time and a quarter'

7. £1·16	8. £1·60	9. £1·44
10. £1·28		

Example 3

An apprentice earns 95p per hour. If his basic week is 40 hours, find his total wage in the week in which he works 47 hours when all overtime is at 'time and a half'.

Basic wage = 95p × 40 = £38
Overtime pay
$$= 95 × 1\tfrac{1}{2}$$
$$= £1·42\tfrac{1}{2} \text{ per hour}$$
No. of hours overtime = 47 − 40 = 7 hours
Total overtime pay
$$= £1·42\tfrac{1}{2} × 7$$
$$= £9·97\tfrac{1}{2}$$
Total wage = £38 + £9·97$\tfrac{1}{2}$ = £47·97$\tfrac{1}{2}$

Exercise 2.6c

Find the total wage if the overtime rate is 'time and a half'.

	Standard rate per hour	Hours in basic week	Actual hours worked
1.	£1·20	40	48
2.	£1·30	40	45
3.	£1·10	40	50
4.	£1·28	42	48
5.	£1·36	42	50

Find the total wage if the overtime rate is 'time and a quarter'.

6.	£1·24	40	44
7.	£1·32	40	42
8.	£1·20	42	50
9.	£1·40	44	46
10.	£1·16	44	54

A salary is the amount earned in a year. It is paid in 12 equal monthly instalments.

Example 4

Find the salary of a person who earns £240 per month.

Salary = £240 × 12 = £2880.

Exercise 2.6d

Find the annual salary for each monthly payment below.

1. £250　　　　2. £320

3. £355　　　　4. £275
5. £316　　　　6. £344
7. £418　　　　8. £293
9. £320·50　　10. £284·50
11. £310·25　　12. £275·25
13. £386·25　　14. £290·75
15. £372·75

Example 5

A man's salary is £3684 per year. How much does he earn per month?

Monthly pay
$$= £3684 ÷ 12$$
$$= £307 \text{ per month.}$$

Exercise 2.6e

Find the monthly earnings from each annual salary below.

1. £3600	2. £4500	3. £3204
4. £3912	5. £3828	6. £3984
7. £3336	8. £3408	9. £4083
10. £4371	11. £3489	12. £4521
13. £2886	14. £4146	15. £4038

Example 6

A man has a choice of two jobs
(a) £320·50 per month
(b) £74 per week

Which is the better rate of pay and by how much a year?

(a) £320·50 per month = £320·50 × 12
　　　　　　　　　　　= £3846 per year
(b) £74 per week　　　= £74 × 52
　　　　　　　　　　　= £3848 per year
∴ £74 per week is better by £2 per year.

Exercise 2.6f

In each case find which job gives the higher pay and by how much per year.

1. £64 per week or £277 per month
2. £55 per week or £239 per month
3. £68 per week or £294·50 per month
4. £80 per week or £346·75 per month
5. £56·50 per week or £244·50 per month
6. £72·50 per week or £313·75 per month
7. £75 per week or £326 per month
8. £66 per week or £286·25 per month

Gas and electricity bills are paid quarterly, i.e. every 13 weeks.

The gas bill is based on the number of therms used each quarter.

The electricity bill is based on the number of units used each quarter. These quantities are found by reading the meter.

Example 1

An electricity meter was read as follows:

Jan. 1st	16943
April 1st	18491
July 1st	19631

Find the number of units used in each quarter.

Units used in quarter ending
$$\text{April 1st} = 18491 - 16943$$
$$= 1548 \text{ units}$$
Units used in quarter ending
$$\text{July 1st} = 19631 - 18491$$
$$= 1140 \text{ units}$$

Exercise 2.7a

Find the number of electricity units used in each quarter from the following meter readings.

1.	Jan. 1st	13562	**6.**	April 1st	18932
	April 1st	15825		July 1st	19559
	July 1st	16499		Oct. 1st	20075
2.	Jan. 1st	17722	**7.**	April 1st	25324
	April 1st	19587		July 1st	26037
	July 1st	20492		Oct. 1st	26466
3.	July 1st	12133	**8.**	Jan. 1st	17532
	Oct. 1st	12588		April 1st	19853
	Jan. 1st	13792		July 1st	20615
4.	July 1st	18576		Oct. 1st	21163
	Oct. 1st	19197	**9.**	July 1st	48382
	Jan. 1st	20739		Oct. 1st	48915
5.	July 1st	25361		Jan. 1st	50662
	Oct. 1st	25895		April 1st	53117
	Jan. 1st	27531	**10.**	Jan. 1st	29234
				April 1st	31127
				July 1st	31942
				Oct. 1st	32691
				Jan. 1st	35777

Electricity bills are usually calculated on a 'two-tier' basis, as shown.

Example 2

A householder uses 1548 units of electricity in a certain quarter. Calculate his bill if the first 72 units cost 5·5p per unit and all remaining units cost 2·25p each.

$$72 @ 5\cdot5\text{p each cost } \pounds3\cdot96$$
$$\text{Remaining units} = 1548 - 72 = 1476$$
$$1476 @ 2\cdot25 \text{ each cost } \pounds33\cdot21$$
$$\therefore \text{ total bill} = \pounds3\cdot96 + \pounds33\cdot21 = \pounds37\cdot17$$

Exercise 2.7b

Calculate the bill for each quarter below if the first 72 units cost 5·5p each and all remaining units cost 2·25p each. Questions 9 to 15 give the meter readings.

1. Jan. 1st to April 1st, 1712 units
2. Jan. 1st to April 1st, 1800 units
3. July 1st to Oct. 1st,   492 units
4. July 1st to Oct. 1st,   720 units
5. April 1st to July 1st, 1032 units
6. April 1st to July 1st, 936 units
7. Oct. 1st to Jan. 1st, 1352 units
8. Oct. 1st to Jan. 1st, 1224 units
9. Jan. 1st: 16362, April 1st: 18194
10. Jan. 1st: 12143, April 1st: 14519
11. April 1st: 14534, July 1st: 15726
12. April 1st: 27781, July 1st: 28637
13. July 1st: 35422, Oct. 1st: 36006
14. Oct. 1st, 23564, Jan. 1st: 25176
15. Oct. 1st: 41254, Jan. 1st: 43270

Gas bills usually consist of a Standing Charge to be paid each quarter in addition to the cost of the gas.

Example 3

In the quarter ending 31 December, a man used 174 therms of gas. In the quarter ending 31 March he used 206 therms of gas. Calculate the total amount he paid for the gas if the gas cost 22p per therm and the Standing Charge was £3·50 per quarter.

$$\text{Total number of therms used} = 174 + 206$$
$$= 380$$
$$\text{Cost of gas} = 22\text{p} \times 380 = \pounds83\cdot60$$
$$\text{Standing Charge (2 quarters)} = \pounds3\cdot50 \times 2$$
$$= \pounds7\cdot00$$
$$\therefore \text{ total bill} = \pounds83\cdot60 + \pounds7\cdot00$$
$$= \pounds90\cdot60$$

Exercise 2.7c

Calculate the total bill for each quarter if the
Standing Charge for each quarter is £3·50 and
the price of gas is 22p per therm. Questions
9 to 15 give the meter readings.

1. Dec. 31st to March 31st, 190 therms
2. March 31st to June 30th, 43 therms
3. June 30th to Sept. 30th, 36 therms
4. Sept. 30th to Dec. 31st, 165 therms
5. Dec. 31st to March 31st, 154 therms
   and March 31st to June 30th, 32 therms
6. March 31st to June 30th, 41 therms and
   June 30th to Sept. 30th, 25 therms

7. Sept. 30th to Dec. 31st, 135 therms and
   Dec. 31st to March 31st, 142 therms
8. Dec. 31st to March 31st, 123 therms and
   March 31st to June 30th, 37 therms
9. Dec. 31st: 234
   March 31st: 386
   June 30th: 429
10. March 31st: 154
    June 30th: 188
    Sept. 30th: 215

11. June 30th: 386
    Sept. 30th: 447
    Dec. 31st: 573
12. Sept. 30th: 058
    Dec. 31st: 195
    March 31st: 406
13. Dec. 31st: 135
    March 31st: 339
    June 30th: 392
    Sept. 30th: 430
    Dec. 31st: 611
14. Dec. 31st: 257
    March 31st: 426
    June 30th: 474
    Sept. 30th: 518
    Dec. 31st: 653
15. Dec. 31st: 008
    March 31st: 202
    June 30th: 251
    Sept. 30th: 290
    Dec. 31st: 448

2.8 24–HOUR CLOCK AND TIMETABLES

Timetables generally use the 24-hour
clock system instead of a.m. and p.m. In
this system, all times are written using four
figures. The first two figures indicate hours
and the last two indicate minutes past the
hour. All p.m. times have 12 hours added to
them.

Example 1

Write these times as 24-hour clock times:
(a) 8 a.m. (b) 10.27 a.m. (c) 1.05 p.m.
(a) 8 a.m. = 08.00 h
(b) 10.27 a.m. = 10.27 h
(c) 1.05 p.m. = 13.05 h

Exercise 2.8a

Write the following as 24-hour clock times.

1. 2.00 p.m.	2. 5.00 p.m.	3. 6.00 p.m.
4. 2.15 p.m.	5. 2.40 p.m.	6. 5.20 p.m.
7. 5.32 p.m.	8. 6.30 p.m.	9. 6.48 p.m.
10. 7.50 p.m.	11. 7.05 p.m.	12. 4.35 p.m.
13. 8.00 p.m.	14. 8.30 p.m.	15. 9.15 p.m.
16. 10.00 p.m.	17. 10.24 p.m.	18. 11.45 p.m.

19. 11.05 p.m.	20. 8.00 a.m.	21. 6.00 a.m.
22. 9.15 a.m.	23. 8.42 a.m.	24. 7.35 a.m.
25. 6.08 a.m.	26. 9.03 a.m.	27. 10.00 a.m.
28. 10.15 a.m.	29. 11.55 a.m.	30. 11.32 a.m.

Example 2

Write these times as 12-hour times
(a) 01.00 h (b) 16.27 h (c) 23.59 h
(a) 01.00 h = 1 a.m.
(b) 16.27 h = 4.27 p.m.
(c) 23.59 h = 11.59 p.m.

Exercise 2.8b

Write the following as 12-hour times

1. 13.00 h	2. 16.00 h	3. 19.00 h
4. 13.30 h	5. 13.50 h	6. 19.35 h
7. 19.18 h	8. 16.25 h	9. 16.05 h
10. 18.40 h	11. 18.12 h	12. 15.55 h
13. 14.36 h	14. 21.00 h	15. 21.45 h
16. 23.00 h	17. 23.30 h	18. 22.45 h
19. 22.06 h	20. 09.00 h	21. 07.00 h
22. 08.45 h	23. 09.53 h	24. 08.56 h
25. 07.02 h	26. 08.09 h	27. 11.00 h
28. 11.25 h	29. 10.50 h	30. 10.36 h

If time is written using the 24-hour clock system, then the difference between two times is easily found by subtraction.

Example 3

A bus leaves Brigg at 14.27 h and arrives in Lincoln at 15.44 h. How long does the journey take?

$$\begin{array}{r} 15.44 \\ -\ 14.27 \\ \hline 1.17 \end{array}$$

∴ time taken is 1 h 17 min.

Exercise 2.8c

Find the time of each journey.

1. A train leaves London (Waterloo) at 09.10 h and arrives in Portsmouth at 10.55 h.
2. A bus leaves Hull at 11.05 h and arrives in Bridlington at 12.40 h.
3. A bus leaves Manchester at 14.15 h and arrives in Preston at 15.30 h.
4. A train leaves Liverpool at 16.20 h and arrives in Crewe at 16.54 h.
5. A bus leaves Darlington at 11.30 h and arrives in Hawes at 13.55 h.
6. A train leaves Glasgow at 12.25 h and arrives in Aberdeen at 15.55 h.
7. A train leaves London (Paddington) at 10.30 h and arrives in Cardiff at 12.20 h.
8. A bus leaves Birmingham at 15.30 h and arrives in Worcester at 17.15 h.
9. A bus leaves Sheffield at 08.40 h and arrives in Chesterfield at 09.25 h.
10. A train leaves London (King's Cross) at 11.20 h and arrives in Leeds at 14.05 h.
11. Find the time taken by the train from London to each station.

London (Waterloo)	(depart)	09.05 h
Southampton	(arrive)	10.17 h
Bournemouth	(arrive)	10.50 h
Poole	(arrive)	11.05 h
Weymouth	(arrive)	11.45 h

12. Find the time taken by the train from London to each station.

London (Euston)	(depart)	17.10 h
Coventry	(arrive)	18.21 h
Birmingham	(arrive)	18.50 h
Wolverhampton	(arrive)	19.15 h
Shrewsbury	(arrive)	19.55 h

13. Find the time taken by the train from London to each station.

London (Paddington)	(depart)	12.30 h
Taunton	(arrive)	14.42 h
Exeter	(arrive)	15.10 h
Plymouth	(arrive)	16.15 h
Penzance	(arrive)	18.30 h

14. Bus timetable

Brigg–Worlaby–Barton					
From Brigg					
Brigg	07.15	08.10	10.30	13.55	21.45
Wrawby	07.20	08.15	10.35	14.00	21.50
Worlaby	07.30	08.25	10.45	14.10	22.00
Ferriby	07.43	08.38	10.58	14.23	22.13
Barton	07.50	08.45	11.05	14.30	22.20
From Barton					
Barton	08.00	09.00	12.15	16.20	22.30
Ferriby	08.07	09.07	12.22	16.27	22.37
Worlaby	08.20	09.20	12.35	16.40	22.50
Wrawby	08.30	09.30	12.45	16.50	23.00
Brigg	08.35	09.35	12.50	16.55	23.05

(a) When (in 12-hour time) does the last bus leave Brigg for Barton? How long does the journey take?

(b) How long does it take to go from Brigg to Worlaby?

(c) A man lives in Wrawby and travels to and from work in Barton by bus.
  (i) If he starts work at 9 a.m. at what time does he catch the bus?
  (ii) At what time does he arrive in Barton?
  (iii) How long does the journey take?
  (iv) If he finishes work at 4.15 p.m. at what time does he catch the bus home?
  (v) At what time does he arrive back in Wrawby?

(d) If the 13.55 bus is 5 minutes late leaving Brigg, at what time will it reach Worlaby? If owing to road works at Ferriby the bus does not arrive in Barton until 14.40, how long has the journey from Worlaby taken?

(e) The 12.15 bus from Barton arrives in Brigg 15 minutes late. At what time does it reach Brigg? How long has a passenger on this bus to wait in Brigg to catch the next train to London which leaves at twenty to five?

15. Bus timetable

Richmond—Barnard Castle			
From Richmond			
Richmond	10.15	12.10	17.30
Gilling	10.25	12.20	17.40
Greta Bridge	11.05	13.00	18.20
Startforth	11.20	13.15	18.35
Barnard Castle	11.25	13.20	18.40
From Barnard Castle			
Barnard Castle	09.00	10.55	16.15
Startforth	09.05	11.00	16.20
Greta Bridge	09.20	11.15	16.35
Gilling	10.00	11.55	17.15
Richmond	10.10	12.05	17.25

(a) I travel on the first bus from Richmond to Barnard Castle, but I alight at Gilling to visit a friend. I travel on to Barnard Castle on the next bus and finally return to Richmond on the last bus of the day.

(i) How long do I have to visit my friend?
(ii) How long after leaving Richmond do I eventually reach Barnard Castle?
(iii) How long do I spend in Barnard Castle before finally returning to Richmond?
(iv) At what time (in 12-hour time) do I arrive back in Richmond?

(b) A lady who lives at Greta Bridge has a dentist's appointment in Richmond from 11.15 to 11.45 a.m.

(i) At what time does she catch the bus to Richmond?
(ii) At what time does she arrive there?
(iii) How long does her journey take?
(iv) How long does she have in Richmond before her appointment?
(v) At what time does she catch the bus home?
(vi) How long does she have after her appointment before catching the bus home?
(vii) At what time, on the 12-hour clock, does she arrive back at Greta Bridge?

(c) I travel from Richmond on the last bus as far as Startforth.

(i) How long does the journey take?
(ii) At what time, on the 12-hour clock do I arrive at Startforth?

16. Bus timetable

Tenbury Wells—Bromyard			
From Tenbury Wells			
Tenbury Wells	08.10	11.45	13.35	17.40
Kyre Magna	08.24	11.59	13.49	17.54
Bank Street	08.28	12.03	13.53	17.58
Collington	08.36	12.11	14.01	18.06
Bromyard	08.50	12.25	14.15	18.20
From Bromyard			
Bromyard	07.20	10.35	14.30	18.25
Collington	07.34	10.49	14.44	18.39
Bank Street	07.42	10.57	14.52	18.47
Kyre Magna	07.46	11.01	14.56	18.51
Tenbury Wells	08.00	11.15	15.10	19.05

(a) When (in 12-hour time) does the first bus after 12 noon leave Tenbury Wells for Bromyard? When does it arrive in Bromyard? How long does the journey take?

(b) How long does it take to go from Kyre Magna to Bromyard?

(c) A man who lives at Bank Street works in Bromyard between 9.30 a.m. and 5.30 p.m.

(i) At what time does he catch the bus in the morning?
(ii) At what time does he arrive in Bromyard?
(iii) How long does his journey take?
(iv) At what time does he catch the bus after work?
(v) How long after he finishes work does the bus depart?

(d) I arrive in Bromyard 35 minutes before the last bus leaves for Tenbury Wells. When do I arrive in Bromyard?

(e) If I arrive at Kyre Magna at 11.15 h, by how many minutes have I missed a bus to Tenbury Wells? How long is it before the next one?

In Britain the pound (£) is the basic unit of currency.
Because £1 = 100 pence, all sums of money can be written as
decimals. Most other countries use a similar system.
  e.g. Germany, 1 mark = 100 pfennigs
In the following exercises, the rates of exchange to be used are shown
in the table below.

Country	Unit of currency	Exchange rate
Belgium	1 franc = 100 centimes	BF 60 = £1
France	1 franc = 100 centimes	Fr. 9·10 = £1
Germany	1 mark = 100 pfennigs	DM 3·84 = £1
Spain	1 peseta = 100 cents	Ptas. 156 = £1
Switzerland	1 franc = 100 centimes	SF 3·60 = £1
U.S.A.	1 dollar = 100 cents	$1·96 = £1

Example 1

How many dollars would a man receive
for £50?
  number of dollars = 1·96 × 50
                    = $98

Exercise 2.9a

Find how much each amount of British money
is worth in the foreign currency.

  1. £200 (dollars)   2. £150 (dollars)
  3. £25 (dollars)   4. £40 (French francs)
  5. £300 (French francs)   6. £140 (French francs)
  7. £75 (marks)   8. £200 (marks)
  9. £250 (marks)   10. £45 (Swiss francs)
 11. £105 (Swiss francs)   12. £16 (pesetas)
 13. £21 (pesetas)   14. £125 (Belgian francs)
 15. £72 (Belgian francs)

Example 2

How many £'s would a man receive for
728 French francs?

  number of £'s = 728 ÷ 9·10
               = 7280 ÷ 91 = £80

Exercise 2.9b

Find how much each amount of foreign currency
is worth in £'s.

  1. $588   2. $490
  3. $147   4. Fr. 546

  5. Fr. 3640   6. Fr. 1456
  7. DM 192   8. DM 1344
  9. DM 480   10. SF 54
 11. SF 126   12. Ptas. 2808
 13. Ptas. 3744   14. BF 6900
 15. BF 3840

Example 3

Petrol in England costs 18p per litre. In
Spain it costs Ptas. 39 per litre. In which
country is petrol cheaper, and by how
much?

  Ptas. 39 = £ (39 ÷ 156)
           = £0·25 = 25p
∴ petrol is cheaper in England by 7p per
litre.

Exercise 2.9c

  1. A typewriter costs £100 in Britain and $245 in
     the United States. Find in terms of British money
     the country where the selling price is the cheaper,
     and by how much.
  2. A motor cycle costs £900 in Britain and Fr. 7826
     in France. Find in terms of British money the
     country where the selling price is the cheaper,
     and by how much.
  3. A kilogram of flour costs 25p in Britain, whereas
     in Switzerland the same kind of flour costs 72
     centimes per kilogram. In which country is it the
     cheaper and by how many pence per kilogram?

4. A kilogram of sugar costs 30p in Britain, whereas in Belgium the same kind of sugar costs 24 francs per kilogram. In which country is it the cheaper and by how many pence per kilogram?

5. In Britain, Spanish oranges cost £1·00 for ten, but in Spain itself the price is Ptas. 117 for the same ten oranges. In which country do the ten oranges cost less and by how many pence?

6. A bottle of whisky costs £4·50 in Britain and $9·80 in the United States. In which country is it cheaper and by how many pence?

7. A bottle of wine costs £1·50 in Britain and DM 4·80 in Germany. In which country is it cheaper and by how many pence?

8. A certain type of car is sold in several countries at the prices given below.

Britain	£2400
U.S.A.	$4900
Switzerland	SF 8820
Spain	Ptas. 343 200
Germany	DM 8640
France	Fr. 24 570
Belgium	BF 156 000

Find in terms of British money the price in each other country, then write a list of the prices in order, beginning with the cheapest.

9. A certain brand of cigarettes is sold in several countries and the price for a packet of twenty is given below.

Britain	60p
Switzerland	SF 2·34
Spain	Ptas. 98·28
France	Fr. 4·55
Belgium	BF 34·20

Find in terms of British money the price in each country, then write a list of the prices in order, beginning with the cheapest.

10. An average rail fare for a journey of 200 kilometres is worked out for several countries and the results are given below.

Britain	£5·00
U.S.A.	$9·31
Switzerland	SF 17·28
Spain	Ptas. 702
Germany	DM 20·16
France	Fr. 47·32
Belgium	BF 294

Find the average fare for each country in terms of British money, then write a list of the average fares in order, beginning with the cheapest.

Area is a measure of the size of a surface; area is measured in square units.
The common metric units are:
Square millimetres (mm²)
Square centimetres (cm²)
Square metres (m²)

$1 \text{ m}^2 = 10\ 000 \text{ cm}^2$
$1 \text{ cm}^2 = 100 \text{ mm}^2$

Perimeter is a measure of the distance round the outline of a shape. It is a measure of length.

Example 1

(a) Change 1·25 m² to cm²
$$1 \cdot 25 \text{ m}^2 = 1 \cdot 25 \times 10\ 000$$
$$= 12\ 500 \text{ cm}^2$$
(b) Change 275 mm² to cm²
$$275 \text{ mm}^2 = 275 \div 100$$
$$= 2 \cdot 75 \text{ cm}^2$$

Exercise 3.1a

Change the following.

1. 3·75 m² to cm²   2. 2·75 cm² to mm²
3. 465 mm² to cm²   4. 12 500 cm² to m²
5. 150 mm² to cm²   6. 0·675 m² to cm²
7. 127 mm² to cm²   8. 4·1 m² to cm²
9. 62 cm² to m²   10. 1764 mm² to m²

Rectangle

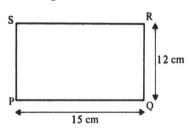

area = length × width
$$= PQ \times QR$$
$$= 15 \text{ cm} \times 12 \text{ cm} = 180 \text{ cm}^2$$
perimeter = PQ + QR + RS + SP
$$= 15 + 12 + 15 + 12 = 54 \text{ cm}$$

Square

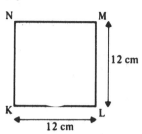

area = length × width
$$= KL \times LM$$
$$= 12 \text{ cm} \times 12 \text{ cm} = 144 \text{ cm}^2$$
perimeter = KL + LM + MN + NK
$$= 12 + 12 + 12 + 12 = 48 \text{ cm}$$

Exercise 3.1b

1. Find (i) the area, (ii) the perimeter of the rectangle PQRS if
   (a) PQ = 16 cm, QR = 9 cm
   (b) PQ = 18 cm, QR = 11 cm
   (c) PQ = 35 mm, QR = 20 mm
   (d) PQ = 40 mm, QR = 24 mm
   (e) PQ = 25 m, QR = 12 m
   (f) PQ = 24 m, QR = 15 m
2. Find (i) the area, (ii) the perimeter of the square KLMN if
   (a) KL = LM = 20 cm
   (b) KL = LM = 16 cm
   (c) KL = LM = 1·4 m
   (d) KL = LM = 1·9 m
   (e) KL = LM = 18 mm
   (f) KL = LM = 25 mm

Parallelogram

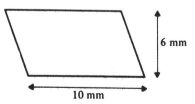

area = base × vertical height
$$= 10 \text{ mm} \times 6 \text{ mm} = 60 \text{ mm}^2$$

Triangle

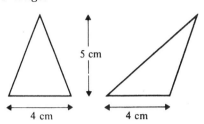

area $= \frac{1}{2}$ base $\times$ vertical height
$$= \frac{1}{2} \times 4 \times 5 = 10 \text{ cm}^2$$

Exercise 3.1c

Find the area of the following parallelograms:

1. base 20 cm, height 5 cm
2. base 4 mm, height 2·5 mm
3. base 1·2 cm, height 1·3 cm
4. base 0·2 m, height 0·3 m
5. base 25 mm, height 1·4 cm
6. base 184 mm, height 15 cm

Find the area of the following triangles:

7. base 20 cm, height 35 cm
8. base 19 cm, height 32 cm
9. base 40 mm, height 15 mm
10. base 25 mm, height 24 mm
11. base 3·6 cm, height 50 mm
12. base 2·5 m, height 120 cm

Find the area of the following right-angled triangles:

13. $XY = 16$ cm, $YZ = 25$ cm, $\hat{Y} = 90°$
14. $PQ = 31$ cm, $QR = 12$ cm, $\hat{Q} = 90°$
15. $RS = 36$ mm, $ST = 15$ mm, $\hat{S} = 90°$
16. $LM = 4·8$ m, $MN = 8$ m, $\hat{M} = 90°$
17. $AB = 5·4$ m, $BC = 3·2$ m, $\hat{B} = 90°$

Example 2

(a) The area of a rectangle is 156 cm². If it is 13 cm long, how wide is it?

   area $=$ length $\times$ width
   $156 = 13 \times$ width
   $\therefore$ width $= 156 \div 13 = 12$ cm

(b) The area of a parallelogram is 1·44 cm². If its base is 0·9 cm long, what is its vertical height?
   area $=$ base $\times$ height
   $1·44 = 0·9 \times$ height

$\therefore$ height $= 1·44 \div 0·9$
$= 14·4 \div 9 = 1·6$ cm

(c) The area of a square is 49 cm². How long is its perimeter?

   length of side
   $= 7$ cm (because $7 \times 7 = 49$)
   $\therefore$ perimeter $= 7 + 7 + 7 + 7 = 28$ cm

Exercise 3.1d

Find the width of the following rectangles.

1. area $= 117$ cm², length $= 9$ cm
2. area $= 136$ cm², length $= 8$ cm
3. area $= 285$ mm², length $= 15$ mm
4. area $= 414$ mm², length $= 18$ mm
5. area $= 16·5$ m², length $= 1·1$ m

Find the vertical height of the following parallelograms:

6. area $= 154$ cm², base length $= 7$ cm
7. area $= 210$ cm², base length $= 15$ cm
8. area $= 432$ mm², base length $= 16$ mm
9. area $= 20·4$ m², base length $= 1·2$ m
10. area $25·2$ m², base length $= 3·6$ m

Find the perimeter of the following squares.

11. area $= 81$ cm²
12. area $= 121$ cm²
13. area $= 225$ cm²
14. area $= 1·44$ m²
15. area $= 2·56$ m²

Example 3

(a) A lawn measuring 10 m by 7·5 m is to be reseeded. A box of grass seed contains 500 g. If 100 g of grass seed covers 1 m², how many boxes will be needed?

   area $= 10$ m $\times 7·5$ m $= 75$ m²
   no. of grams of seed needed
   $= 75 \times 100$
   $= 7500$ g
   no. of boxes $= 7500 \div 500$
   $= 15$

(b) A floor measuring 2 m by 1·5 m is to be covered with tiles measuring 25 cm square. How many tiles will be needed?

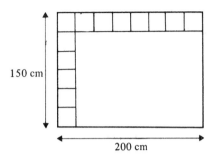

150 cm

200 cm

no. of tiles along length = $200 \div 25 = 8$
no. of tiles along width = $150 \div 25 = 6$
$\therefore$ tiles needed = 8×6 = 48

Exercise 3.1e

1. A metal plate measuring 32 cm by 12 cm is cut up into small squares of side length 4 cm. How many squares will be cut?
2. A lawn 5 m by 4 m is to be covered by pieces of square turf, each of side length 50 cm. How many pieces will be required?
3. A wall space in a bathroom measuring 2 m by 60 cm is to be covered with square tiles of side length 20 cm. How many tiles will be required? Find the cost of surfacing if the price of the tiles is £2·80 for ten.

4. A ceiling measures 6 m by 4 m. It is to be covered with square tiles of side length 40 cm. How many tiles will be required? Find the cost of the covering if the price of the tiles is £1·44 for twelve.
5. A yard measuring 10 m by 5 m is to be surfaced with concrete the dry mixture for which is supplied in 100-kg bags. If 80 kg can suitably surface 1 m², how many bags will be required?
6. A pathway 20 m long and 3 m wide is to be covered with shingle which is supplied in 50-kg bags. If 40 kg of shingle can suitably cover 1 m², how many bags will be required?
7. A small garden plot measuring 5 m by 4 m is to be covered with a special kind of soil which is supplied in 25-kg bags. If 30 kg of soil can suitably cover 1 m², how many bags will be required?
8. A lawn measuring 12 m by 10 m is to be covered with pieces of turf each 60 cm by 40 cm. How many pieces will be required?
9. A pavement of length 600 m and width 1 m is to be laid by using slabs of dimensions 80 cm by 50 cm. How many slabs are required?
10. A pavement of length 500 m and width 1·5 m is to be laid by using slabs of dimensions 75 cm by 40 cm. How many slabs are required?

3.2 SIMPLE AREA II

Rhombus

6 cm   8 cm

9·6 cm

8 cm   6 cm

10 cm

(a)   area = base $\times$ vertical height
      = $10 \times 9·6 = 96$ cm²
(b)   area = $\frac{1}{2}$ (product of diagonals)
      = $\frac{1}{2}[(8 + 8) \times (6 + 6)]$
      = $\frac{1}{2}(16 \times 12) = 96$ cm²

Kite

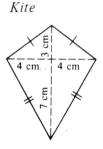

3 cm

4 cm   4 cm

7 cm

area = $\frac{1}{2}$ (product of diagonals)
   = $\frac{1}{2}[(4 + 4) \times (3 + 7)]$
   = $\frac{1}{2}(8 \times 10) = 40$ cm²

Exercise 3.2a

Questions 1 to 10 refer to the rhombus ABCD.
Find the area in each case
(a) from the base and vertical height,
(b) from the diagonals.

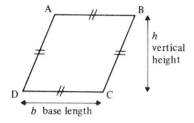

1. $b = 12$ cm, $h = 5\cdot5$ cm, AC = 6 cm, BD = 22 cm
2. $b = 20$ cm, $h = 16\cdot2$ cm, AC = 18 cm, BD = 36 cm
3. $b = 50$ cm, $h = 48$ cm, AC = 60 cm, BD = 80 cm
4. $b = 0\cdot6$ m, $h = 0\cdot275$ m, AC = 0·3 m, BD = 1·1 m
5. $b = 16$ m, $h = 15$ m, AC = 20 m, BD = 24 m
6. $b = 0\cdot072$ m, $h = 0\cdot06$ m, AC = 0·072 m, BD = 0·12 m
7. $b = 20$ mm, $h = 11\cdot4$ mm, AC = 12 mm, BD = 38 mm
8. $b = 30$ mm, $h = 23\cdot4$ mm, AC = 26 mm, BD = 54 mm
9. $b = 10$ cm, $h = 3\cdot92$ cm, AC = 4 cm, BD = 19·6 cm
10. $b = 5$ cm, $h = 3\cdot68$ cm, AC = 4 cm, BD = 9·2 cm

Questions 11 to 20 refer to the kites illustrated.
Find the area in each case.

11.

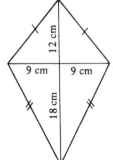

12.

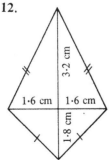

13.

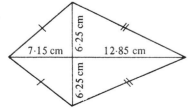

14.

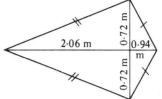

15.

16.

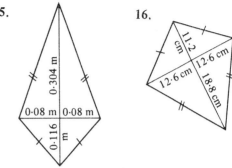

17.

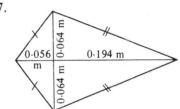

18.

19.

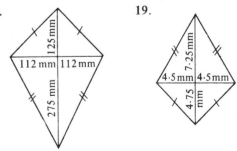

20.

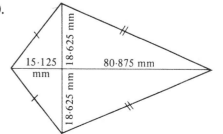

Trapezium

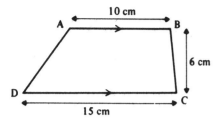

area = $\frac{1}{2}$ (sum of parallel sides) × height

$= \frac{1}{2}$ (AB + DC) × 6

$= \frac{1}{2}$ (10 + 15) × 6

$= 12\frac{1}{2}$ × 6 = 75 cm²

Exercise 3.2b

Questions 1 to 10 refer to the trapezium PQRS. Find the area in each case.

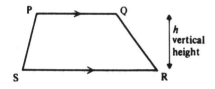

1. PQ = 12 cm, RS = 18 cm, h = 8 cm
2. PQ = 26 cm, RS = 34 cm, h = 12 cm
3. PQ = 42 cm, RS = 54 cm, h = 15 cm
4. PQ = 2·7 cm, RS = 4·3 cm, h = 1·6 cm
5. PQ = 0·36 m, RS = 0·92 m, h = 0·5 m
6. PQ = 0·28 m, RS = 0·52 m, h = 0·19 m
7. PQ = 0·021 m, RS = 0·039 m, h = 0·018 m
8. PQ = 324 mm, RS = 676 mm, h = 250 mm
9. PQ = 146·7 mm, RS = 153·3 mm, h = 180 mm
10. PQ = 21·78 mm, RS = 78·22 mm, h = 33·2 mm

To find the area of an irregular shape, divide it up into shapes whose area can be found.

Example 1

Find the area of the following shape.

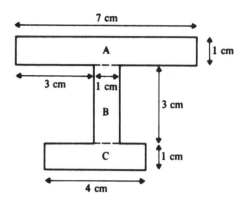

Divide the figure into three rectangles A, B, and C as shown.

area of A = 7 × 1 = 7 cm²
area of B = 3 × 1 = 3 cm²
area of C = 4 × 1 = 4 cm²
∴ area of figure = 7 + 3 + 4 = 14 cm²

Exercise 3.2c

Find the area of each shape.

1.

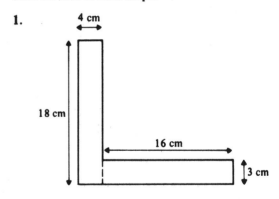

2.

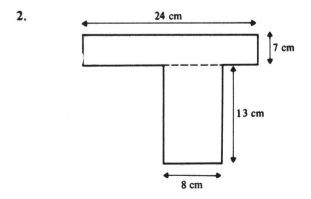

3.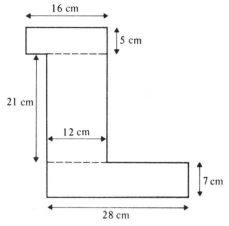

16 cm
5 cm
21 cm
12 cm
7 cm
28 cm

7.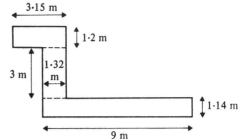

3·15 m
1·2 m
3 m
1·32 m
1·14 m
9 m

4.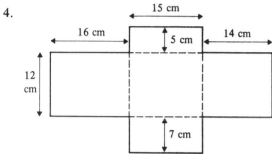

15 cm
16 cm
5 cm
14 cm
12 cm
7 cm

8.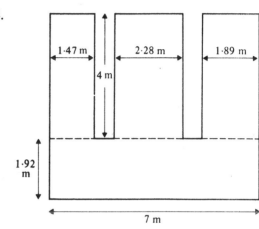

1·47 m
2·28 m
1·89 m
4 m
1·92 m
7 m

9. Find the area of the shaded part.

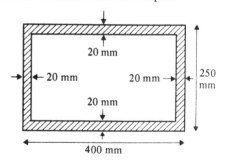

20 mm
20 mm
20 mm
20 mm
250 mm
400 mm

5.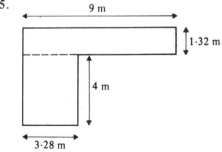

9 m
1·32 m
4 m
3·28 m

6.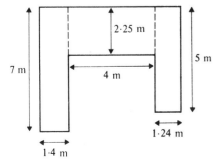

2·25 m
7 m
5 m
4 m
1·24 m
1·4 m

10. Find the area of the shaded part.

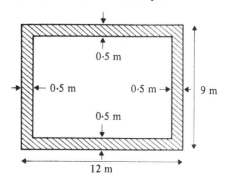

0·5 m
0·5 m
0·5 m
0·5 m
9 m
12 m

Example 2

Find the area of the following shape.

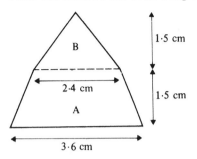

Divide the figure into the trapezium A
and the triangle B.

area of A = $\frac{1}{2}$ (2·4 + 3·6) × 1·5
= 3 × 1·5 = 4·5 cm²

area of B = $\frac{1}{2}$ (2·4 × 1·5)
= $\frac{1}{2}$ × 3·6 = 1·8 cm²

∴ area of figure = 4·5 + 1·8 = 6·3 cm²

Exercise 3.2d

Find the area of each shape.

1.

2.

3.

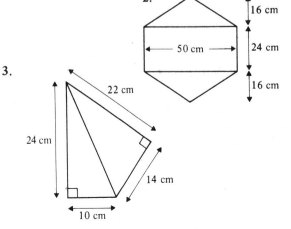

4.

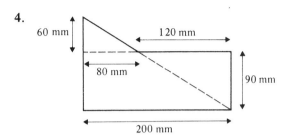

5.

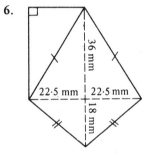

6.

7.

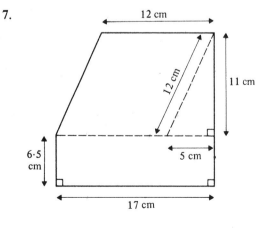

8. Find the area of the shaded part.

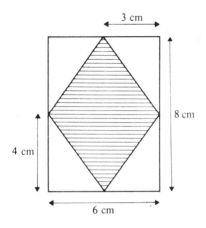

9. Find the area of the shaded part

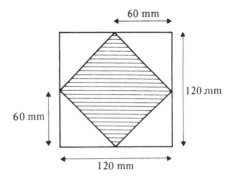

3.3 THE CIRCLE

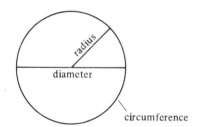

The length of the diameter D is twice the length of the radius r.

$$D = 2r$$
$$\text{circumference} = \pi D \text{ or } 2\pi r$$
$$\text{area} = \pi r^2$$
$$\text{diameter} = \text{circumference} \div \pi$$
$$\text{radius} = \sqrt{(\text{area} \div \pi)}$$

Example 1

Find the circumference of a circle when

(a) the radius is 14 cm, using $\pi = 3\frac{1}{7}$
(b) the diameter is 20 cm, using $\pi = 3\cdot14$

(a) circumference $= 2\pi r$

$$= \frac{2}{1} \times \frac{\overset{}{22}}{\underset{1}{\cancel{7}}} \times \frac{\overset{2}{\cancel{14}}}{1} = 88 \text{ cm}$$

(b) circumference $= \pi D$
$$= 20 \times 3\cdot14 = 62\cdot8 \text{ cm}$$

Exercise 3.3a

Find the circumference of each circle.

In questions 1 to 10 assume $\pi = 3\frac{1}{7}$ or $\frac{22}{7}$.

1. diameter $= 21$ cm   **2.** diameter $= 49$ cm
3. radius $= 35$ cm   **4.** radius $= 7$ cm
5. radius $= 2\cdot1$ m   **6.** radius $= 2\cdot8$ m
7. diameter $= 8\cdot4$ m   **8.** diameter $= 147$ mm
9. diameter $= 105$ mm   **10.** diameter $= 91$ mm

In questions 11 to 20 assume $\pi = 3\cdot14$.

11. diameter $= 5$ cm   **12.** diameter $= 3$ cm
13. diameter $= 11$ cm   **14.** radius $= 4$ cm
15. radius $= 15$ cm   **16.** radius $= 200$ mm
17. radius $= 300$ mm   **18.** radius $= 450$ mm
19. diameter $= 1\cdot5$ m **20.** diameter $= 2\cdot5$ m

In questions 21 to 30 assume $\pi = 3\cdot142$.

21. diameter $= 100$ mm   **22.** diameter $= 20$ cm
23. diameter $= 5$ m   **24.** diameter $= 8$ cm
25. radius $= 5$ cm   **26.** radius $= 50$ cm
27. radius $= 200$ mm   **28.** radius $= 25$ cm
29. diameter $= 150$ mm   **30.** radius $= 150$ mm

Example 2

Find (a) the diameter, (b) the radius of a circle whose circumference is 132 cm. (Take $\pi = 3\frac{1}{7}$).

(a) circumference $= \pi D$

$$132 = 3\tfrac{1}{7} D$$

$\therefore$ diameter $D = 132 \div 3\frac{1}{7}$

$$= \frac{\overset{6}{\cancel{132}}}{1} \times \frac{7}{\underset{1}{\cancel{22}}} = 42 \text{ cm}$$

(b) radius $=$ (diameter) $\div 2$
$$= 42 \div 2 = 21 \text{ cm}$$

Exercise 3.3b

Find (a) the diameter, (b) the radius of each circle, assuming $\pi = 3\frac{1}{7}$ or $\frac{22}{7}$.

1. circumference $= 176$ cm
2. circumference $= 264$ cm
3. circumference $= 352$ mm
4. circumference $= 308$ mm
5. circumference $= 396$ mm
6. circumference $= 484$ mm
7. circumference $= 13\cdot2$ m
8. circumference $= 4\cdot4$ m
9. circumference $= 57\cdot2$ cm
10. circumference $= 61\cdot6$ cm

Example 3

Find the area of a circle when

(a) the radius is 14 cm, using $\pi = 3\frac{1}{7}$
(b) the diameter is 20 cm, using $\pi = 3\cdot14$

(a) area $= \pi r^2 = 3\frac{1}{7} \times 14 \times 14$

$$= \frac{22}{\underset{1}{\cancel{7}}} \times \frac{\overset{2}{\cancel{14}}}{1} \times \frac{14}{1} = 616 \text{ cm}^2$$

(b) diameter $= 20$ cm, so radius $= 10$ cm
$$\begin{aligned} \text{area} = \pi r^2 &= 3\cdot14 \times 10 \times 10 \\ &= 3\cdot14 \times 100 \\ &= 314 \text{ cm}^2 \end{aligned}$$

Exercise 3.3c

Find the area of each circle.
In questions 1 to 10 assume $\pi = 3\frac{1}{7}$ or $\frac{22}{7}$.

1. radius $= 21$ cm
2. radius $= 7$ cm
3. diameter $= 70$ mm
4. diameter $= 56$ mm
5. radius $= 0\cdot7$ m
6. radius $= 1\cdot05$ m
7. radius $= 3\frac{1}{2}$ cm
8. radius $= 1\frac{3}{4}$ cm
9. radius $= \frac{7}{20}$ m
10. diameter $= 2\frac{4}{5}$ m

In questions 11 to 20 assume $\pi = 3\cdot14$

11. radius $= 2$ cm
12. radius $= 3$ cm
13. diameter $= 10$ cm
14. diameter $= 60$ mm
15. diameter $= 40$ mm
16. diameter $= 30$ mm
17. diameter $= 50$ mm
18. radius $= 1\cdot5$ m
19. radius $= 0\cdot5$ m
20. radius $= 2\cdot5$ m

Example 4

How many revolutions does a cycle wheel of diameter 63 cm complete in travelling 396 m? (Take $\pi = \frac{22}{7}$)

circumference of wheel $= \pi D$

$$= \tfrac{22}{7} \times \tfrac{63}{1} = 198 \text{ cm}$$

distance to travel $= 396$ m
$$= 396 \times 100 = 39\,600 \text{ cm}$$
$\therefore$ no. of revolutions $= 39\,600 \div 198 = 200$

Example 5

A circular pond has an area of 154 m². How much would it cost to fence round the pond if 1 m of fencing costs £1·50? (Take $\pi = 3\frac{1}{7}$).

$$\text{area} = \pi r^2; \therefore r = \sqrt{(\text{area} \div \pi)}$$
$$= \sqrt{(154 \div 3\tfrac{1}{7})}$$
$$= \sqrt{(\tfrac{154}{1} \times \tfrac{7}{22})}$$
$$= \sqrt{49} = 7 \text{ cm}$$

circumference $= 2\pi r = \tfrac{2}{1} \times \tfrac{22}{7} \times \tfrac{7}{1}$
$$= 44 \text{ m}$$

$\therefore$ cost $= £1\cdot50 \times 44 = £66$

Exercise 3.3d

Assume that $\pi = 3\frac{1}{7}$ or $\frac{22}{7}$

1. The pulley on the jib of a crane has a diameter of 56 cm. Find
   (a) the circumference of the pulley,
   (b) the number of revolutions turned by the pulley when the crane raises a load through a vertical height of 44 m.
2. The pulley above a lift cage has a diameter of 14 cm. Find
   (a) the circumference of the pulley,
   (b) the number of revolutions it makes while the cage ascends 33 m.
3. At Cannock colliery the cage pulley has a diameter of 3·5 m. Find
   (a) the circumference of the pulley,
   (b) the number of revolutions it makes while the miner's cage descends to the coal seam 495 m below the ground.
4. A bowling-green roller has a diameter of 1·4 m. Find
   (a) the circumference of the roller,
   (b) the number of revolutions it makes while travelling the length of a 30·8 m green.
5. A glass cutter has a cutting wheel of diameter 9·8 mm. Find
   (a) the circumference of the wheel,
   (b) the number of revolutions it makes while cutting a pane of width 154 cm.
6. A wall tin-opener has a cutting wheel of diameter 1·68 cm. How many revolutions will it make while opening a tin of diameter 15·12 cm?
7. A sheet of metal has dimensions 56 cm by 33 cm. It is melted down and recast into discs of the same thickness and radius 7 cm. How many discs will be cast?

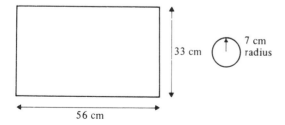

8. Find the area of the shaded part in the diagram.

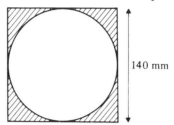

9. A circular pond has a radius of 10·5 m and its base is to be covered with a concrete layer. If 80 kg of dry mixture is sufficient to cover 1 m^2 and it is supplied in 120-kg bags, how many bags are required?
10. A 10p coin has an area of 6·16 cm^2.
    Find (a) its radius
         (b) its diameter
         (c) its circumference
         (d) the number of grooves on its milled edge circumference if the distance between two successive grooves is 0·55 mm.
11. Find the area of the shaded part in the diagram.

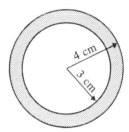

12. The diagram shows a point X on the circumference of a gramophone record and a point Y on the circumference of the label.

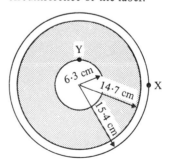

    (a) What is the distance travelled by the point X in 30 minutes when the record is played at $33\frac{1}{3}$ r.p.m.?
    (b) What is the speed of the point Y in km/h when the record is played at $33\frac{1}{3}$ r.p.m.?
    (c) What is the area of the playing surface (shaded portion) of the record?

Volume is a measure of the space occupied by a solid; volume is measured in cubic units, e.g. m^3, cm^3, mm^3.

Capacity is the volume of a liquid or a gas, measured in litres which a vessel of the same shape and size as the solid can hold, 1 litre = 1000 cm^3 or millilitres

Prism

A prism is a solid with a uniform cross section, as shown in the examples below.

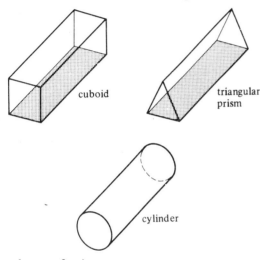

cuboid

triangular prism

cylinder

volume of prism
   = (area of cross section) × length
   = (area of end) × length

Example 1

A cuboid measures 5 cm by 12 cm by 10 cm. What is its volume?

area of end = 5 × 12 cm^2
   volume = 5 × 12 × 10 cm^3
          = 600 cm^3

Example 2

A rectangular tank measures 2 m by 3 m by 4 m and is full of water. What is its capacity in litres?

area of end = 200 × 300 cm^2
   volume = 200 × 300 × 400 cm^3

$$capacity = \frac{\overset{1}{\cancel{200}} \times \overset{60}{\cancel{300}} \times 400}{\underset{5}{\cancel{1000}}} \text{ litres}$$

   = 24 000 litres

Exercise 3.4a

Find the volume of each cuboid.

  1. height = 5 cm, length = 11 cm, width = 6 cm.
  2. height = 3 cm, length = 20 cm, width = 12 cm.
  3. height = 5 cm, length = 16 cm, width = 15 cm.
  4. height = 1·2 m, length = 5 m, width = 1·5 m.
  5. height = 2·5 m, length = 3·5 m, width = 3·2 m.
  6. height = 1·25 m, length = 6·4 m, width = 4·5 m.

Find the volumes of the following cubes.

  7. side-length = 9 cm.   8. side-length = 12 cm.
  9. side-length = 60 mm.   10. side-length = 80 mm.

Find (a) the volume of each of the following tanks and (b) the capacity of each tank in litres.

 11. height = 3 m, length = 6 m, width = 4 m.
 12. height = 2 m, length = 7 m, width = 6 m.
 13. height = 1·6 m, length = 5 m, width = 3·5 m.
 14. height = 2·4 m, length = 4·5 m, width = 2·5 m.
 15. height = 1·5 m, length = 3·75 m, width = 1·6 m.
 16. height = 0·15 m, length = 0·5 m, width = 0·32 m.
 17. height = 0·16 m, length = 0·25 m, width = 0·2 m.
 18. height = 25 cm, length = 30 cm, width = 16 cm.
 19. height = 36 cm, length = 50 cm, width = 15 cm.
 20. height = 24 cm, length = 35 cm, width = 25 cm.

Example 3

Find the volume of the triangular prism shown.

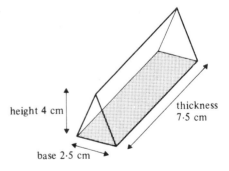

height 4 cm

thickness 7·5 cm

base 2·5 cm

area of end = $\frac{1}{2}$ (2·5 × 4) cm^2

$$volume = \frac{2 \cdot 5 \times \overset{2}{\cancel{4}}}{\underset{1}{\cancel{2}}} \times 7 \cdot 5 \ cm^3$$

   = 5 × 7·5 = 37·5 cm^3

Example 4

A cylinder has a radius of $3\frac{1}{2}$ cm and is 14 cm long. Find its volume. (Take $\pi = \frac{22}{7}$).

area of end $= \pi r^2 = \frac{22}{7} \times \frac{7}{2} \times \frac{7}{2}$ cm^2

$$\text{volume} = \frac{\overset{11}{\cancel{22}}}{\cancel{7}_1} \times \frac{\overset{1}{\cancel{7}}}{\cancel{2}_1} \times \frac{7}{\cancel{2}_1} \times \frac{\overset{7}{\cancel{14}}}{1}$$

$$= 539 \text{ cm}^3$$

Exercise 3.4b

Find the volume of each triangular prism.

1. base = 6 cm
   height = 5 cm
   thickness = 12 cm
2. base = 7 cm
   height = 4 cm
   thickness = 11 cm
3. base = 8 cm
   height = 9 cm
   thickness = 15 cm
4. base = 12 cm
   height = 10 cm
   thickness = 25 cm
5. base = 40 mm
   height = 50 mm
   thickness = 75 mm
6. base = 30 mm
   height = 48 mm
   thickness = 25 mm
7. base = 50 mm
   height = 32 mm
   thickness = 15 mm
8. base = 18 mm
   height = 25 mm
   thickness = 24 mm
9. base = 0·8 m
   height = 1·5 m
   thickness = 5 m
10. base = 4·8 m
    height = 2·5 m
    thickness = 1·5 m

Find the volume of each cylinder (take $\pi = 3\frac{1}{7}$ or $\frac{22}{7}$).

11. radius = 7 cm
    length = 12 cm
12. radius = 3 cm
    length = 21 cm
13. radius = 5 cm
    length = 14 cm
14. radius = 8 cm
    length = 35 cm
15. radius = 10 mm
    length = 28 mm
16. radius = 14 mm
    length = 15 mm
17. radius = 20 mm
    length = 56 mm
18. radius = 15 mm
    length = 70 mm
19. radius = 0·7 m
    length = 2 m
20. radius = 0·5 m
    length = 1·4 m

Example 5

A machine for making ice freezes 5·76 litres of water into ice cubes measuring 4 cm by 3 cm by 2 cm. How many bricks will be made?

volume of 1 ice brick $= 4 \times 3 \times 2 = 24$ cm^3
volume of water $= 5·76$ litres
$= 5·76 \times 1000$ cm^3
$= 5760$ cm^3
$\therefore$ no. of bricks made $= 5760 \div 24 = 240$

Example 6

A storage tank at a petrol station is a cuboid measuring 4 m long by 3 m wide by 2 m deep. During one day the garage sells 7200 litres of petrol from the tank. What is the fall in the level of the petrol in the tank?

volume of petrol sold $= 7200 \times 1000$ cm^3
area of cross section of tank
$= 400 \times 300$ cm^2

$$\therefore \text{depth fallen} = \frac{\overset{24}{\cancel{7200}} \times \overset{10}{\cancel{1000}}}{\underset{4}{\cancel{400}} \times \underset{1}{\cancel{300}}}$$

$$= \frac{240}{4} = 60 \text{ cm}$$

Exercise 3.4c

1. A rectangular block of metal has dimensions equal to 24, 12 and 8 cm. It is melted down and recast into cubes of edge length 4 cm. How many cubes will be cast?
2. A child has a number of toy bricks, all of which are cubes of edge length 6 cm. He also has a wooden box which is 30 cm long, 24 cm wide and 18 cm deep. If all of the bricks pack exactly into the box, how many bricks does he have?
3. An ice machine freezes 6 litres of water and then forms ice bricks of dimensions 5, 4 and 2 cm. How many ice bricks will be formed?
4. The platform at the base of a statue is to be built to a height of 50 cm on a square base measuring 150 cm by 150 cm. It is to be constructed from bricks of dimensions 30, 15 and 10 cm. How many bricks will be required?
5. A wall, 18 m long, 120 cm high and 45 cm thick, is to be constructed using bricks of dimensions 30, 15 and 10 cm. How many bricks will be required?

6. A coffee urn has a square base measuring 30 cm by 30 cm and it is 50 cm in height. It is used to serve coffee into cups of capacity 0.15 litre. By how many centimetres will the level of the coffee in the urn drop after 120 cups have been served?

7. The dimensions of a greenhouse are shown in the diagram. Calculate the total volume enclosed by the structure.

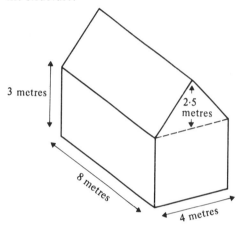

3 metres

2·5 metres

8 metres

4 metres

8. A drinking trough for cattle takes the shape of an inverted triangular prism. It is 3 m long, 40 cm wide and its maximum depth is 30 cm. Find its capacity in litres. If it is filled by using a bucket of capacity 15 litres, how many buckets full of water are required?

9. A beer cask has a radius of 20 cm and is 70 cm high. Find its capacity in litres. How many glasses full of beer can it serve if the capacity of each glass is 0·55 litres? $(\pi = \frac{22}{7})$.

10. A railway tunnel 140 m in length is to be bored with a circular cross section of radius 4 metres. What volume of spoil has to be excavated? If the spoil is to be taken away in wagons of capacity 88 m^3, how many wagon loads will be moved? $(\pi = \frac{22}{7})$.

3.5 VOLUME II

Pyramid

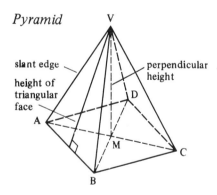

slant edge

perpendicular height

height of triangular face

A

V

D

M

C

B

volume of a pyramid

$$= \frac{\text{(area of base)} \times \text{(perpendicular height)}}{3}$$

Example 1

Find the volume of a square-based pyramid VABCD in which AB = BC = CD = AD = 8 cm, and vertical height VM = 9cm.

$$\text{volume} = \frac{8 \times 8 \times \overset{3}{\cancel{9}}}{\underset{1}{\cancel{3}}} = 192 \text{ cm}^3$$

Exercise 3.5a

Find the volume of each square-based pyramid.

1. base dimension = 5 cm, perpendicular height
= 6 cm.
2. base dimension = 9 cm, perpendicular height
= 7 cm.
3. base dimension = 12 cm, perpendicular height
= 10 cm.
4. base dimension = 15 cm, perpendicular height
= 16 cm.
5. base dimension = 40 mm, perpendicular height
= 60 mm.
6. base dimension = 30 mm, perpendicular height
= 45 mm.
7. base dimension = 25 mm, perpendicular height
= 18 mm.
8. base dimension = 0·6 m, perpendicular height
= 0·5 m.
9. base dimension = 0·8 m, perpendicular height
= 1·5 m.
10. base dimension = 1·2 m, perpendicular height
= 0·75 m.

Cone

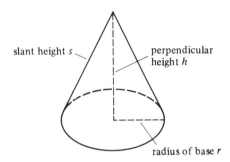

slant height *s*

perpendicular height *h*

radius of base *r*

volume of a cone

$$= \frac{(\text{area of base}) \times (\text{perpendicular height})}{3}$$

$$= \frac{\pi r^2 h}{3}$$

Example 2

Find the volume of a cone, with base radius 6 cm, and perpendicular height 5 cm (take $\pi = 3\cdot14$).

$$\text{volume} = \frac{\pi r^2 h}{3} = \frac{3\cdot14 \times \overset{2}{6} \times 6 \times 5}{\underset{1}{3}} \text{ cm}^3$$

$$= 3\cdot14 \times 60 = 188\cdot4 \text{ cm}^3$$

Exercise 3.5b

Find the volumes of the following cones.

1. radius = 3 cm, perpendicular height = 7 cm. $(\pi = \frac{22}{7})$
2. radius = 9 cm, perpendicular height = 35 cm. $(\pi = \frac{22}{7})$
3. radius = 10 cm, perpendicular height = 15 cm. $(\pi = 3\cdot14)$
4. radius = 15 cm, perpendicular height = 40 cm. $(\pi = 3\cdot14)$
5. radius = 60 mm, perpendicular height = 105 mm. $(\pi = \frac{22}{7})$
6. radius = 45 mm, perpendicular height = 70 mm. $(\pi = \frac{22}{7})$
7. radius = 20 mm, perpendicular height = 60 mm. $(\pi = 3\cdot14)$
8. radius = 0·7 m, perpendicular height = 1·5 m. $(\pi = \frac{22}{7})$
9. radius = 0·35 m, perpendicular height = 1·2 m. $(\pi = \frac{22}{7})$
10. radius = 0·4 m, perpendicular height = 0·75 m. $(\pi = 3\cdot14)$

Sphere

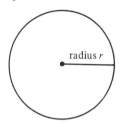

radius *r*

$$\text{volume of a sphere} = \frac{4\pi r^3}{3}$$

Example 3

Find the volume of a sphere with radius 7 mm (Take $\pi = 3\frac{1}{7}$).

$$\text{volume} = \frac{4\pi r^3}{3}$$

$$= \frac{4 \times 22 \times 7 \times 7 \times \overset{1}{\cancel{7}}}{3 \times \underset{1}{\cancel{7}}}$$

$$= \frac{88 \times 49}{3}$$

$$= \frac{4312}{3} = 1437\frac{1}{3} \text{ mm}^3$$

Exercise 3.5c

Find the volume of each sphere.

1. radius = 3 cm $(\pi = 3\cdot14)$
2. radius = 21 mm $(\pi = \frac{22}{7})$
3. radius = 90 mm $(\pi = 3\cdot14)$
4. diameter = 12 cm $(\pi = 3\cdot14)$
5. diameter = 3 cm $(\pi = 3\cdot14)$
6. radius = 1·05 cm $(\pi = \frac{22}{7})$
7. radius = $5\frac{1}{4}$ cm $(\pi = \frac{22}{7})$
8. radius = $4\frac{1}{2}$ cm $(\pi = 3\cdot14)$
9. radius = 2·25 cm $(\pi = 3\cdot14)$
10. radius = $\frac{3}{4}$ cm $(\pi = 3\cdot14)$

Example 4

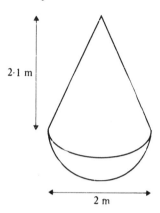

2·1 m

2 m

A navigational buoy consists of a hemispherical base of diameter 2 m with a cone
2·1 m high on top of it.
Find the volume of the buoy.
(Take $\pi = 3\cdot14$)

volume of hemispherical base

$$= \frac{2\pi r^3}{3} = \frac{2 \times 3\cdot14 \times 1 \times 1 \times 1}{3}$$

$$= 2\cdot093 \text{ m}^3$$

volume of cone $= \dfrac{\pi r^2 h}{3}$

$$= \frac{3\cdot14 \times 1 \times 1 \times 2\cdot1}{3}$$

$$= 2\cdot198 \text{ m}^3$$

volume of buoy $= 2\cdot093 + 2\cdot198$
$$= 4\cdot291 \text{ m}^3$$

Exercise 3.5d

1.

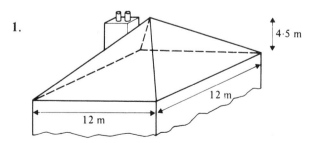

4·5 m

12 m

12 m

The roof dimensions of a detached house are
shown in the diagram. Find the volume enclosed
by the loft.

2. The dimensions of
a gate post are
shown in the
diagram.
Find its volume.

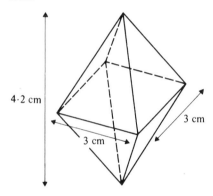

6 cm

20 cm

20 cm

1 m

3. A research chemist has grown a large crystal of
alum. The crystal is an octahadron, i.e. two
identical square-based pyramids with a common
base.

4·2 cm

3 cm

3 cm

Find the volume of the crystal from the dimensions
shown. If 1 cm^3 of alum weighs 1·5 g, how much
does this crystal weigh?

4. A conical fire extinguisher has a radius of 20 cm
and a perpendicular height of 54 cm. How many
litres of pressurized foam does it contain when
full? ($\pi = 3\cdot14$).

5. A large conical flask has a radius of 10 cm and a
perpendicular height of 30 cm. On the side it is
marked 'Approx. 3 litres'. How much does this
figure differ from its true capacity? (Take $\pi = 3\cdot14$).

6. A wooden cone has an outer radius of 30 cm and
an inner radius of 25 cm. The outer and inner
heights are 20 cm and 18 cm respectively. Find
the volume of wood in the cone. ($\pi = 3\cdot14$).

7. Oil fills an inverted metal cone to a depth of
20 cm. If the radius of the surface of the oil is
15 cm, find the volume of oil.
It is then poured into a rectangular can of base
25 cm by 15·7 cm. Find the depth of oil in the
can. Assume that $\pi = 3\cdot14$.

8. A steel cuboid measuring 62·8 mm by 18 mm by
8 mm is melted down and cast into ball bearings
of radius 3 mm. How many ball bearings are
cast? (Take $\pi = 3\cdot14$).

The length of an arc is proportional to the angle it subtends at the centre of the circle.

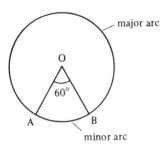

In the diagram
length of minor arc AB
$= \frac{60}{360}$ of circumference.

Example 1

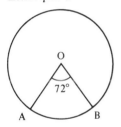

A circle has a circumference of 35 cm. What is the length of the arc AB if the angle it subtends at the centre is 72°?

length of arc AB

$= \frac{72}{360}$ of circumference

$= \frac{\overset{1}{\cancel{72}}}{\underset{\cancel{1}}{\cancel{360}}} \times \frac{\overset{7}{\cancel{35}}}{1} = 7$ cm

Exercise 3.6a

Find the length of each arc.

	Circumference of circle	Angle subtended
1.	18 cm	60°
2.	32 cm	45°
3.	15 cm	120°
4.	24 cm	270°
5.	36 cm	210°

	Circumference of circle	Angle subtended
6.	20 mm	90°
7.	50 mm	108°
8.	80 mm	135°
9.	40 mm	225°
10.	70 mm	324°
11.	4·5 cm	240°
12.	9·6 cm	150°
13.	14·4 cm	75°
14.	13·5 cm	160°
15.	3·75 cm	192°

Example 2

An arc of a circle of circumference 70 cm is 42 cm long. What is the angle subtended by this arc at the centre of the circle?

angle required $= \frac{42}{70}$ of 360°

$= \frac{\overset{6}{\cancel{42}}}{\underset{1}{\cancel{70}}} \times \frac{\overset{36}{\cancel{360}}}{1}$

$= 216°$

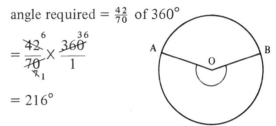

Exercise 3.6b

Find the angle subtended by the arc at the centre of the circle.

	Length of arc	Circumference of circle
1.	24 cm	96 cm
2.	30 cm	75 cm
3.	18 cm	216 cm
4.	105 mm	126 mm
5.	180 mm	225 mm
6.	385 mm	840 mm
7.	54 mm	486 mm
8.	7·7 cm	26·4 cm
9.	4·5 cm	16·2 cm
10.	22·4 cm	25·6 cm
11.	14·4 cm	27 cm
12.	18 cm	32·4 cm
13.	0·168 m	0·840 m
14.	0·196 m	0·280 m
15.	0·352 m	38·4 cm

Example 3

A circle has a radius of $3\frac{1}{2}$ cm.
Find (a) the angle subtended at the
    centre by an arc $5\frac{1}{2}$ cm long,
   (b) the length of an arc which
    subtends an angle of $108°$.

circumference of circle $= \pi D$

$$= \tfrac{22}{7} \times \tfrac{7}{1} = 22 \text{ cm}$$

(a) angle subtended $= \dfrac{5\frac{1}{2}}{22}$ of $360°$

$$= \tfrac{11}{44} \times \tfrac{360}{1} = 90°$$

(b) length of arc $= \tfrac{108}{360}$ of 22

$$= \tfrac{108}{360} \times \tfrac{22}{1}$$

$$= \tfrac{33}{5} = 6\cdot6 \text{ cm}$$

Exercise 3.6c

For each circle find the angle subtended by the
arc. $(\pi = \tfrac{22}{7})$

	Radius of circle	Arc length
1.	21 mm	44 mm
2.	70 mm	66 mm
3.	10·5 mm	55 mm
4.	31·5 mm	33 mm
5.	3·5 cm	6·6 cm
6.	2·1 cm	3·3 cm
7.	2·8 cm	2·2 cm
8.	0·42 m	0·33 m
9.	0·28 m	110 cm
10.	0·35 m	0·88 m

For each circle find the length of arc subtended
by the given angle. $(\pi = \tfrac{22}{7})$

	Radius of circle	Size of angle subtended
11.	21 mm	150°
12.	70 mm	72°
13.	10·5 mm	60°
14.	31·5 mm	140°
15.	3·5 cm	144°
16.	2·1 cm	120°
17.	2·8 cm	216°
18.	0·42 m	60°
19.	0·28 m	135°
20.	0·35 m	54°

A circle of longitude is a great circle,
i.e. a circle with a radius which is the
radius of the Earth.
The distance between two places on the
same line of longitude on the Earth is the
length of the minor arc between them.

Example 4

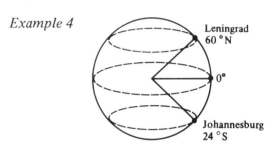

Leningrad
60 °N

0°

Johannesburg
24 °S

Leningrad and Johannesburg lie on the
same circle of longitude, 29 °E. If their
latitudes are 60 °N and 24 °S respectively,
find the distance between them. Take the
radius of the Earth to be 6300 km and
$\pi = \tfrac{22}{7}$.

angle subtended at centre $= 60° + 24°$
$$= 84°$$
circumference of circle of longitude
$$= 2 \times \tfrac{22}{7} \times 6300 \text{ km}$$
distance between the two places

$$= \tfrac{84}{360} \times 2 \times \tfrac{22}{7} \times \tfrac{6300}{1}$$

$$= 9240 \text{ km}$$

Exercise 3.6d

Find the distance between each pair of towns
from their latitude north or south of the equator.
Take the radius of the Earth to be 6300 km and
$\pi = \tfrac{22}{7}$.

1. Budapest (47 °N 19 °E);
   Cape Town (33 °S, 19 °E).
2. Berlin (52 °N, 13 °E);
   Luanda (8 °S, 13 °E).
3. Jerusalem (31 °N, 35 °E);
   Beira (19 °S, 35 °E).
4. Tokyo (36 °N, 139 °E);
   Adelaide (34 °S, 139 °E).
5. Cairo (30 °N, 31 °E);
   Salisbury, Rhodesia (18 °S, 31 °E).
6. Archangel (65 °N, 40°E);
   Dar es Salaam (7 °S, 40 °E).

7. Buffalo (42 °N, 78 °W);
   Lima (12 °S, 78 °W).
8. Moscow (55 °N, 37 °E);
   Nairobi (1 °S, 37 °E).
9. Manila (13 °N, 121 °E);
   Kalgoorlie (31 °S, 121 °E).
10. Boston (42 °N, 71 °W);
   Santiago (33 °S, 71 °W).

Example 5

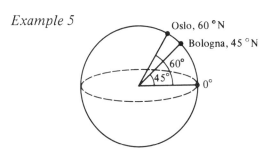

Oslo and Bologna lie on the same circle of longitude, 11 °E. If their latitudes are 60 °N and 45 °N respectively, find the distance between them. Take the radius of the Earth to be 6300 km and $\pi = \frac{22}{7}$.

angle subtended at centre $= 60° - 45°$
$\qquad\qquad\qquad\qquad\quad = 15°$
circumference of circle of longitude
$\qquad\qquad\qquad = 2 \times \frac{22}{7} \times 6300$ km
distance between the two places

$= \frac{15}{360} \times \frac{2}{1} \times \frac{22}{7} \times \frac{6300}{1}$

$= 1650$ km

Exercise 3.6e

Find the distance between each pair of towns from their latitude. Take the radius of the Earth to be 6300 km and $\pi = \frac{22}{7}$.

1. Ankora (39 °N, 32 °E);
   Khartoum (15 °N, 32 °E).
2. Winnipeg (49 °N, 98 °W);
   Mexico City (19 °N, 98 °W).
3. Port Arthur (48 °N, 90 °W);
   New Orleans (30 °N, 90 °W).
4. Omsk (54 °N, 73 °E);
   Bombay (18 °N, 73 °E).
5. Shanghai (31 °N, 121 °E);
   Manila (13 °N, 121°E).
6. Prague (50 °N, 14 °E);
   Naples (41 °N, 14 °E).
7. London (51 °N, 0°);
   Accra (6 °N, 0°).
8. Peace River (57 °N, 118 °W);
   Los Angeles (35 °N, 118 °W);
9. Durban (30 °S, 31 °E);
   Salisbury, Rhodesia (18 °S, 31 °E).
10. South Orkney Islands (62 °S, 44 °W);
   Rio de Janeiro (22 °S, 44 °W).

The number which is:

> 4 more than $x = x + 4$
> y more than $2 = 2 + y$
> 3 less than $b = b - 3$
> m less than $5 = 5 - m$

Exercise 4.1a

Write down the number which is:

1. 3 more than x
2. 7 more than y
3. 5 more than t
4. 8 more than u
5. 11 more than v
6. x more than 9
7. y more than 4
8. a more than 6
9. b more than 12
10. c more than 1
11. 4 less than x
12. 8 less than y
13. 10 less than l
14. 2 less than m
15. 6 less than n
16. x less than 5
17. y less than 3
18. p less than 11
19. q less than 7
20. r less than 9

The number which is:

> 2 times $x = 2x$
> 4 times $b = 4b$
> $\frac{1}{4}$ of $a = \frac{1}{4}a$ or $\frac{a}{4}$
> $\frac{2}{3}$ of $b = \frac{2}{3}b$ or $\frac{2b}{3}$
> x divided by $2 = \frac{x}{2}$
> 3 divided by $y = \frac{3}{y}$

Exercise 4.1b

Write down the numbers which are,

1. 3 times x
2. 9 times y
3. 4 times z
4. 7 times a
5. 12 times b
6. $\frac{1}{5}$ of x
7. $\frac{1}{8}$ of y
8. $\frac{1}{3}$ of z
9. $\frac{3}{4}$ of l
10. $\frac{5}{6}$ of m
11. $\frac{2}{5}$ of n
12. x divided by 6
13. y divided by 9
14. z divided by 12
15. t divided by 10
16. 5 divided by x
17. 7 divided by y
18. 11 divided by z
19. 8 divided by u
20. 4 divided by v

The number which is:

> y more than $x = x + y$
> y less than $x = x - y$
> x times $y = xy$
> x times $x = x^2$
> x divided by $y = \frac{x}{y}$

Exercise 4.1c

Write down the numbers which are,

1. b more than x
2. c more than y
3. x more than m
4. y more than n
5. u less than x
6. v less than y
7. x less than p
8. y less than q
9. a times b
10. m times n
11. u times v
12. p times q
13. a times a
14. b times b
15. t times t
16. x divided by a
17. y divided by b
18. t divided by z
19. m divided by x
20. n divided by y

Example 1

How many grams are there in

(a) 1 kg?   (b) 8 kg?   (c) x kg?

In 1 kg there are 1000 g.
$\therefore$ in 8 kg there are $1000 \times 8 = 8000$ g
$\therefore$ in x kg there are $1000 \times x = 1000x$ g

Example 2

How many £'s are there in

(a) 200 pence?   (b) $8y$ pence?

100 pence = £1

$\therefore$ 200 pence $= £\frac{200}{100} = £2$

$\therefore$ $8y$ pence $= £\frac{8y}{100} = £\frac{2y}{25}$

Exercise 4.1d

1. How many metres are there in
   (a) 1 km?   (b) 5 km?   (c) z km?
2. How many centimetres are there in
   (a) 1 m?   (b) 3 m?   (c) s m?
3. How many millilitres are there in
   (a) 1 litre?   (b) 8 litres?   (c) a litres?
4. How many pence are there in
   (a) £1?   (b) £6?   (c) £b?
5. How many seconds are there in
   (a) 1 minute? (b) 4 minutes? (c) t minutes?
6. How many millimetres are there in
   (a) 1 cm?   (b) 9 cm?   (c) p cm?
7. How many kilograms are there in
   (a) 1 tonne? (b) 7 tonnes? (c) w tonnes?
8. How many minutes are there in
   (a) 1 hour? (b) 12 hours? (c) r hours?

9. How many hours are there in
   (a) 1 day? (b) 5 days? (c) q days?
10. How many litres are there in
   (a) 1 m³? (b) 11 m³? (c) v m³?
11. How many litres are there in
   (a) 3000 ml? (b) 400a ml?
12. How many kilometres are there in
   (a) 4000 m? (b) 750l m?
13. How many £'s are there in
   (a) 900 p? (b) 60c p?
14. How many metres are there in
   (a) 600 cm? (b) 80d cm?

15. How many days are there in
   (a) 72 hours? (b) 16e hours?
16. How many minutes are there in
   (a) 300 seconds? (b) 15f seconds?
17. How many hours are there in
   (a) 480 minutes? (b) 25g minutes?
18. How many centimetres are there in
   (a) 70 mm? (b) 2h mm?
19. How many cubic metres are there in
   (a) 12 000 litres? (b) 625i litres?
20. How many tonnes are there in
   (a) 10 000 kilograms? (b) 375j kilograms?

4.2 DIRECTED NUMBERS

To add directed numbers with the *same* sign, put down their sign together with their sum.

(a) $+2 + +4 = +2 + 4 = +6$
(b) $-2 + -4 = -2 - 4 = -6$
(c) $-3 + -4 + -5 = -3 - 4 - 5 = -12$

To add two directed numbers with *different* signs, put down the sign of the larger together with their difference.

(a) $+2 + -4 = +2 - 4 = -2$
(b) $-2 + +4 = -2 + 4 = +2$
(c) $(+2 + -3) + +4 = -1 + 4 = +3$

Exercise 4.2a

1. $+6 + +8$	2. $+9 + +12$
3. $+24 + +11$	4. $+45 + +12$
5. $+13 + +62$	6. $+3 + +7 + +8$
7. $+8 + +5 + +9$	8. $+6 + +12 + +10$
9. $+11 + +4 + +18$	10. $+15 + +9 + +12$
11. $-9 + -11$	12. $-8 + -16$
13. $-15 + -12$	14. $-24 + -21$
15. $-35 + -13$	16. $-3 + -8 + -4$
17. $-7 + -6 + -12$	18. $-9 + -10 + -11$
19. $-15 + -4 + -16$	20. $-21 + -15 + -14$
21. $+15 + -11$	22. $+24 + -6$
23. $+28 + -12$	24. $-6 + +21$
25. $-7 + +32$	26. $-25 + +10$
27. $-36 + +27$	28. $-17 + +9$
29. $+15 + -36$	30. $+8 + -23$
31. $+5 + +7 + -8$	32. $+3 + +16 + -5$
33. $+21 + +4 + -17$	34. $+25 + -12 + -9$
35. $+33 + -16 + -5$	36. $+7 + +8 + -22$
37. $-35 + +19 + +13$	38. $+17 + -26 + +5$
39. $+12 + -8 + -9$	40. $-18 + -6 + +15$

To subtract a directed number, change its sign and add.

(a) $+2 - +3 = +2 - 3 = -1$
(b) $+2 - -3 = +2 + 3 = +5$
(c) $-2 - +3 = -2 - 3 = -5$
(d) $-2 - -3 = -2 + 3 = +1$

Exercise 4.2b

1. $+27 - +15$	2. $+19 - +6$	3. $+13 - +4$
4. $+32 - +17$	5. $+34 - +16$	6. $+12 - +27$
7. $+18 - +29$	8. $+10 - +15$	9. $+9 - +17$
10. $+25 - +34$	11. $+13 - -2$	12. $+17 - -6$
13. $+19 - -8$	14. $+11 - -14$	15. $+16 - -17$
16. $+23 - -12$	17. $+15 - -18$	18. $+12 - -24$
19. $+18 - -14$	20. $+29 - -16$	21. $-7 - +12$
22. $-16 - +5$	23. $-12 - +18$	24. $-13 - +19$
25. $-22 - +13$	26. $-26 - +14$	27. $-12 - +19$
28. $-16 - +17$	29. $-24 - +18$	30. $-19 - +25$
31. $-7 - -12$	32. $-6 - -18$	33. $-11 - -26$
34. $-8 - -22$	35. $-15 - -31$	36. $-19 - -14$
37. $-16 - -10$	38. $-23 - -8$	39. $-24 - -16$
40. $-32 - -17$		

When two directed numbers with the *same* sign are multiplied together, the answer is positive.

(a) $+2 \times +4 = +8$
(b) $-2 \times -4 = +8$
(c) $-2\frac{1}{2} \times -4 = +10$

When two directed numbers with *different* signs are multiplied together, the answer is negative.

(a) $+2 \times -4 = -8$
(b) $-2 \times +4 = -8$
(c) $-2 \times -3 \times -4 = +6 \times -4 = -24$

Exercise 4.2c

1. $+8 \times +3$
2. $+15 \times +4$
3. $+16 \times +5$
4. $+14 \times +3$
5. $+18 \times +4$
6. $+1\frac{1}{2} \times +6$
7. $+2\frac{1}{2} \times +8$
8. $+1\frac{1}{4} \times +12$
9. $+3{\cdot}5 \times +4$
10. $+2{\cdot}2 \times +5$
11. -8×-4
12. -15×-5
13. -18×-3
14. $+3 \times -4 \times -11$
15. $-8 \times +2 \times -3$
16. -8^2
17. -11^2
18. $-1\frac{1}{2} \times -8$
19. $-1\frac{1}{4} \times -4$
20. $-2{\cdot}5 \times -6$
21. $+20 \times -4$
22. $+25 \times -3$
23. $+40 \times -5$
24. $-2 \times -5 \times -6$
25. $-4 \times -3 \times -3$
26. $+7 \times -7$
27. $+12 \times -12$
28. $+2\frac{1}{2} \times -10$
29. $+1\frac{1}{2} \times -12$
30. $+1\frac{1}{4} \times -8$
31. $-30 \times +3$
32. $-50 \times +4$
33. $-25 \times +6$
34. $-5 \times +3 \times +4$
35. $-4 \times +6 \times +5$
36. $-6 \times +6$
37. $-9 \times +9$
38. $-2\frac{1}{2} \times +12$
39. $-1\frac{1}{2} \times +4$
40. $-3{\cdot}5 \times +6$

When dividing directed numbers, the same rules are used as in multiplication:
(i) like signs give a positive answer;
(ii) unlike signs give a negative answer.

(a) $+16 \div +8 = +2$

(b) $-16 \div -4 = +4$

(c) $+16 \div -12 = -1\frac{1}{3}$

(d) $-12 \div +16 = -\frac{3}{4}$

Exercise 4.2d

1. $+60 \div +5$
2. $+108 \div +12$
3. $+72 \div +9$
4. $+132 \div +11$
5. $+140 \div +20$
6. $+6 \div +30$
7. $+4 \div +36$
8. $+9 \div +15$
9. $+60 \div +48$
10. $+28 \div +20$
11. $-144 \div -9$
12. $-156 \div -12$
13. $-121 \div -11$
14. $-128 \div -8$
15. $-132 \div -6$
16. $-7 \div -56$
17. $-5 \div -45$
18. $-14 \div -70$
19. $-56 \div -32$
20. $-40 \div -25$
21. $+135 \div -9$
22. $+144 \div -8$
23. $+133 \div -7$
24. $+126 \div -6$
25. $+168 \div -12$
26. $+12 \div -84$
27. $+9 \div -108$
28. $+28 \div -35$
29. $+24 \div -16$
30. $+36 \div -30$
31. $-176 \div +8$
32. $-216 \div +9$
33. $-228 \div +12$
34. $-91 \div +7$
35. $-102 \div +6$
36. $-8 \div +32$
37. $-11 \div +110$
38. $-16 \div +40$
39. $-25 \div +20$
40. $-27 \div +15$

4.3 SUBSTITUTION

Points to note:

1. $3a = a + a + a = 3 \times a$
2. a cubed $= a^3 = a \times a \times a$
3. $2a^2 = 2 \times a^2 = 2 \times a \times a$
4. $ab = a$ times $b = a \times b$
5. $2ab = 2 \times ab = 2 \times a \times b$
6. $2ab^2 = 2 \times ab^2 = 2 \times a \times b^2$
   $\qquad = 2 \times a \times b \times b$
7. $5a^2 b = 5 \times a^2 b = 5 \times a^2 \times b$
   $\qquad = 5 \times a \times a \times b$
8. $(4b)^2 = 4b \times 4b$
   $\qquad = 4 \times b \times 4 \times b$
   $\qquad = 4 \times 4 \times b \times b = 16b^2$
9. $a^2 \div 3 = \frac{1}{3}$ of $a^2 = \dfrac{a^2}{3}$

Example 1

If $x = 5$, find the value of:

(a) $3x$ (b) x^2 (c) $2x^3$

If $x = 5$ then

(a) $3x = 3 \times x$
$\qquad = 3 \times 5 = 15$
(b) $x^2 = x \times x$
$\qquad = 5 \times 5 = 25$
(c) $2x^3 = 2 \times x \times x \times x$
$\qquad = 2 \times 5 \times 5 \times 5$
$\qquad = 250$

Exercise 4.3a

Find the value of each of the following when
the letter assumes the given value.

1. (a) $4p$ (b) p^2 (c) p^3 if $p = 6$
2. (a) $5q$ (b) $3q^2$ (c) $4q^3$ if $q = 5$
3. (a) $7r$ (b) $8r^2$ (c) $3r^3$ if $r = 4$
4. (a) $12t$ (b) $3t^2$ (c) $2t^3$ if $t = 8$
5. (a) $9u$ (b) $5u^2$ (c) u^3 if $u = 7$
6. (a) $11v$ (b) $6v^2$ (c) $5v^3$ if $v = 3$
7. (a) $36x$ (b) $20x^2$ (c) $12x^3$ if $x = 2$
8. (a) $120y$ (b) $54y^2$ (c) $48y^3$ if $y = 1$
9. (a) $20z$ (b) $8z^2$ (c) z^3 if $z = 9$
10. (a) $100l$ (b) $7l^2$ (c) $6l^3$ if $l = 10$
11. (a) $12m$ (b) $5m^2$ (c) $3m^3$ if $m = 4$
12. (a) $11n$ (b) $9n^2$ (c) $5n^3$ if $n = 6$
13. (a) $30a$ (b) $6a^2$ (c) $3a^3$ if $a = 8$
14. (a) $16b$ (b) $8b^2$ (c) $10b^3$ if $b = 5$
15. (a) $12c$ (b) $4c^2$ (c) $2c^3$ if $c = 7$
16. (a) $32d$ (b) $15d^2$ (c) $25d^3$ if $d = 2$
17. (a) $50x$ (b) $12x^2$ (c) $4x^3$ if $x = 3$
18. (a) $40y$ (b) $6y^2$ (c) $2y^3$ if $y = 9$
19. (a) $1000z$ (b) $100z^2$ (c) $10z^3$ if $z = 10$
20. (a) $15t$ (b) $50t^2$ (c) $5t^3$ if $t = 20$

Example 2

If $x = 2$ and $y = 3$, find the value of:

(a) xy (b) $5x^2y$ (c) $(5y)^2$

(a) $xy = x \times y$
$= 2 \times 3 = 6$
(b) $5x^2y = 5 \times x \times x \times y$
$= 5 \times 2 \times 2 \times 3 = 60$
(c) $(5y)^2 = 5 \times y \times 5 \times y$
$= 5 \times 3 \times 5 \times 3 = 225$

Exercise 4.3b

Find the value of the following when the letters
assume the given values.

1. If $a = 5$ and $b = 6$:
   (a) $3ab$,   (b) $6a^2b$,   (c) $(2b)^2$
2. If $m = 3$ and $n = 5$:
   (a) $5mn$,   (b) $8m^2n$,   (c) $(4n)^2$
3. If $p = 8$ and $q = 2$:
   (a) $9pq$,   (b) $7p^2q$,   (c) $(5q)^2$
4. If $u = 4$ and $v = 3$:
   (a) $12uv$,   (b) $4u^2v$,   (c) $(3v)^2$

5. If $x = 5$ and $y = 4$:
   (a) $8xy$,   (b) $5x^2y$,   (c) $(2y)^2$
6. If $b = 3$ and $c = 6$:
   (a) $7bc$,   (b) $3b^2c$,   (c) $(5c)^2$
7. If $l = 2$ and $m = 3$:
   (a) $4lm$,   (b) $9l^2m$,   (c) $(4m)^2$
8. If $q = 10$ and $r = 5$:
   (a) $5qr$,   (b) $6q^2r$,   (c) $(2r)^2$
9. If $t = 8$ and $u = 4$:
   (a) $9tu$,   (b) $5t^2u$,   (c) $(5u)^2$
10. If $y = 5$ and $z = 2$:
    (a) $100yz$,   (b) $7y^2z$,   (c) $(3z)^2$
11. If $a = 5$ and $b = 8$:
    (a) $5ab$,   (b) $3ab^2$,   (c) $(6a)^2$
12. If $m = 2$ and $n = 3$:
    (a) $3mn$,   (b) $4mn^2$,   (c) $(6m)^2$
13. If $p = 2$ and $q = 7$:
    (a) $7pq$,   (b) $2pq^2$,   (c) $(4p)^2$
14. If $u = 4$ and $v = 9$:
    (a) $4uv$,   (b) $5uv^2$,   (c) $(3u)^2$
15. If $x = 2$ and $y = 10$:
    (a) $30xy$,   (b) $50xy^2$,   (c) $(10x)^2$
16. If $b = 3$ and $c = 8$:
    (a) $6bc$,   (b) $3bc^2$,   (c) $(2b)^2$
17. If $l = 3$ and $m = 7$:
    (a) $12lm$,   (b) $5lm^2$,   (c) $(10l)^2$
18. If $q = 10$ and $r = 6$:
    (a) $7qr$,   (b) $2qr^2$,   (c) $(2q)^2$
19. If $t = 2$ and $u = 4$:
    (a) $11tu$,   (b) $4tu^2$,   (c) $(2t)^2$
20. If $y = 10$ and $z = 3$:
    (a) $9yz$,   (b) $6yz^2$,   (c) $(3y)^2$

Example 3

If $y = 6$, find the value of:

(a) $5y - 6$ (b) $3y^2 + 2y$ (c) $\dfrac{y^3}{4}$

(a) $5y - 6 = (5 \times y) - 6$
$= (5 \times 6) - 6$
$= 30 - 6 = 24$
(b) $3y^2 + 2y = (3 \times y \times y) + (2 \times y)$
$= (3 \times 6 \times 6) + (2 \times 6)$
$= 108 + 12 = 120$

(c) $\dfrac{y^3}{4} = \dfrac{y \times y \times y}{4}$
$= \dfrac{6 \times 6 \times 6}{4} = \dfrac{216}{4} = 54$

Exercise 4.3c

Find the value of each of the following when the letter assumes the given value.

1. If $p = 4$:

   (a) $3p + 8$,   (b) $p^2 - 2p$,   (c) $\dfrac{p^2}{2}$

2. If $q = 10$:

   (a) $4q + 5$,   (b) $3q^2 + 5q$,   (c) $\dfrac{q^2}{4}$

3. If $r = 6$:

   (a) $5r - 6$,   (b) $2r^2 + 3r$,   (c) $\dfrac{r^2}{2}$

4. If $t = 10$:

   (a) $9t + 10$,   (b) $5t^2 - 2t$,   (c) $\dfrac{t^2}{5}$

5. If $u = 8$:

   (a) $7u - 6$,   (b) $2u^2 - 3u$,   (c) $\dfrac{u^2}{4}$

6. If $v = 6$:

   (a) $9v - 14$,   (b) $3v^2 + 2v$,   (c) $\dfrac{v^2}{3}$

7. If $x = 4$:

   (a) $12x + 7$,   (b) $5x^2 + 4x$,   (c) $\dfrac{x^2}{8}$

8. If $y = 10$:

   (a) $11y + 15$,   (b) $2y^2 + 15y$,   (c) $\dfrac{y^2}{20}$

9. If $z = 8$:

   (a) $6z - 13$,   (b) $z^2 - 7z$,   (c) $\dfrac{z^2}{16}$

10. If $l = 10$:

    (a) $20l - 31$,   (b) $5l^2 - 14l$,   (c) $\dfrac{l^2}{50}$

11. If $m = 2$:

    (a) $16m - 7$,   (b) $15m^2 - 16m$,   (c) $\dfrac{m^3}{4}$

12. If $n = 3$:

    (a) $20n + 21$,   (b) $11n^2 + 2n$,   (c) $\dfrac{n^3}{9}$

13. If $a = 6$:

    (a) $12a - 23$,   (b) $3a^2 - 4a$,   (c) $\dfrac{a^3}{12}$

14. If $b = 8$:

    (a) $7b + 8$,   (b) $2b^2 + 5b$,   (c) $\dfrac{b^3}{4}$

15. If $c = 4$:

    (a) $6c + 11$,   (b) $4c^2 - 3c$,   (c) $\dfrac{c^3}{8}$

16. If $d = 6$:

    (a) $8d + 7$,   (b) $d^2 + 9d$,   (c) $\dfrac{d^3}{9}$

17. If $x = 10$:

    (a) $100x + 225$, (b) $6x^2 - 25x$, (c) $\dfrac{x^3}{50}$

18. If $y = 6$:

    (a) $11y + 25$, (b) $4y^2 - 5y$, (c) $\dfrac{y^3}{8}$

19. If $z = 4$:

    (a) $7z + 17$,   (b) $3z^2 - 10z$, (c) $\dfrac{z^3}{16}$

20. If $t = 10$:

    (a) $12t - 29$, (b) $4t^2 + 11t$, (c) $\dfrac{t^3}{200}$

When negative numbers are substituted, extra care is required.

Example 4

If $x = -5$, find the value of:

(a) $3x$ (b) x^2 (c) $2x^3$

(a) $3x = 3 \times x$
$\qquad = 3 \times -5 = -15$

(b) $x^2 = x \times x$
$\qquad = -5 \times -5 = +25$

(c) $2x^3 = 2 \times x \times x \times x$
$\qquad = 2 \times -5 \times -5 \times -5 = -250$

Exercise 4.3d

Find the value of each of the following when the letter assumes the given value.

1. If $a = -10$:
   (a) $5a$,   (b) a^2,   (c) a^3
2. If $b = -2$:
   (a) $12b$,   (b) $5b^2$,   (c) $9b^3$
3. If $c = -4$:
   (a) $8c$,   (b) $3c^2$,   (c) $5c^3$
4. If $d = -6$:
   (a) $11d$,   (b) $4d^2$,   (c) $2d^3$
5. If $l = -5$:
   (a) $9l$,   (b) $8l^2$,   (c) $4l^3$
6. If $m = -7$:
   (a) $12m$,   (b) $2m^2$,   (c) $2m^3$
7. If $n = -3$:
   (a) $30n$,   (b) $11n^2$,   (c) $6n^3$
8. If $p = -1$:
   (a) $150p$,   (b) $180p^2$,   (c) $81p^3$
9. If $q = -9$:
   (a) $6q$,   (b) $5q^2$,   (c) q^3
10. If $r = -8$:
    (a) $9r$,   (b) $10r^2$,   (c) $2r^3$

11. If $t = -4$:
   (a) $7t$ (b) $6t^2$,   (c) $4t^3$
12. If $u = -5$:
   (a) $8u$,   (b) $4u^2$,   (c) $3u^3$
13. If $v = -10$:
   (a) $100v$,   (b) $400v^2$,   (c) $5v^3$
14. If $x = -2$:
   (a) $25x$,   (b) $15x^2$,   (c) $12x^3$
15. If $y = -6$:
   (a) $12y$,   (b) $5y^2$,   (c) $3y^3$
16. If $z = -9$:
   (a) $9z$,   (b) $4z^2$,   (c) $2z^3$
17. If $a = -3$:
   (a) $40a$,   (b) $20a^2$,   (c) $4a^3$
18. If $b = -7$:
   (a) $6b$,   (b) $3b^2$,   (c) b^3
19. If $c = -8$:
   (a) $11c$,   (b) $5c^2$,   (c) c^3
20. If $d = -20$:
   (a) $25d$,   (b) $20d^2$,   (c) $3d^3$

9. If $b = -8$ and $c = -2$:
   (a) $6bc$,   (b) $3b^2c$ (c) $(5c)^2$
10. If $l = -10$ and $m = -6$:
   (a) $7lm$,   (b) $5l^2m$,   (c) $(2m)^2$
11. If $p = -3$ and $q = -6$:
   (a) $8pq$,   (b) $3pq^2$,   (c) $(2p)^2$
12. If $u = -6$ and $v = -5$:
   (a) $5uv$,   (b) $4uv^2$ (c) $(5u)^2$
13. If $x = -6$ and $y = -10$:
   (a) $9xy$,   (b) $3xy^2$,   (c) $(2x)^2$
14. If $a = -3$ and $b = -5$:
   (a) $8ab$,   (b) $4ab^2$,   (c) $(10a)^2$
15. If $m = -5$ and $n = -8$:
   (a) $12mn$,   (b) $2mn^2$,   (c) $(4m)^2$
16. If $q = -2$ and $r = -12$:
   (a) $6qr$,   (b) $5qr^2$,   (c) $(3q)^2$
17. If $t = +2$ and $u = -10$:
   (a) $12tu$,   (b) $8tu^2$,   (c) $(10t)^2$
18. If $y = +4$ and $z = -5$:
   (a) $11yz$,   (b) $12yz^2$,   (c) $(2y)^2$
19. If $b = +5$ and $c = -9$:
   (a) $3bc$,   (b) $2bc^2$,   (c) $(6b)^2$
20. If $l = +2$ and $m = -3$:
   (a) $7lm$,   (b) $5lm^2$,   (c) $(4l)^2$

Example 5

If $x = +2$ and $y = -3$, find the value of:

(a) xy (b) $5x^2y$ (c) $(5y)^2$

(a) xy $\quad = x \times y$
$\qquad = 2 \times -3 = -6$
(b) $5x^2y$ $\quad = 5 \times x \times x \times y$
$\qquad = 5 \times 2 \times 2 \times -3 = -60$
(c) $(5y)^2$ $\quad = 5 \times y \times 5 \times y$
$\qquad = 5 \times -3 \times 5 \times -3 = 225$

Exercise 4.3e

Find the value of the following when the letters assume the given values.

1. If $p = +2$ and $q = -5$:
   (a) $12pq$,   (b) $3p^2q$,   (c) $(2q)^2$
2. If $u = +4$ and $v = -3$:
   (a) $7uv$,   (b) $2u^2v$,   (c) $(3v)^2$
3. If $x = +6$ and $y = -2$:
   (a) $9xy$,   (b) $4x^2y$,   (c) $(6y)^2$
4. If $a = +2$ and $b = -4$:
   (a) $11ab$,   (b) $5a^2b$,   (c) $(5b)^2$
5. If $m = +5$ and $n = -10$:
   (a) $5mn$,   (b) $6m^2n$,   (c) $(3n)^2$
6. If $q = +9$ and $r = -2$:
   (a) $4qr$,   (b) $2q^2r$,   (c) $(2r)^2$
7. If $t = -8$ and $u = -3$:
   (a) $4tu$ (b) $3t^2u$,   (c) $(4u)^2$
8. If $y = -6$ and $z = -4$:
   (a) $3yz$,   (b) $2y^2z$,   (c) $(3z)^2$

Example 6

If $y = -6$, find the value of:

(a) $5y - 6$ $\quad$ (b) $3y^2 + 2y$ $\quad$ (c) $\dfrac{y^3}{4}$

(a) $5y - 6$ $\quad = (5 \times y) - 6$
$\qquad = (5 \times -6) - 6$
$\qquad = -30 - 6 = -36$
(b) $3y^2 + 2y = (3 \times y \times y) + (2 \times y)$
$\qquad = (3 \times -6 \times -6) + (2 \times -6)$
$\qquad = 108 - 12 = 96$

(c) $\dfrac{y^3}{4} = \dfrac{y \times y \times y}{4}$

$\qquad = \dfrac{-6 \times -6 \times -6}{4} = \dfrac{-216}{4} = -54$

Exercise 4.3f

Find the value of the following when the letter assumes the given value.

1. If $a = -10$:
   (a) $3a - 2$,   (b) $a^2 + 8a$,   (c) $\dfrac{a^2}{5}$
2. If $b = -4$:
   (a) $7b - 3$,   (b) $3b^2 + 5b$,   (c) $\dfrac{b^2}{2}$

3. If $c = -10$:

  (a) $5c - 6$,   (b) $2c^2 + 9c$,   (c) $\dfrac{c^2}{4}$

4. If $d = -6$:

  (a) $4d - 9$,   (b) $3d^2 + 8d$ (c) $\dfrac{d^2}{2}$

5. If $l = -10$:

  (a) $3l + 32$,   (b) $l^2 + 12l$,   (c) $\dfrac{l^2}{20}$

6. If $m = -8$:

  (a) $2m + 21$,   (b) $m^2 + 9m$,   (c) $\dfrac{m^2}{4}$

7. If $n = -4$:

  (a) $7n + 30$,   (b) $2n^2 + 11n$,   (c) $\dfrac{n^2}{8}$

8. If $p = -10$:

  (a) $6p + 50$,   (b) $3p^2 - 20p$,   (c) $\dfrac{p^2}{50}$

9. If $q = -6$:

  (a) $5q + 24$,   (b) $5q^2 - 3q$,   (c) $\dfrac{q^2}{3}$

10. If $r = -8$:

  (a) $4r + 21$,   (b) $2r^2 - 5r$,   (c) $\dfrac{r^2}{16}$

11. If $t = -3$:

  (a) $4t - 9$,   (b) $10t^2 + 9t$,   (c) $\dfrac{t^3}{9}$

12. If $u = -2$:

  (a) $5u - 8$,   (b) $12u^2 + 11u$,   (c) $\dfrac{u^3}{4}$

13. If $v = -6$:

  (a) $3v - 7$,   (b) $2v^2 + 7v$,   (c) $\dfrac{v^3}{9}$

14. If $x = -4$:

  (a) $8x + 35$,   (b) $2x^2 + 9x$,   (c) $\dfrac{x^3}{8}$

15. If $y = -10$:

  (a) $4y + 49$,   (b) $y^2 + 11y$,   (c) $\dfrac{y^3}{50}$

16. If $z = -8$:

  (a) $6z + 53$,   (b) $z^2 + 12z$,   (c) $\dfrac{z^3}{4}$

17. If $a = -6$:

  (a) $7a + 36$,   (b) $4a^2 - 3a$,   (c) $\dfrac{a^3}{12}$

18. If $b = -10$:

  (a) $5b + 42$,   (b) $2b^2 - 5b$,   (c) $\dfrac{b^3}{200}$

19. If $c = -4$:

  (a) $8c + 27$,   (b) $3c^2 - 4c$,   (c) $\dfrac{c^3}{16}$

20. If $d = -6$:

  (a) $11d + 60$,   (b) $d^2 - 9d$,   (c) $\dfrac{d^3}{8}$

4.4 LIKE TERMS AND BRACKETS

Terms are all the parts of an expression which are connected by $+$ or $-$ signs. Like terms are collected together and written as a single term.

Example 1

(a) $3x + 4x = 7x$
(b) $3x + 4x - 6x = x$ (Note: not $1x$).

(c) $-2a - 3a = -5a$
(d) $-2a - 3a + 5a = 0$ (Note: not $0a$).

Unlike terms cannot be collected together.

Example 2

(a) $5a + 4b$ is in its simplest form.
(b) $4x + 2x^2$ cannot be written as a single term.

Exercise 4.4a

Simplify the following by collecting like terms.

1. $5a + 4a$ **2.** $7b + 8b$
3. $12c - 9c$ **4.** $4d - 8d$
5. $-3l - 7l$ **6.** $-6m - 12m$
7. $5n + 6n + n$ **8.** $3p + 9p - 4p$
9. $6q + 3q - 8q$ **10.** $2r + 6r - 10r$
11. $5t + 10t - 16t$ **12.** $12u - 3u - 4u$
13. $10v - 2v - 7v$ **14.** $15x - 6x - 9x$
15. $9y - 4y - y$ **16.** $3a + 4b + 7a + 5b$
17. $5m + 8n + 10m + 6n$ **18.** $7p + 9q + p + 3q$
19. $5z + 2z^2 + 7z + 3z^2$ **20.** $4u - 3v + 11u - 5v$
21. $5x - 6y + 4x - 4y$ **22.** $12b - c + 3b - 7c$
23. $9a - a^2 + 7a - 3a^2$ **24.** $6l + 11m + 9l - 3m$
25. $7q + 12r + 3q - 11r$ **26.** $3t + 2u + 13t - 7u$
27. $4y + z + 5y - 5z$ **28.** $11a - 2b + 5a + 9b$
29. $6m - 3n + 12m + 8n$ **30.** $8p - q + 7p + 3q$
31. $5u - 8v + 9u + 5v$ **32.** $3x - 11y + 8x + 4y$
33. $4b - 7c + 12b + 6c$ **34.** $11l - 7m - 8l + 12m$

35. $10b - 2b^2 - 5b + 11b^2$
36. $16c - 7c^2 - 15c + 3c^2$
37. $12d - 9d^2 - d + 10d^2$
38. $12q - 6r - 8q - 3r$
39. $8t - 11u - t - 5u$ **40.** $20y - z - 19y - 9z$

When collecting like terms, remember:
(i) the term xy is the same as the term yx.
(ii) $3ab$ is only one term.

Example 3

Simplify by collecting like terms:

(a) $x^2 + xy + yx + y^2 = x^2 + 2xy + y^2$
(b) $2abc - 3bca - 4acb = -5abc$

Exercise 4.4b

Simplify the following by collecting like terms.

1. $a^2 + 3ab + 2ba + b^2$
2. $m^2 + mn + 5nm + n^2$
3. $3p^2 + 8pq - 3qp - 3q^2$
4. $2u^2 + 4uv - vu - 2v^2$
5. $x^2 + 2xy - 6yx + y^2$
6. $5b^2 - 3bc - 7cb - 5c^2$
7. $2l^2 - lm - 4ml - 2m^2$
8. $2abc + 3cab + 6bca$
9. $3xy^2 + 5yx^2 + 2y^2x$
10. $4a^2b + ba^2 + 6ab^2$
11. $3lmn + 7nlm - 8mnl$
12. $5pq^2 + 4p^2q - 2q^2p$
13. $8x^2y + yx^2 - 8y^2x$
14. $5pqr - 7rpq + 3qrp$
15. $9bc^2 - 5cb^2 + 3c^2b$
16. $7l^2m - 6ml^2 + 12lm^2$
17. $10xyz - 8zxy - 4yzx$
18. $8qr^2 - 5q^2r - 3r^2q$
19. $12u^2v - 3vu^2 - v^2u$
20. $-3abc - 4cab - 5bca$

When removing brackets, remember:
(i) A term outside a bracket multiplies *each* of the terms inside the bracket.
(ii) A negative sign in front of a bracket changes *all* the signs inside the bracket.
(iii) After the brackets have been removed, collect all like terms together.

Example 4

Simplify by removing brackets and collecting like terms.

(a) $3x + 2(x + 7) = 3x + 2x + 14$
$$= 5x + 14$$
(b) $4 + 3(3x - 2) + 8x = 4 + 9x - 6 + 8x$
$$= 17x - 2$$

Exercise 4.4c

Simplify by removing brackets and collecting like terms.

1. $2a + 3(a + 4)$ 2. $3b + 4(b + 2)$
3. $c + 5(c + 3)$ 4. $4l + 2(l - 5)$
5. $3m + 4(m - 3)$ 6. $2n + 5(2n + 3)$
7. $4p + 2(5p + 1)$ 8. $3q + 4(2q - 3)$
9. $5r + 3(3r - 2)$ 10. $2t + 5(4t - 5)$
11. $5u + 2(2u + 3) + 4$ 12. $v + 3(5v + 4) + 6$
13. $4x + 5(4x + 3) - 5$ 14. $3y + 3(2y - 6) + 20$
15. $2z + 4(3z - 2) + 6$ 16. $5a + 2(5a - 1) - 7$
17. $7 + 5(2b + 1) + 5b$ 18. $5 + 2(3c + 2) - 4c$
19. $12 + 3(4l - 3) + 3l$ 20. $8 + 3(5m - 2) - 12m$

Example 5

Simplify by removing brackets and collecting like terms.

(a) $3x - 2(x + 7) = 3x - 2x - 14$
$$= x - 14$$
(b) $4 - 3(3x - 2) + 8x = 4 - 9x + 6 + 8x$
$$= 10 - x$$

Exercise 4.4d

Simplify by removing brackets and collecting like terms.

1. $5p - 2(p + 3)$ 2. $8q - 3(q + 2)$
3. $6r - 4(r + 5)$ 4. $5t - 2(t - 4)$
5. $7u - 3(u - 2)$ 6. $3v - 5(v - 3)$
7. $12x - 3(3x + 1)$ 8. $9y - 2(2y + 3)$
9. $11z - 2(4z - 5)$ 10. $15a - 4(3a - 2)$
11. $9b - 3(2b + 3) + 12$ 12. $15c - 4(3c + 2) + 5$
13. $12l - 2(5l + 1) - 6$ 14. $10m - 4(2m - 5) + 4$
15. $14n - 3(4n - 3) - 5$ 16. $13t - 5(2t - 1) - 8$
17. $15 - 3(2u + 4) + 9u$ 18. $16 - 2(4v + 5) - v$
19. $7 - 4(3x - 2) + 18x$ 20. $5 - 5(2y - 3) - 6y$

Example 6

Simplify:

(a) $3x + 2x(x - 6) = 3x + 2x^2 - 12x$
$$= 2x^2 - 9x$$
(b) $3x - 2x(6 - 3x) = 3x - 12x + 6x^2$
$$= 6x^2 - 9x$$

Note: $-2x \times -3x = -2 \times x \times -3 \times x$
$$= -2 \times -3 \times x \times x$$
$$= 6x^2$$

Exercise 4.4e

Simplify the following.

1. $2l + 4l(l + 3)$
2. $8m + 2m(3m - 1)$
3. $5n + 3n(4n - 3)$
4. $10t + 4t(2 - t)$
5. $9u - 2u(u + 3)$
6. $10v - 3v(3v + 2)$
7. $2x - 4x(2x - 5)$
8. $y - 5y(4y - 1)$
9. $8a - 2a(3 - a)$
10. $11b - 3b(2 - 3b)$

When multiplying a bracket by a bracket, each term in the second bracket is multiplied separately by each term in the first bracket.

Example 7

Expand:

(a) $(x + 2)(x + 3) = x(x + 3) + 2(x + 3)$
$$= x^2 + 3x + 2x + 6$$
$$= x^2 + 5x + 6$$

Exercise 4.4f

Expand the following.

1. $(a + 3)(a + 2)$
2. $(b + 2)(b + 6)$
3. $(c + 8)(c + 4)$
4. $(d + 1)(d + 12)$
5. $(x + 4y)(x + 3y)$
6. $(l - 2)(l + 5)$
7. $(m - 4)(m + 6)$
8. $(n - 1)(n + 8)$
9. $(p - 4)(p + 5)$
10. $(a - 4b)(a + 7b)$
11. $(q - 5)(q + 2)$
12. $(r - 8)(r + 4)$
13. $(t - 3)(t + 1)$
14. $(u - 7)(u + 6)$
15. $(m - 6n)(m + 4n)$
16. $(v - 2)(v - 5)$
17. $(x - 3)(x - 9)$
18. $(x - 5)(x - 6)$
19. $(x - 1)(x - 8)$
20. $(p - 2q)(p - 4q)$
21. $(2y + 3)(5y + 2)$
22. $(3z + 4)(2z + 1)$
23. $(2a + 5)(a + 3)$
24. $(4m + 3n)(m + 2n)$
25. $(3p + 5q)(4p + 3q)$
26. $(3b - 2)(4b + 5)$

27. $(5c - 1)(2c + 3)$
28. $(2d - 3)(3d + 5)$
29. $(5l - 2)(l + 1)$
30. $(2u - 3v)(u + 5v)$
31. $(4x - 3)(9x + 5)$
32. $(2y - 5)(4y + 7)$
33. $(5z - 3)(7z + 4)$
34. $(x - 2y)(4x + 3y)$
35. $(b - 4c)(3b + c)$
36. $(3a - 4)(2a - 3)$
37. $(2b - 7)(3b - 1)$
38. $(5c - 3)(2c - 5)$
39. $(8l - m)(5l - 4m)$
40. $(3q - 8r)(q - 2r)$

Example 8

Expand:

(a) $(3x + 2)(3x - 2)$
$$= 3x(3x - 2) + 2(3x - 2)$$
$$= 9x^2 - 6x + 6x - 4$$
$$= 9x^2 - 4$$
(b) $(3x + 2)^2 = (3x + 2)(3x + 2)$
$$= 9x^2 + 6x + 6x + 4$$
$$= 9x^2 + 12x + 4$$
(c) $(3x - 2)^2 = (3x - 2)(3x - 2)$
$$= 9x^2 - 6x - 6x + 4$$
$$= 9x^2 - 12x + 4$$

Exercise 4.4g

Expand the following.

1. $(x - 2)(x + 2)$
2. $(y - 4)(y + 4)$
3. $(z - 3)(z + 3)$
4. $(a - 10)(a + 10)$
5. $(b - 1)(b + 1)$
6. $(4c - 5)(4c + 5)$
7. $(2l - 9)(2l + 9)$
8. $(7m - 2)(7m + 2)$
9. $(2n - 1)(2n + 1)$
10. $(4 - 3y)(4 + 3y)$
11. $(x + 2)^2$
12. $(p + 5)^2$
13. $(q + 1)^2$
14. $(r + 12)^2$
15. $(u + 10v)^2$
16. $(2t + 5)^2$
17. $(4a + 1)^2$
18. $(2b + 3)^2$
19. $(5c + 2d)^2$
20. $(4x + 3y)^2$
21. $(x - 1)^2$
22. $(y - 7)^2$
23. $(z - 3)^2$
24. $(l - 12)^2$
25. $(m - 5)^2$
26. $(2p - 5)^2$
27. $(5q - 1)^2$
28. $(2t - \frac{1}{2})^2$
29. $(4z - \frac{1}{2})^2$
30. $(8a - \frac{1}{2})^2$
31. $(5b - 3c)^2$
32. $(6 - 7y)^2$
33. $(a + \frac{1}{a})^2$
34. $(x - \frac{1}{x})^2$
35. $(x + 2)(x^2 + 3x + 2)$
36. $(x - 2)(x^2 + 3x + 2)$
37. $(x + 1)^3$
38. $(a - 1)^3$
39. $(x + 1)(x - 1)(x^2 + 1)$
40. $(a - b)(a + b)(a^2 + b^2)$

A simple equation is solved by finding the number which replaces the letter so that both sides of the equation remain equal.

Example 1

(a) $x + 3 = 6$
To solve, take 3 from both sides.
$x + 3 - 3 = 6 - 3$
$\therefore x = 3$

(b) $a + 7 = 4$
To solve, take 7 from both sides.
$a + 7 - 7 = 4 - 7$
$\therefore a = -3$

Exercise 4.5a

Solve the following equations.

1. $x + 3 = 8$	2. $x + 4 = 5$
3. $x + 5 = 5$	4. $a + 6 = 5$
5. $x + 27 = 53$	6. $c + 18 = 29$
7. $y + 46 = 83$	8. $p + 21 = 19$
9. $36 + x = 17$	10. $27 = x + 4$
11. $20 = a + 21$	12. $4\frac{1}{2} = x + 2\frac{1}{2}$
13. $3 \cdot 5 + b = 4 \cdot 6$	14. $x + 12 \cdot 3 = 16 \cdot 7$
15. $y + 3 \cdot 5 = 2 \cdot 1$	

Example 2

(a) $x - 3 = 6$
To solve, add 3 to both sides
$x - 3 + 3 = 6 + 3$
$\therefore x = 9$

(b) $a - 7 = -4$
To solve, add 7 to both sides
$a - 7 + 7 = -4 + 7$
$\therefore a = 3$

Exercise 4.5b

Solve the following equations.

1. $x - 3 = 3$	2. $x - 5 = 2$
3. $x - 7 = 3$	4. $a - 9 = 5$
5. $x - 2 = -2$	6. $x - 3 = -2$
7. $x - 4 = -6$	8. $c - 12 = -2$
9. $36 = x - 12$	10. $41 = x - 16$
11. $-22 = b - 19$	12. $-18 = p - 32$
13. $x - 2\frac{1}{4} = 4$	14. $a - \frac{1}{4} = -2$
15. $c - 0 \cdot 6 = -2 \cdot 4$	

Example 3

(a) $\qquad 3x = 6$
To solve, divide both sides by 3

$$\frac{3x}{3} = \frac{6}{3}$$

$\therefore x = 2$

(b) $\qquad \frac{a}{4} = 2$

To solve, multiply both sides by 4.

$$\frac{a}{4} \times 4 = 2 \times 4$$

$\therefore a = 8$

Exercise 4.5c

Solve the following equations.

1. $2x = 4$	2. $3y = 12$
3. $4y = 20$	4. $5x = -30$
5. $7x = -28$	6. $8 = 4x$
7. $36 = 9a$	8. $2b = 11$
9. $3a = -10$	10. $4c = 18$
11. $\frac{a}{2} = 4$	12. $\frac{b}{3} = 6$
13. $\frac{x}{3} = -2$	14. $\frac{b}{7} = 9$
15. $\frac{c}{11} = 8$	16. $\frac{a}{4} = 2\frac{1}{2}$
17. $\frac{x}{6} = -2$	18. $\frac{x}{3} = -2\frac{1}{2}$
19. $\frac{x}{10} = 1 \cdot 5$	20. $\frac{c}{100} = -0 \cdot 62$

Example 4

$3x + 6 = 12$
$\therefore 3x = 6 \qquad$ (see example 1a)
$\quad x = 2 \qquad$ (see example 3a)

Example 5

$$\frac{a}{5} - 4 = 3$$

$\therefore \frac{a}{5} = 7 \qquad$ (see example 1b)

$\quad a = 35 \qquad$ (see example 3b)

Exercise 4.5d

Solve the following equations.

1. $3x + 5 = 11$ 2. $5x + 5 = 10$
3. $7x + 1 = 8$ 4. $3x + 15 = 27$
5. $9x - 5 = 22$ 6. $2x - 18 = 12$
7. $10a - 19 = 1$ 8. $7 = 3a - 2$
9. $5a - 26 = 54$ 10. $11x - 19 = 102$

11. $\frac{x}{2} + 1 = 2$ 12. $\frac{c}{5} + 2 = 3$

13. $\frac{l}{6} - 3 = 9$ 14. $\frac{x}{6} - 4 = 10$

15. $\frac{a}{4} + 1 = 2\frac{1}{2}$ 16. $\frac{a}{7} + 4 = 1$

17. $\frac{x}{2} - 4 = 0$ 18. $\frac{c}{3} - 1\frac{1}{2} = 4\frac{1}{2}$

19. $\frac{x}{5} - 1 = 2$ 20. $\frac{b}{2} + 0.5 = 1$

Example 6

$$\frac{5x}{4} = 15$$

$\therefore 5x = 60$ (see example 3b)
$x = 12$ (see example 3a)

Example 7

$3\frac{1}{2}x - 2 = 5$

$\therefore 3\frac{1}{2}x = 7$ (see example 2a)

$\frac{7}{2}x = 7$

$7x = 14$ (see example 3b)
$\therefore x = 2$ (see example 3a)

Exercise 4.5e

Solve the following equations.

1. $\frac{2x}{3} = 4$ 2. $\frac{3x}{2} = 12$

3. $\frac{5a}{3} = 10$ 4. $\frac{3c}{4} = 1\frac{1}{2}$

5. $\frac{7b}{3} = 7$ 6. $\frac{2x}{3} = -2$

7. $\frac{5a}{3} = -10$ 8. $7 = \frac{2a}{3}$

9. $12 = 1\frac{1}{2}x$ 10. $3\frac{1}{2}a = -7$

11. $1\frac{1}{2}x - 1 = 2$ 12. $4\frac{1}{2}x + 5 = 14$

13. $5\frac{1}{2}b - 1 = 21$ 14. $1\frac{1}{4}x - 2 = 3$

15. $7\frac{1}{2}x - 4 = 26$ 16. $1\frac{1}{5}a - 1 = 5$

17. $2\frac{1}{2}x - 9 = 6$ 18. $1\frac{1}{3}a - \frac{1}{2} = -4\frac{1}{2}$

19. $2\frac{1}{2}x + 2.2 = 22.2$ 20. $2.5x - 0.5 = 4.5$

Example 8

$5x + 4 = 16 - 3x$
To solve
(a) add $3x$ to both sides
$8x + 4 = 16$
(b) take 4 from both sides
$8x = 12$
$x = 1\frac{1}{2}$ (see example 3a).

Example 9

$5 - 3(x - 4) = -7$
$5 - 3x + 12 = -7$
$5 + 12 = -7 + 3x$
$5 + 12 + 7 = 3x$
$24 = 3x$
$8 = x$

Exercise 4.5f

Solve the following equations.

1. $5x + 3 = 2x + 15$
2. $6x + 5 = 3x + 14$
3. $10x + 3 = 4x + 21$
4. $9x + 1 = 5x + 9$
5. $5x - 3 = 2x + 9$
6. $9x - 11 = 3x + 13$
7. $6x - 5 = 4x + 7$
8. $11x - 26 = 10x + 18$
9. $8x - 12 = 5x - 3$
10. $12x - 16 = 7x - 1$
11. $9x - 19 = 2x - 5$
12. $5x - 4 = 2x + 5$
13. $23x - 27 = 16x - 13$
14. $16x - 35 = 7x - 8$
15. $11x - 51 = 6x - 1$
16. $17x - 40 = 8x - 13$
17. $8(x + 1) = 2(x + 16)$
18. $10(x + 1) = 7(x + 4)$
19. $5(x + 4) = 3(x + 12)$
20. $5(x - 3) = 4(x + 2)$
21. $5(x - 4) = 2(x - 7)$
22. $8(x - 9) = 3(x - 4)$
23. $6(x - 5) = 5(x - 4)$
24. $9(x - 2) = 8(x - 1)$
25. $a + 2(a + 1) = 8$
26. $3b + 4(b + 6) = -4$
27. $4(x - 6) = 3(x - 1)$
28. $2(x - 1) = 33 - 3(2x + 1)$
29. $5(x + 2) - 9(x - 2) = 0$
30. $8(3 - x) - 5 = -3(3 - 2x)$

Example 10

$$\frac{x+2}{3} = \frac{x+12}{8}$$

Multiply both sides by 24, the common denominator of 3 and 8.

$$\frac{\overset{8}{\cancel{24}}(x+2)}{\underset{1}{\cancel{3}}} = \frac{\overset{3}{\cancel{24}}(x+12)}{\underset{1}{\cancel{8}}}$$

Remove brackets.

$$8x + 16 = 3x + 36$$
$$8x - 3x = 36 - 16$$
$$5x = 20$$
$$\therefore x = 4$$

Exercise 4.5g

Solve the following equations.

1. $\dfrac{x+2}{5} = \dfrac{x+4}{7}$ 2. $\dfrac{x+2}{4} = \dfrac{x+12}{9}$

3. $\dfrac{x+2}{9} = \dfrac{x+4}{11}$ 4. $\dfrac{x+1}{7} = \dfrac{x+6}{12}$

5. $\dfrac{x-1}{3} = \dfrac{x+4}{8}$ 6. $\dfrac{x-3}{3} = \dfrac{x+3}{5}$

7. $\dfrac{x-4}{8} = \dfrac{x-1}{11}$ 8. $\dfrac{x-4}{3} = \dfrac{x-11}{10}$

9. $\dfrac{x-8}{5} = \dfrac{x-12}{9}$ 10. $\dfrac{2x+3}{5} = \dfrac{4x+9}{11}$

11. $\dfrac{3x+2}{2} = \dfrac{6x+11}{5}$ 12. $\dfrac{4x+3}{3} = \dfrac{6x+7}{5}$

13. $\dfrac{5x-1}{7} = \dfrac{7x+3}{12}$ 14. $\dfrac{9x-5}{7} = \dfrac{6x+2}{5}$

15. $\dfrac{7x-10}{5} = \dfrac{8x-5}{7}$

Exercise 4.5h

In all the questions below find the equation with the 'Odd solution out.'

1. (a) $\frac{3}{4}x = 6$
   (b) $3x + 40 = 61$
   (c) $15x + 22 = 6x + 85$

2. (a) $\frac{5}{3}x = 15$
   (b) $12x - 50 = 46$
   (c) $24x - 27 = 13x + 72$

3. (a) $\frac{x}{4} + 19 = 28$
   (b) $\frac{x}{2} - 5 = 12$
   (c) $13x - 15 = 12x + 19$

4. (a) $\frac{8}{x} + 9 = 11$
   (b) $\frac{12}{x} - 2 = 1$
   (c) $13x - 42 = 5x - 18$

5. (a) $2x + 12 = x + 19$
   (b) $7x - 14 = 2x + 11$
   (c) $11(x + 1) = 8(x + 4)$

6. (a) $9x - 18 = x + 14$
   (b) $20x - 51 = 9x - 7$
   (c) $7(x - 3) = 2(x + 2)$

7. (a) $\dfrac{x+2}{4} = \dfrac{x+12}{9}$
   (b) $\dfrac{x+2}{3} = \dfrac{x+12}{8}$
   (c) $9(x - 2) = 4(x + 3)$

8. (a) $11(2x + 3) = 5(4x + 9)$
   (b) $9(4x - 5) = 5(5x + 2)$
   (c) $\dfrac{x-1}{5} = \dfrac{x+6}{12}$

9. (a) $7(3x - 8) = 5(3x + 2)$
   (b) $5(8x + 3) = 9(4x + 7)$
   (c) $\dfrac{x-5}{7} = \dfrac{x-3}{9}$

10. (a) $\dfrac{x-2}{5} = \dfrac{x+4}{11}$
    (b) $\dfrac{x-5}{5} = \dfrac{x-2}{8}$
    (c) $\dfrac{x-3}{4} = \dfrac{x+2}{9}$

11. (a) $\dfrac{3x+2}{4} = \dfrac{4x+11}{7}$
    (b) $\dfrac{2x-1}{9} = \dfrac{2x+1}{11}$
    (c) $\dfrac{3x-5}{5} = \dfrac{5x-1}{12}$

12. (a) $\dfrac{11x-2}{6} = \dfrac{12x+1}{7}$
    (b) $\dfrac{3x-8}{4} = \dfrac{2x-3}{5}$
    (c) $\dfrac{4x-5}{7} = \dfrac{5x-4}{11}$

Simultaneous equations are solved by
finding the numbers which replace the
letters so that both sides of each equation
remain equal.

Example 1

(a) Solve $x + y = 4$ (i)
              $x - y = 2$ (ii)
   Eliminate y by adding (i) to (ii)
         Then $2x = 6$
                 $x = 3$
   Replace x by 3 in (i)
         Then $3 + y = 4$
                 $y = 1$ $\therefore x = 3, y = 1$
   Check by putting $x = 3, y = 1$ in (ii)
         So $3 - 1 = 2$

(b) Solve $3x + y = 7$ (i)
            $2x + y = 6$ (ii)
   Eliminate y by subtracting (ii) from (i).
         Then $x = 1$
   Substitute $x = 1$ in (ii)
         Then $2 + y = 6$
                 $y = 4$ $\therefore x = 1, y = 4$
   Check by putting $x = 1, y = 4$ in (i).
         So $3 + 4 = 7$

Exercise 4.6a

Solve the following pairs of simultaneous
equations.

1. $x + y = 10$ 2. $x + y = 11$
   $x - y = 4$ $x - y = 7$
3. $a + b = 15$ 4. $2x + y = 21$
   $a - b = 9$ $x - y = 6$
5. $5x + y = 28$ 6. $3m + n = 25$
   $x - y = 2$ $m - n = 3$
7. $x + y = 7$ 8. $x + y = 11$
   $2x - y = 2$ $5x - y = 25$
9. $p + q = 6$ 10. $u + v = 8$
   $3p - q = 10$ $4u - v = 17$
11. $5x + y = 19$ 12. $2x + y = 11$
   $2x - y = 2$ $3x - y = 4$
13. $4x + y = 18$ 14. $2b + c = 13$
   $7x - y = 15$ $6b - c = 3$
15. $7l + m = 31$ 16. $6x + y = 20$
   $2l - m = 5$ $3x + y = 11$
17. $5x + y = 21$ 18. $9q + r = 13$
   $2x + y = 12$ $3q + r = 7$

19. $6t + u = 16$ 20. $4x + y = 19$
   $2t + u = 8$ $x + y = 10$
21. $5a + b = 27$ 22. $4x + y = 23$
   $a + b = 7$ $3x + y = 18$
23. $6m + n = 16$ 24. $x + 7y = 13$
   $5m + n = 14$ $x + 3y = 9$
25. $x + 6y = 17$ 26. $x + 5y = 25$
   $x + 2y = 9$ $x + 3y = 17$
27. $p + 7q = 28$ 28. $x + 6y = 31$
   $p + 2q = 13$ $x + y = 6$
29. $u + 8v = 26$ 30. $x + 5y = 11$
   $u + v = 5$ $x + 4y = 10$

Example 2

(a) Solve $x + y = 6$ (i)
                     $2x - 3y = 2$ (ii)
   Multiply (i) by 3 to make the coefficients
   of y the same.
         Then $3x + 3y = 18$ (iii)
               $2x - 3y = 2$ (iv)
   Eliminate y by adding (iii) to (iv)
         Then $5x = 20$
                 $x = 4$
   Substitute $x = 4$ in (i).
         Then $4 + y = 6$
                 $y = 2$
   Check by substituting $x = 4, y = 2$ in (ii).
       $(2 \times 4) - (3 \times 2)$ $= 8 - 6 = 2$
                 $\therefore x = 4, y = 2$

(b) Solve $5x + 2y = 7$ (i)
                       $x + 3y = 4$ (ii)
   Multiply (ii) by 5 to make the cofficients
   of x the same.
         Then $5x + 2y = 7$ (iii)
               $5x + 15y = 20$ (iv)
   Eliminate x by subtracting (iii) from (iv).
         Then $13y = 13$
                 $y = 1$
   Substitute $y = 1$ in (i)
         Then $5x + (2 \times 1) = 7$
               $5x + 2 = 7$
                 $x = 1$
   Check by putting $x = 1, y = 1$ in (ii)
         $1 + (3 \times 1) = 1 + 3 = 4$
                 $\therefore x = 1, y = 1$

Exercise 4.6b

Solve the following pairs of simultaneous equations.

1. $3x + y = 13$
    $5x - 2y = 7$

2. $4x + y = 6$
    $9x - 4y = 1$

3. $2x + y = 11$
    $3x - 2y = 6$

4. $x + y = 7$
    $11x - 4y = 2$

5. $6a + b = 20$
    $4a - 3b = 6$

6. $3m + n = 10$
    $5m - 2n = 2$

7. $5p + q = 16$
    $3p - 2q = 7$

8. $4u + v = 22$
    $5u - 4v = 17$

9. $3x + 2y = 13$
    $2x - y = 4$

10. $4x + 3y = 17$
    $3x - y = 3$

11. $3x + 2y = 16$
    $x - y = 2$

12. $10x + 3y = 19$
    $5x - y = 2$

13. $3b + 4c = 10$
    $2b - c = 3$

14. $11l + 2m = 34$
    $7l - m = 8$

15. $2q + 5r = 33$
    $5q - r = 15$

16. $3x + 5y = 11$
    $x + 2y = 4$

17. $2x + 5y = 31$
    $x + 6y = 33$

18. $3x + 2y = 17$
    $x + 3y = 15$

19. $2x + 5y = 16$
    $x + 3y = 9$

20. $3t + 4u = 24$
    $t + 5u = 19$

21. $5a + 3b = 14$
    $a + 2b = 7$

22. $3m + 2n = 16$
    $m + 3n = 17$

23. $3p + 4q = 25$
    $p + 2q = 11$

24. $5x + 2y = 16$
    $4x + y = 11$

25. $6x + 5y = 23$
    $2x + y = 7$

26. $3x + 4y = 20$
    $2x + y = 10$

27. $7x + 3y = 13$
    $5x + y = 7$

28. $3u + 2v = 16$
    $2u + v = 9$

29. $5b + 4c = 23$
    $3b + c = 11$

30. $6l + 5m = 32$
    $3l + m = 10$

Example 3

(a) Solve $\qquad 2x + 5y = 16 \qquad$ (i)
$\qquad\qquad\qquad 3x - 2y = 5 \qquad$ (ii)

Multiply (i) by 2 and (ii) by 5 to make the coefficients of y the same.
$$4x + 10y = 32 \qquad \text{(iii)}$$
$$15x - 10y = 25 \qquad \text{(iv)}$$
Eliminate y by adding (iii) to (iv).
$$19x = 57$$
$$x = 3$$
Substitute $x = 3$ in (i)
$$\text{Then } 6 + 5y = 16$$
$$5y = 10$$
$$y = 2$$
Check by substituting $x = 3$, $y = 2$ in (ii)
$$(3 \times 15) - (2 \times 10) = 45 - 20 = 25$$
$$\therefore x = 3, y = 2$$

(b) Solve $\qquad 2a + 3b = 5 \qquad$ (i)
$\qquad\qquad\qquad 3a + 2b = 0 \qquad$ (ii)

Multiply (i) by 3 and (ii) by 2 to make the coefficients of a the same.
$$6a + 9b = 15 \qquad \text{(iii)}$$
$$6a + 4b = 0 \qquad \text{(iv)}$$
Eliminate a by subtracting (iv) from (iii).
$$5b = 15$$
$$b = 3$$
Substitute $b = 3$ in (i)
$$2a + 9 = 5$$
$$2a = -4$$
$$a = -2$$
Check by substituting $a = -2$, $b = 3$ in (ii)
$$(3 \times -2) + (2 \times 3) = -6 + 6 = 0$$
$$\therefore a = -2, b = 3$$

Exercise 4.6c

Solve the following pairs of simultaneous equations.

1. $3x + 2y = 13$
    $5x - 3y = 9$

2. $2x + 3y = 18$
    $9x - 4y = 11$

3. $3x + 2y = 14$
    $13x - 5y = 6$

4. $5x + 3y = 11$
    $7x - 2y = 3$

5. $4x + 3y = 15$
    $5x - 4y = 11$

6. $2q + 5r = 19$
    $7q - 2r = 8$

7. $4t + 3u = 14$
    $5t - 2u = 6$

8. $2a + 3b = 11$
    $16a - 5b = 1$

9. $6m + 5n = 17$
    $5m - 4n = 6$

10. $3p + 2q = 15$
    $7p - 5q = 6$

11. $5u + 2v = 13$
    $16u - 3v = 4$

12. $4a + 3b = 22$
    $3a - 4b = 4$

13. $7m + 2n = 20$
    $8m - 5n = 1$

14. $9p + 3q = 30$
    $5p - 2q = 2$

15. $2u + 5v = 31$
    $7u - 2v = 11$

16. $5x + 3y = 19$
    $7x + 2y = 20$

17. $2x + 5y = 31$
    $3x + 2y = 19$

18. $4x + 3y = 32$
    $5x + 2y = 33$

19. $2x + 5y = 16$
    $3x + 4y = 17$

20. $2x + 7y = 17$
    $3x + 5y = 20$

21. $3a + 4b = 25$
    $5a + 3b = 27$

22. $2m + 5n = 23$
    $3m + 4n = 24$

23. $3p + 4q = 16$
    $4p + 5q = 21$

24. $2u + 3v = 16$
    $3u + 4v = 23$

25. $3b + 5a = 25$
    $2b + 3a = 16$

26. $5l + 6m = 17$
    $3l + 5m = 13$

27. $2q + 3r = 9$
    $5q + 4r = 19$

28. $5t + 7u = 24$
    $2t + 3u = 10$

29. $3a + 5b = 22$
    $2a + 7b = 22$

30. $4m + 3n = 23$
    $3m + 2n = 16$

Sometimes it is necessary to rewrite a formula in order to find a new subject.

Example 1

(a) Find l when $A = lb$

Divide both sides by b,

then $\dfrac{A}{b} = l$

(b) Find C when $\pi = \dfrac{C}{D}$

Multiply both sides by D,

then $\pi D = C$

Exercise 4.7a

1. Find V when $P = VI$
2. Find a when $v = at$
3. Find (a) f, (b) l when $v = lf$
4. Find (a) m, (b) a when $F = ma$
5. Find D when $C = \pi D$
6. Find (a) b, (b) h when $A = bh$
7. Find (a) f, (b) t when $a = ft$
8. Find (a) V, (b) T when $S = \dfrac{V}{T}$
9. Find (a) m, (b) V when $D = \dfrac{m}{V}$
10. Find (a) s, (b) t when $v = \dfrac{s}{t}$
11. Find (a) V, (b) I when $R = V/I$
12. Find (a) x, (b) y when $n = x/y$

Example 2

(a) Find t when $v = u + at$

Subtract u from both sides,
then $v - u = at$
Divide both sides by a,

then $\dfrac{v - u}{a} = t$

(b) Find Y when $I = \dfrac{PRY}{100}$

Multiply both sides by 100.
then $100I = PRY$.
Divide both sides by PR,

then $\dfrac{100I}{PR} = Y$

Exercise 4.7b

1. Find $\hat{Y}$ when $\hat{X} + \hat{Y} + \hat{Z} = 180°$
2. Find l when $p = m + 2l$
3. Find x when $3x + 2y = 540°$
4. Find t when $P = p + at$
5. Find (a) T, (b) b when $H = h + bT$
6. Find (a) t, (b) v when $s = vt - k$
7. Find (a) t, (b) c when $L = l - ct$
8. Find f when $m = \dfrac{v}{f} - 1$
9. Find A when $n = \dfrac{360}{A} - 1$
10. Find t when $C = rPt$
11. Find (a) h, (b) D when $P = hDg$
12. Find (a) l, (b) h when $V = lbh$
13. Find (a) k, (b) A when $R = k\dfrac{l}{A}$
14. Find (a) U, (b) ΔT when $s = \dfrac{U}{m\Delta T}$
15. Find (a) P, (b) R when $PV = RT$

Example 3

(a) Find b when $p = 2(l + b)$
    Remove the brackets,
          then $p = 2l + 2b$
    Take $2l$ from both sides,
          then $p - 2l = 2b$
    Divide both sides by 2
          then $\dfrac{p - 2l}{2} = b$

(b) Find F when $C = \dfrac{5}{9}(F - 32)$

    Multiply both sides by 9 to remove the fraction, then $9C = 5(F - 32)$
    Remove the brackets to give:
          $9C = 5F - 160$
    Add 160 to both sides,
          then $9C + 160 = 5F$
    Divide both sides by 5

          then $\dfrac{9C + 160}{5} = F$

Exercise 4.7c

1. Find b when $a = 3(b + c)$
2. Find n when $m = 5(n + 2)$
3. Find r when $p = 4(q - r)$
4. Find v when $u = 2(v - 5)$
5. Find y when $x = 3(5 - y)$

6. Find c when $a = b(c + d)$
7. Find n when $A = 180(n - 2)$
8. Find r when $E = (R + r)I$
9. Find u when $s = \dfrac{t(u + v)}{2}$

10. Find m when $A = \dfrac{h}{2}(l - m)$

11. Find a when $L = 4(a + b + c)$
12. Find (a) m, (b) n when $l = m(n + 4)$
13. Find (a) r, (b) s when $p = q(r - s)$
14. Find (a) u, (b) v when $t = u(v - 3)$
15. Find (a) y, (b) z when $x = y(5 - z)$

Example 4

(a) Find r when $A = \pi r^2$
   Divide both sides by π,
   then $\dfrac{A}{\pi} = r^2$

   Take the square root of both sides,
   then $\sqrt{\dfrac{A}{\pi}} = r$

(b) Find t when $A = 2\sqrt{lt}$
   Square both sides,
   then $A^2 = 4lt$
   Divide both sides by $4l$,
   then $\dfrac{A^2}{4l} = t$

Exercise 4.7d

1. Find l when $A = l^2$
2. Find a when $V = a^2 b$
3. Find r when $V = \pi r^2 l$
4. Find t when $s = \frac{1}{2}at^2$
5. Find l when $V = \frac{1}{3}l^2 h$
6. Find r when $V = \frac{1}{3}\pi r^2 h$
7. Find l when $V = l^3$
8. Find r when $V = \frac{4}{3}\pi r^3$
9. Find a when $x = \sqrt{ab}$
10. Find n when $y = 5\sqrt{mn}$
11. Find c when $z = \sqrt{\dfrac{c}{a}}$
12. Find n when $t = 3\sqrt{\dfrac{n}{l}}$
13. Find g when $T = 2\pi\sqrt{\dfrac{l}{g}}$

14. Find a when $x = \sqrt[3]{ab}$
15. Find c when $y = \sqrt[3]{\dfrac{c}{a}}$

Example 5

If $y = mx + c$, find the value of m
when $y = 8$, $x = 2$, and $c = 2$.
With these values $y = mx + c$
becomes $8 = 2m + 2$
$6 = 2m$
So $3 = m$

Exercise 4.7e

1. If $y = mx + c$, find the value of m when $y = 15, x = 4, c = 3$.
2. If $z = mt - c$, find the value of m when $z = 13, t = 5, c = 7$.
3. If $y = m(x + b)$, find the value of b when $y = 14, m = 2, x = 4$.
4. If $z = m(t - b)$, find the value of b when $z = 9, m = 3, t = 7$.
5. If $p = abc$, find the value of c when $p = 36, a = 2, b = 6$.
6. If $q = \dfrac{l}{mn}$, find the value of n when $q = 2$, $l = 48, m = 3$.
7. If $A = 360(n - 2)$, find the value of n when $A = 3240$.
8. If $s = \dfrac{u}{mt}$, find t when $s = 2, u = 9000, m = 150$.
9. If $V = abc$, find the value of c when $V = 105$, $a = 3, b = 7$.
10. If $P = p + at$, find the value of a when $P = 82$, $p = 76, t = 20$. For the same values of p and a, find the value of P when $t = 100$.
11. If $x = \sqrt{ab}$, find the value of b when $x = 6$, $a = 4$.
12. If $y = 3\sqrt{mn}$, find the value of n when $y = 30, m = 20$.
13. If $z = \sqrt{\dfrac{p}{q}}$, find the value of q when $z = 3, p = 45$.
14. If $t = 5\sqrt{\dfrac{u}{v}}$, find the value of v when $t = 20, u = 48$.
15. If $x = \sqrt[3]{\dfrac{a}{b}}$, find the value of b when $x = 2, a = 24$.

If $(x - 4)$ is multiplied by 3, the answer
is $3x - 12$.
3 and $(x - 4)$ are the *factors* of $3x - 12$.
Because both $3x$ and -12 can be divided
by 3, then 3 is the *common factor* of
$3x - 12$.
When factorising, common factors are
always taken out first.

Example 1

(a) Factorise $2a + 4b$
    $2a + 4b = 2(a + 2b)$
    The common factor is 2.

(b) Factorise $3xy + 6yz$
    $3xy + 6yz = 3y(x + 2z)$
    The common factor is $3y$

(c) Factorise $24t^3 + 8t^2$
    $24t^3 + 8t^2 = 8t^2(3t + 1)$
    The common factor is $8t^2$

(d) Factorise $10p^3 - 25pq - 15p^2$
    $10p^3 - 25pq - 15p^2$
    $\qquad\qquad = 5p(2p^2 - 5q - 3p)$
    The common factor is $5p$.

Exercise 4.8a

Factorise the following.

1. $3a + 9b$	2. $4m - 20n$
3. $15p + 5q$	4. $24u - 6v$
5. $10x + 15y$	6. $9b - 15c$
7. $24l + 16m$	8. $8q - 6r$
9. $3tu + 3tv$	10. $xy - 9xz$
11. $4ab + 12a^2$	12. $16mn + 8m^2$
13. $8pq - 10p^2$	14. $25uv + 15u^2$
15. $10xy + 9x^2$	16. $9b^2 - 54ab$
17. $45q^2 - 5pq$	18. $16v^2 + 40uv$
19. $27y^2 - 18xy$	20. $5c^2 - 6bc$
21. $30t^3 + 6t^2$	22. $7a^3 + 21a^2$
23. $24m^3 - 12m^2$	24. $9c^3 - 36c^2$
25. $16p^3q + 15p^2q$	26. $15n^3 + 20n^2$
27. $45u^3v + 36u^2v$	28. $24x^3y - 30x^2y$
29. $18z^3 - 12z^2$	30. $9t^3u - 16t^2u$

Sometimes the common factor is more
than one term; it is usually written in a
bracket.

Example 2

(a) Factorise $a(x + y) + b(x + y)$
    $a(x + y) + b(x + y) = (x + y)(a + b)$
    The common factor is $(x + y)$

(b) Factorise $3x(a + b) + 2y(b + a)$
    $3x(a + b) + 2y(b + a)$
    $\qquad\qquad = (a + b)(3x + 2y)$
    As $(a + b) = (b + a)$, the common
    factor is $(a + b)$.

Exercise 4.8b

Factorise the following.

1. $u(l + m) + v(l + m)$
2. $a(q - r) + b(q - r)$
3. $m(t + u) + 3(t + u)$
4. $5(y + z) + n(y + z)$
5. $4(b - c) + 5(b - c)$
6. $x(l - m) - y(l - m)$
7. $a(q + r) - b(q + r)$
8. $p(t + u) - 2(t + u)$
9. $7(y - z) - q(y - z)$
10. $8(b + c) - 3(b + c)$
11. $3b(p + q) + 4c(q + p)$
12. $5l(u - v) + 2m(u - v)$
13. $3q(x - y) + 2(x - y)$
14. $8(a + b) + 7r(b + a)$
15. $2t(m + n) + 3t(n + m)$
16. $2y(p - q) - 7z(p - q)$
17. $b(u - v) - 3c(u - v)$
18. $5l(x + y) - 3(x + y)$
19. $9(a + b) - 4m(a + b)$
20. $10u(m - n) - 3u(m - n)$

To factorise an expression with four terms
(a tetranomial), arrange the expression
into two pairs so that each pair has a
common factor.

Example 3

(a) Factorise $3ac + 2bc + 3ad + 2bd$

$3ac + 2bc + 3ad + 2bd$

$= c(3a + 2b) + d(3a + 2b)$ \qquad (i)
$= (3a + 2b)(c + d)$ \qquad\qquad (ii)
In (i), c is a common factor of the
first pair, and d of the second pair.
In (ii), $(3a + 2b)$ is the common
factor.

(b) Factorise $3mn + 2pn + 3mq + 2pq$.
Rearrange the whole expression to
give
$3mn + 3mq + 2pn + 2pq$
$= 3m(n + q) + 2p(n + q)$ (i)
$= (n + q)(3m + 2p)$ (ii)
In (i), $3m$ is a common factor of the
first pair, and $2p$ of the second pair.
In (ii), $(n + q)$ is the common factor.

Exercise 4.8c

Factorise the following.

1. $4pr + 3qr + 4ps + 3qs$
2. $2xz - 5yz + 2xt - 5yt$
3. $9ac + 6bc + 12ad + 8bd$
4. $20km + 15lm + 20kn + 15ln$
5. $8pr + 4qr + 6ps + 3qs$
6. $6xz - 18yz + 6xt - 18yt$
7. $3ac + 7bc - 3ad - 7bd$
8. $4km + 6lm - 6kn - 9ln$
9. $9pr - 5qr - 9ps + 5qs$
10. $10xz - 6yz - 10xt + 6yt$
11. $5ab + 2cb + 5ad + 2cd$
12. $12kl + 5ml + 12kn + 5mn$
13. $2pq + 3rq - 2ps - 3rs$
14. $8xy + 5zy - 8xt - 5zt$
15. $4ab - 3cb + 4ad - 3cd$
16. $9kl - 12ml + 9kn - 12mn$
17. $7pq - 5rq - 7ps + 5rs$
18. $3xy - 6zy - 3xt + 6zt$
19. $2ab + 5cb + 5cd + 2ad$
20. $8kl + 12ml - 12mn - 8kn$

The product of $(x + 3)(x + 4)$ is
$x^2 + 7x + 12$; this expression is called a
trinomial because it has three terms.
$(x + 3)$ and $(x + 4)$ are the factors of this
trinomial.

To factorise any trinomial:

1. Find the product of the coefficient of
   the first term and the constant term.
2. Find the factors (of this product) whose
   sum is the coefficient of the middle term.
3. Rewrite the trinomial as a tetranomial,
   using the factors.
4. Factorise this tetranomial.

Example 4

(a) Factorise $x^2 + 8x + 15$.
     product $= +1 \times +15 = +15$
        sum $= +8$
  factors are $+3, +5$
  $\therefore x^2 + 8x + 15 = x^2 + 3x + 5x + 15$
          $= x(x + 3) + 5(x + 3)$
          $= (x + 3)(x + 5)$

(b) Factorise $x^2 - 7x + 12$
     product $= +1 \times +12 = +12$
        sum $= -7$
  factors are $-3, -4$
  $\therefore x^2 - 7x + 12 = x^2 - 3x - 4x + 12$
          $= x(x - 3) - 4(x - 3)$
          $= (x - 3)(x - 4)$

(c) Factorise $x^2 - x - 42$
     product $= +1 \times -42 = -42$
        sum $= -1$
  factors are $-7, +6$
  $\therefore x^2 - x - 42 = x^2 - 7x + 6x - 42$
          $= x(x - 7) + 6(x - 7)$
          $= (x - 7)(x + 6)$

Exercise 4.8d

Factorise the following.

1. $x^2 + 7x + 12$
2. $x^2 + 8x + 12$
3. $x^2 + 10x + 24$
4. $x^2 + 14x + 24$
5. $a^2 + 11a + 18$
6. $b^2 + 9b + 18$
7. $l^2 + 10l + 21$
8. $m^2 + 22m + 21$
9. $n^2 + 20n + 96$
10. $x^2 - 12x + 32$
11. $x^2 - 18x + 32$
12. $y^2 - 15y + 36$
13. $z^2 - 13z + 36$
14. $p^2 - 16p + 48$
15. $q^2 - 14q + 48$
16. $u^2 - 18u + 72$
17. $v^2 - 17v + 72$
18. $x^2 + 2x - 24$
19. $x^2 + 5x - 24$
20. $y^2 + 3y - 28$
21. $z^2 + 12z - 28$
22. $a^2 + 3a - 54$
23. $b^2 + 4b - 12$
24. $c^2 + 11c - 12$
25. $l^2 + 7l - 30$
26. $m^2 + 13m - 30$
27. $x^2 + x - 12$
28. $x^2 + x - 42$
29. $y^2 + y - 30$
30. $x^2 - 7x - 44$
31. $x^2 - 20x - 44$
32. $y^2 - 14y - 32$
33. $z^2 - 4z - 32$
34. $q^2 - 2q - 63$
35. $r^2 - 18r - 63$
36. $t^2 - 10t - 56$
37. $u^2 - 26u - 56$
38. $x^2 - x - 90$
39. $y^2 - y - 132$
40. $z^2 - z - 72$

Example 5

(a) Factorise $2x^2 + 3x + 1$

$$\text{product} = +2 \times +1 = +2$$
$$\text{sum} = +3$$
$$\text{factors are } +2, +1$$
$$\therefore 2x^2 + 3x + 1 = 2x^2 + 2x + x + 1$$
$$= 2x(x + 1) + 1(x + 1)$$
$$= (x + 1)(2x + 1)$$

(b) Factorise $3x^2 - 5x + 2$

$$\text{product} = +3 \times +2 = +6$$
$$\text{sum} = -5$$
$$\text{factors are } -3, -2$$
$$\therefore 3x^2 - 5x + 2 = 3x^2 - 3x - 2x + 2$$
$$= 3x(x - 1) - 2(x - 1)$$
$$= (x - 1)(3x - 2)$$

(c) Factorise $9x^2 + 17x - 2$

$$\text{product} = +9 \times -2 = -18$$
$$\text{sum} = +17$$
$$\text{factors are } +18, -1$$
$$\therefore 9x^2 + 17x - 2 = 9x^2 + 18x - x - 2$$
$$= 9x(x + 2) - 1(x + 2)$$
$$= (x + 2)(9x - 1)$$

Exercise 4.8e

Factorise the following.

1. $3x^2 + 11x + 6$	2. $5x^2 + 23x + 12$
3. $3x^2 + 10x + 8$	4. $6x^2 + 17x + 12$
5. $4a^2 + 16a + 15$	6. $6b^2 + 19b + 15$
7. $4m^2 + 20m + 9$	8. $8n^2 + 18n + 9$
9. $5p^2 + 11p + 2$	10. $5x^2 - 38x + 21$
11. $8x^2 - 34x + 21$	12. $10y^2 - 23y + 9$
13. $5z^2 - 46z + 9$	14. $3q^2 - 29q + 18$
15. $6r^2 - 31r + 18$	16. $4t^2 - 24t + 35$
17. $8u^2 - 38u + 35$	18. $5x^2 + 7x - 6$
19. $4x^2 + 5x - 6$	20. $4y^2 + 9y - 9$
21. $6z^2 + 25z - 9$	22. $8b^2 + 6b - 5$
23. $8c^2 + 18c - 5$	24. $6l^2 + 7l - 10$
25. $9m^2 + 9m - 10$	26. $5n^2 + 2n - 7$
27. $10x^2 + x - 21$	28. $6x^2 + x - 12$
29. $3y^2 + y - 2$	30. $7x^2 - 13x - 2$
31. $12x^2 - 5x - 2$	32. $8y^2 - 6y - 9$
33. $4z^2 - 9z - 9$	34. $6p^2 - 13p - 15$
35. $12q^2 - 8q - 15$	36. $6u^2 - 5u - 21.$
37. $4v^2 - 25v - 21$	38. $12x^2 - x - 20$
39. $6y^2 - y - 15$	40. $3z^2 - z - 24$

Example 6

(a) Factorise $x^2 + 12x + 36$

$$\text{product} = +1 \times +36 = +36$$
$$\text{sum} = +12$$
$$\text{factors are } +6, +6$$
$$\therefore x^2 + 12x + 36 = x^2 + 6x + 6x + 36$$
$$= x(x + 6) + 6(x + 6)$$
$$= (x + 6)(x + 6)$$
$$= (x + 6)^2$$

(b) Factorise $9x^2 - 12x + 4$

$$\text{product} = +9 \times +4 = +36$$
$$\text{sum} = -12$$
$$\text{factors are } -6, -6$$
$$\therefore 9x^2 - 12x + 4 = 9x^2 - 6x - 6x + 4$$
$$= 3x(3x - 2) - 2(3x - 2)$$
$$= (3x - 2)(3x - 2)$$
$$= (3x - 2)^2$$

Exercise 4.8f

Factorise the following.

1. $x^2 + 6x + 9$	2. $y^2 + 16y + 64$
3. $z^2 + 2z + 1$	4. $a^2 + 20a + 100$
5. $b^2 + 14b + 49$	6. $x^2 - 4x + 4$
7. $y^2 - 10y + 25$	8. $z^2 - 8z + 16$
9. $m^2 - 18m + 81$	10. $n^2 - 24n + 144$
11. $4x^2 + 12x + 9$	12. $9y^2 + 24y + 16$
13. $16z^2 + 8z + 1$	14. $100p^2 + 60p + 9$
15. $25q^2 + 20q + 4$	16. $9x^2 - 6x + 1$
17. $4y^2 - 20y + 25$	18. $16z^2 - 24z + 9$
19. $100u^2 - 20u + 1$	20. $25v^2 - 40v + 16$

The expression $a^2 - b^2$ is known as the difference between two squares; the factors of this expression are $(a + b)$ and $(a - b)$.

Example 7

(a) Factorise $x^2 - 4$

$$x^2 - 4 = x^2 - (2)^2$$
$$= (x + 2)(x - 2)$$

(b) Factorise $a^2 - \frac{1}{9}$

$$a^2 - \frac{1}{9} = a^2 - (\tfrac{1}{3})^2$$
$$= (a + \tfrac{1}{3})(a - \tfrac{1}{3})$$

(c) Factorise $4x^2 - 1$

$$4x^2 - 1 = (2x)^2 - 1$$
$$= (2x + 1)(2x - 1)$$

(d) Factorise $9a^2 - 16b^2$
$$9a^2 - 16b^2 = (3a)^2 - (4b)^2$$
$$= (3a + 4b)(3a - 4b)$$

(c) Find the value of $(0 \cdot 6)^2 - (0 \cdot 4)^2$
$$(0 \cdot 6)^2 - (0 \cdot 4)^2 = (0 \cdot 6 + 0 \cdot 4)(0 \cdot 6 - 0 \cdot 4)$$
$$= 1 \cdot 0 \times 0 \cdot 2 = 0 \cdot 2$$

Exercise 4.8g

1. $x^2 - 16$
2. $y^2 - 25$
3. $z^2 - 81$
4. $a^2 - 1$
5. $b^2 - 36$
6. $c^2 - 144$
7. $4x^2 - 25$
8. $16y^2 - 9$
9. $9z^2 - 64$
10. $4l^2 - 81$
11. $25m^2 - 16$
12. $4q^2 - 49$

13. $9r^2 - 25$
14. $x^2 - \frac{1}{4}$

15. $y^2 - \frac{1}{25}$
16. $z^2 - \frac{1}{16}$

17. $t^2 - \frac{1}{100}$
18. $a^2 - \frac{4}{}$

19. $b^2 - \frac{16}{25}$
20. $c^2 - \frac{9}{16}$

Exercise 4.8h

Evaluate the following.

1. $32^2 - 28^2$
2. $25^2 - 15^2$
3. $21^2 - 19^2$
4. $23^2 - 7^2$
5. $54^2 - 36^2$
6. $95^2 - 5^2$
7. $36^2 - 24^2$
8. $17^2 - 3^2$
9. $38^2 - 2^2$
10. $29^2 - 11^2$

11. $(8\frac{3}{5})^2 - (1\frac{2}{5})^2$
12. $(7\frac{4}{5})^2 - (2\frac{1}{5})^2$

13. $(5\frac{7}{10})^2 - (4\frac{3}{10})^2$
14. $(3\frac{9}{10})^2 - (1\frac{1}{10})^2$

15. $(6\frac{2}{3})^2 - (2\frac{1}{3})^2$
16. $(4\frac{5}{6})^2 - (1\frac{1}{6})^2$

17. $(3\frac{7}{12})^2 - (2\frac{5}{12})^2$
18. $(5\frac{5}{9})^2 - (3\frac{4}{9})^2$

19. $(6\frac{5}{8})^2 - (3\frac{3}{8})^2$
20. $(5\frac{7}{8})^2 - (4\frac{1}{8})^2$

21. $(0 \cdot 7)^2 - (0 \cdot 3)^2$
22. $(0 \cdot 9)^2 - (0 \cdot 1)^2$

23. $(3 \cdot 5)^2 - (0 \cdot 5)^2$
24. $(5 \cdot 5)^2 - (1 \cdot 5)^2$

25. $(6 \cdot 5)^2 - (3 \cdot 5)^2$
26. $(4 \cdot 1)^2 - (0 \cdot 9)^2$

27. $(3 \cdot 2)^2 - (1 \cdot 8)^2$
28. $(6 \cdot 4)^2 - (3 \cdot 6)^2$

29. $(1 \cdot 4)^2 - (0 \cdot 6)^2$
30. $(2 \cdot 8)^2 - (1 \cdot 2)^2$

The method of factorising the difference between squares can be used to simplify arithmetical calculations.

Example 8

(a) Find the value of $51^2 - 49^2$
$$51^2 - 49^2 = (51 + 49)(51 - 49)$$
$$= 100 \times 2 = 200$$
(b) Find the value of $(6\frac{3}{4})^2 - (3\frac{1}{4})^2$
$$(6\frac{3}{4})^2 - (3\frac{1}{4})^2 = (6\frac{3}{4} + 3\frac{1}{4})(6\frac{3}{4} - 3\frac{1}{4})$$
$$= 10 \times 3\frac{1}{2} = 35$$

4.9 ALGEBRAIC FRACTIONS

An algebraic fraction (like a common fraction) is not changed in value when the numerator *and* the denominator are multiplied or divided by the same quantity.

Example 1

(a) $\dfrac{x}{y} = \dfrac{x \times 2}{y \times 2} = \dfrac{2x}{2y}$

(b) $\dfrac{3a}{4b} = \dfrac{3a \times a}{4b \times a} = \dfrac{3a^2}{4ab}$

(c) $\dfrac{2x}{3y} = \dfrac{2x \times 2x}{3y \times 2x} = \dfrac{4x^2}{6xy}$

Exercise 4.9a

Copy and complete the following.

1. $\dfrac{a}{b} = \dfrac{3a}{?} = \dfrac{au}{?} = \dfrac{a^2}{?} = \dfrac{8au}{?} = \dfrac{12a^2}{?}$

2. $\dfrac{m}{n} = \dfrac{?}{5n} = \dfrac{?}{nv} = \dfrac{?}{n^2} = \dfrac{?}{7nv} = \dfrac{?}{3n^2}$

3. $\dfrac{2p}{3q} = \dfrac{8p}{?} = \dfrac{2px}{?} = \dfrac{2p^2}{?} = \dfrac{10px}{?} = \dfrac{6p^2}{?}$

4. $\dfrac{5b}{6c} = \dfrac{25b}{?} = \dfrac{5by}{?} = \dfrac{5b^2}{?} = \dfrac{45by}{?} = \dfrac{(5b)^2}{?}$

5. $\dfrac{3l}{4m} = \dfrac{?}{20m} = \dfrac{?}{4mq} = \dfrac{?}{4m^2} = \dfrac{?}{12mq} = \dfrac{?}{24m^2}$

Example 2

(a) Simplify $\dfrac{4a}{4b}$

$$\frac{4a}{4b} = \frac{4a \div 4}{4b \div 4} = \frac{a}{b}$$

(b) Simplify $\dfrac{2x^2}{3xy}$

$$\frac{2x^2}{3xy} = \frac{2x^2 \div x}{3xy \div x} = \frac{2x}{3y}$$

(c) Simplify $\dfrac{4a^2 b}{10ab^2}$

$$\frac{4a^2 b}{10ab^2} = \frac{4a^2 b \div 2ab}{10ab^2 \div 2ab}$$
$$= \frac{2a}{5b}$$

Exercise 4.9b

Simplify the following.

1. $\dfrac{3a}{3b}$ 　　 2. $\dfrac{15m}{20n}$ 　　 3. $\dfrac{4p}{12q}$

4. $\dfrac{5u^2}{6uv}$ 　　 5. $\dfrac{8x^2}{12xy}$ 　　 6. $\dfrac{9b^2}{36bc}$

7. $\dfrac{8lm}{15m^2}$ 　　 8. $\dfrac{16qr}{24r^2}$ 　　 9. $\dfrac{9t^2 u}{14tu^2}$

10. $\dfrac{6y^2 z}{30yz^2}$ 　　 11. $\dfrac{10ab^2}{24a^2 b}$ 　　 12. $\dfrac{9mn^2}{15m^2 n}$

13. $\dfrac{8p^2}{21p^2 q}$ 　　 14. $\dfrac{3u^3}{18u^3 v}$ 　　 15. $\dfrac{9x^3}{12xy^2}$

16. $\dfrac{15b^3}{24bc^2}$ 　　 17. $\dfrac{4l^2 m}{28l^3}$ 　　 18. $\dfrac{16q^2 r}{25q^3}$

19. $\dfrac{30tu^2}{36t^3}$ 　　 20. $\dfrac{21yz^2}{36y^3}$

To multiply algebraic fractions:

1. Factorise the numerator and the denominator where possible.
2. Cancel down and then multiply out.

Example 3

(a) $\dfrac{x}{y} \times \dfrac{3y^2}{x} = \dfrac{\cancel{x}^1}{\cancel{y}_1} \times \dfrac{\cancel{3y^2}^{3y}}{\cancel{x}_1}$

$$= 3y$$

(b) $\dfrac{a^2 - b^2}{4a} \times \dfrac{2a^2}{a-b}$

$$= \frac{(a + b)\cancel{(a-b)}^1}{\cancel{4a}_2} \times \frac{\cancel{2a^2}^{a}}{\cancel{(a-b)}_1}$$

$$= \frac{a(a + b)}{2}$$

Exercise 4.9c

Simplify the following.

1. $\dfrac{4a}{b} \times \dfrac{b^2}{a}$ 　　 2. $\dfrac{6m}{n} \times \dfrac{n^2}{2m^2}$

3. $\dfrac{3q^2}{10r} \times \dfrac{5r^2}{12q}$ 　　 4. $\dfrac{30v^2}{u^2} \times \dfrac{u^3}{6v}$

5. $\dfrac{4x^2}{21y} \times \dfrac{7y^2}{16x^3}$ 　　 6. $\dfrac{b^3}{12c} \times \dfrac{10c^3}{b}$

7. $\dfrac{l}{15m} \times \dfrac{20m^3}{l^2}$ 　　 8. $\dfrac{9t^3}{30u} \times \dfrac{15u^2}{16t}$

9. $\dfrac{36z^3}{y} \times \dfrac{y^3}{9z}$ 　　 10. $\dfrac{16a}{15b} \times \dfrac{5b^3}{8a^3}$

11. $\dfrac{m}{30n^2} \times \dfrac{12n^3}{m^3}$ 　　 12. $\dfrac{p^3}{24q^2} \times \dfrac{40q}{p^2}$

13. $\dfrac{28u^2}{15v^2} \times \dfrac{5v}{7u^3}$ 　　 14. $\dfrac{8x^3}{y^2} \times \dfrac{y^3}{21x^2}$

15. $\dfrac{8b^2}{27c^2} \times \dfrac{9c^3}{40b^3}$ 　　 16. $\dfrac{2a + 6b}{5c} \times \dfrac{15c^2}{a + 3b}$

17. $\dfrac{4l - 8m}{30n^2} \times \dfrac{6n}{l - 2m}$ 　　 18. $\dfrac{p^2 - q^2}{24r^2} \times \dfrac{9r^3}{p - q}$

19. $\dfrac{t^2 - u^2}{27v} \times \dfrac{36v^3}{t + u}$ 　　 20. $\dfrac{15x + 5y}{20x^3} \times \dfrac{32x^2}{3x + y}$

To divide algebraic fractions:

1. Factorise the numerator and the denominator where possible.
2. Multiply by the inverse of the divisor.

Example 4

(a) $\dfrac{3b}{d} \div \dfrac{6}{d^2} = \dfrac{3b}{\cancel{d}_1} \times \dfrac{\cancel{d^2}^{d}}{\cancel{6}_2}$

$$= \frac{bd}{2}$$

(b) $\dfrac{3a + 6b}{4} \div \dfrac{a + 2b}{12}$

$= \dfrac{3(a + 2b)}{4} \div \dfrac{(a + 2b)}{12}$

$= \dfrac{3\cancel{(a + 2b)}^1}{\cancel{4}_1} \times \dfrac{\cancel{12}^3}{\cancel{(a + 2b)}_1} = 9$

Exercise 4.9d

Simplify the following.

1. $\dfrac{a}{b} \div \dfrac{4}{b^2}$

2. $\dfrac{4b}{c} \div \dfrac{8}{5c^2}$

3. $\dfrac{m}{n^2} \div \dfrac{6}{n^3}$

4. $\dfrac{3l^2}{m} \div \dfrac{9l^3}{2}$

5. $\dfrac{p}{4q} \div \dfrac{3}{16q^3}$

6. $\dfrac{2q}{3r} \div \dfrac{8q^3}{21}$

7. $\dfrac{5u^2}{8v^2} \div \dfrac{15u}{2v^3}$

8. $\dfrac{7t^2}{30u^2} \div \dfrac{21t^3}{10u}$

9. $\dfrac{27x^2}{16y} \div \dfrac{9x}{4y^3}$

10. $\dfrac{5y}{2z^2} \div \dfrac{25y^3}{6z}$

11. $\dfrac{8a^3}{27b^2} \div \dfrac{40a}{9b^3}$

12. $\dfrac{3b^2}{5c^3} \div \dfrac{12b^3}{35c}$

13. $\dfrac{6m^2}{7n} \div \dfrac{18}{49n^3}$

14. $\dfrac{16l}{9m^2} \div \dfrac{20l^3}{3}$

15. $\dfrac{8p}{7q^2} \div \dfrac{12p^2}{35}$

16. $\dfrac{4a + 16b}{8} \div \dfrac{a + 4b}{6}$

17. $\dfrac{2m - 10n}{16} \div \dfrac{m - 5n}{12}$

18. $\dfrac{p^2 - q^2}{45} \div \dfrac{p + q}{27}$

19. $\dfrac{u^2 - v^2}{36} \div \dfrac{u - v}{32}$

20. $\dfrac{10x - 5y}{48} \div \dfrac{2x - y}{36}$

To add or to subtract algebraic fractions:

1. Find the L.C.M. of the denominators.
2. Write each fraction with this L.C.M.
3. Simplify the numerator.

Example 5

(a) $\dfrac{a}{3} + \dfrac{b}{2}$

The L.C.M. of the denominator is 6.

so $\dfrac{a}{3} + \dfrac{b}{2} = \dfrac{2a}{6} + \dfrac{3b}{6} = \dfrac{2a + 3b}{6}$

(b) $\dfrac{3}{2x} - \dfrac{4}{3y}$

The L.C.M. of the denominator is $6xy$.

so $\dfrac{3}{2x} - \dfrac{4}{3y} = \dfrac{9y}{6xy} - \dfrac{8x}{6xy}$

$= \dfrac{9y - 8x}{6xy}$

Exercise 4.9e

Simplify the following.

1. $\dfrac{a}{4} + \dfrac{b}{12}$

2. $\dfrac{m}{6} + \dfrac{n}{10}$

3. $\dfrac{3p}{10} + \dfrac{7q}{30}$

4. $\dfrac{5u}{12} + \dfrac{3v}{20}$

5. $\dfrac{x}{4} + \dfrac{5y}{12}$

6. $\dfrac{b}{4} - \dfrac{c}{8}$

7. $\dfrac{l}{9} - \dfrac{m}{12}$

8. $\dfrac{4p}{5} - \dfrac{3r}{10}$

9. $\dfrac{9t}{10} - \dfrac{4u}{15}$

10. $\dfrac{y}{2} - \dfrac{3z}{14}$

11. $\dfrac{5a}{12} - \dfrac{b}{15}$

12. $\dfrac{1}{p} + \dfrac{1}{4q}$

13. $\dfrac{1}{2u} + \dfrac{1}{6v}$

14. $\dfrac{8}{x} + \dfrac{5}{y}$

15. $\dfrac{7}{2b} + \dfrac{9}{4c}$

16. $\dfrac{3}{4l} + \dfrac{5}{6m}$

17. $\dfrac{1}{6q} - \dfrac{8}{r}$

18. $\dfrac{10}{3t} - \dfrac{11}{6u}$

19. $\dfrac{3}{2y} - \dfrac{8}{5z}$

20. $\dfrac{7}{6a} - \dfrac{4}{3b}$

21. $\dfrac{a}{3} + \dfrac{a}{4} + \dfrac{a}{6}$

22. $\dfrac{a}{3} + \dfrac{a}{4} - \dfrac{a}{12}$

23. $\dfrac{1}{a} + \dfrac{1}{2a} + \dfrac{1}{3a}$

24. $\dfrac{1}{a} + \dfrac{1}{2a} - \dfrac{1}{6a}$

Example 6

(a) $\dfrac{a+b}{2} + \dfrac{a-b}{3}$

The L.C.M. of the denominator is 6.

so $\dfrac{a+b}{2} + \dfrac{a-b}{3}$

$= \dfrac{3(a+b)}{6} + \dfrac{2(a-b)}{6}$

$= \dfrac{3(a+b)+2(a-b)}{6}$

$= \dfrac{3a+3b+2a-2b}{6}$

$= \dfrac{5a+b}{6}$

(b) $\dfrac{a-b}{3a} - \dfrac{a-2b}{4a}$

The L.C.M. of the denominator is $12a$.

so $\dfrac{(a-b)}{3a} - \dfrac{(a-2b)}{4a}$

$= \dfrac{4(a-b)}{12a} - \dfrac{3(a-2b)}{12a}$

$= \dfrac{4(a-b)-3(a-2b)}{12a}$

$= \dfrac{4a-4b-3a+6b}{12a} = \dfrac{a+2b}{12a}$

Exercise 4.9f

Simplify the following.

1. $\dfrac{x+4}{4} + \dfrac{x+3}{6}$

2. $\dfrac{y+6}{5} + \dfrac{y-2}{2}$

3. $\dfrac{z-3}{4} + \dfrac{z+2}{3}$

4. $\dfrac{t-8}{4} + \dfrac{t-1}{12}$

5. $\dfrac{a+b}{6} + \dfrac{a-b}{8}$

6. $\dfrac{m-n}{3} + \dfrac{m+n}{12}$

7. $\dfrac{p+2q}{3} + \dfrac{p-2q}{6}$

8. $\dfrac{u+3v}{6} + \dfrac{u-4v}{2}$

9. $\dfrac{b-2c}{3} + \dfrac{b-5c}{2}$

10. $\dfrac{2l+m}{4} + \dfrac{2l-m}{3}$

11. $\dfrac{x+2}{2} - \dfrac{x+1}{5}$

12. $\dfrac{y-2}{4} - \dfrac{y+3}{8}$

13. $\dfrac{z+5}{2} - \dfrac{z-3}{3}$

14. $\dfrac{t-1}{3} - \dfrac{t-4}{6}$

15. $\dfrac{q+r}{2} - \dfrac{q-r}{5}$

16. $\dfrac{t-u}{3} - \dfrac{t+u}{4}$

17. $\dfrac{a+3b}{2} - \dfrac{a-5b}{4}$

18. $\dfrac{m-2n}{2} - \dfrac{m-3n}{3}$

19. $\dfrac{3p-q}{3} - \dfrac{2p-q}{5}$

20. $\dfrac{2u+v}{2} - \dfrac{4u-v}{4}$

4.10 INDICES

Expressions of the form x^2, 5^3 are called powers of x, powers of 5. The smaller figures 2, 3, are called *indices*.

Index notation is used to avoid repetition of terms,

so $x \times x = x^2$
   $5 \times 5 \times 5 = 5^3$

Example 1

(a) Find the value of 3^3
    $3^3 = 3 \times 3 \times 3 = 27$

(b) Find the value of 10^4
    $10^4 = 10 \times 10 \times 10 \times 10 = 10\ 000$

(c) $2^4 \times 3^2 = 2 \times 2 \times 2 \times 2 \times 3 \times 3$
    $= 16 \times 9 = 144$

Example 2

Express in index form:

(a) $4 \times 4 \times 4$,   (b) $x \times x \times x \times x \times x$

(a) $4 \times 4 \times 4 = 4^3$

(b) $x \times x \times x \times x \times x = x^5$

Exercise 4.10a

Find the value of:

1. 3^5
2. 6^5
3. 7^4
4. 9^4
5. 5^6
6. 4^7
7. 11^3
8. 3^9
9. 2^{10}
10. 8^4
11. $2^3 \times 5^2$
12. $2^4 \times 4^2$
13. $2^5 \times 3^2$
14. $3^2 \times 4^3$
15. $3^3 \times 4^2$

Express in index form.

16. $4 \times 4 \times 4 \times 4 \times 4$
17. $5 \times 5 \times 5 \times 5$
18. $8 \times 8 \times 8 \times 8 \times 8 \times 8$
19. $3 \times 3 \times 3 \times 3 \times 3 \times 3 \times 3$
20. $12 \times 12 \times 12$
21. $2 \times 2 \times 2 \times 2 \times 2 \times 2 \times 2 \times 2 \times 2$
22. $6 \times 6 \times 6 \times 6 \times 6 \times 6 \times 6 \times 6$
23. $7 \times 7 \times 7 \times 7 \times 7 \times 7 \times 7 \times 7 \times 7 \times 7$
24. $x \times x \times x \times x \times x \times x \times x \times x$
25. $y \times y \times y \times y \times y \times y \times y \times y$
26. $z \times z \times z \times z \times z \times z \times z \times z \times z$
27. $a \times a \times a \times a \times a \times a$
28. $b \times b \times b \times b \times b \times b \times b \times b \times b \times b$
29. $c \times c \times c \times c \times c$
30. $t \times t \times t \times t \times t \times t \times t \times t \times t \times t$

Rule 1 To multiply the *same* quantity raised to different powers, add the powers.

e.g. $a^2 \times a^3 = a \times a \times a \times a \times a = a^5$

$\therefore a^2 \times a^3 = a^{2+3} = a^5$

Example 3

(a) Simplify $x^4 \times x^5$

$x^4 \times x^5 = x^{4+5} = x^9$

(b) Simplify $2a^2 \times 5a^3$

$2a^2 \times 5a^3 = 2 \times 5 \times a^2 \times a^3$

$= 10 \times a^{2+3} = 10a^5$

Exercise 4.10b

Simplify the following.

1. $x^2 \times x^6$
2. $y^3 \times y^4$
3. $z^4 \times z^6$
4. $p^5 \times p^7$
5. $q^3 \times q^6$
6. $2^2 \times 2^3$
7. $2^4 \times 2^5$
8. $3^2 \times 3^4$
9. 3×3^3
10. $4^2 \times 4^3$
11. $3a^3 \times 4a^5$
12. $5b^4 \times 3b^6$
13. $2c^5 \times 15c^7$
14. $4l^3 \times 6l^{12}$
15. $6m \times 9m^8$
16. $a^2 x^2 \times a^5 x^3$

17. $b^3 y^2 \times b^7 y^4$
18. $cz^4 \times c^3 z^5$
19. $m^3 u \times m^4 u^7$
20. $n^2 v^6 \times n^2 v^4$

Rule 2 To divide the *same* quantity raised to different powers, subtract the powers.

e.g. $x^5 \div x^2 = \dfrac{\cancel{x}^1 \times \cancel{x}^1 \times x \times x \times x}{{}_1\cancel{x} \times {}_1\cancel{x}} = x^3$

$\therefore x^5 \div x^2 = x^{5-2} = x^3$

Example 4

(a) Simplify $x^4 \div x^3$

$x^4 \div x^3 = x^{4-3} = x^1 = x$

(b) Simplify $a^2 \div a^2$

$a^2 \div a^2 = a^{2-2} = a^0 = 1$

Note: $\dfrac{a \times a}{a \times a} = 1$

(c) Simplify $12b^5 \div 3b^2$

$12b^5 \div 3b^2 = \dfrac{12b^5}{3b^2} = 4 \times b^{5-2}$

$= 4b^3$

Exercise 4.10c

Simplify the following.

1. $x^5 \div x^3$
2. $y^9 \div y^4$
3. $z^{12} \div z^3$
4. $m^{11} \div m^{10}$
5. $n^{15} \div n^{15}$
6. $2^7 \div 2^3$
7. $2^8 \div 2^2$
8. $3^9 \div 3^6$
9. $5^7 \div 5^4$
10. $5^{12} \div 5^8$
11. $15p^7 \div 3p^4$
12. $36q^8 \div 4q^3$
13. $72r^{10} \div 12r^2$
14. $7u^{12} \div 35u^8$
15. $8v^{16} \div 56v^{16}$
16. $p^5 t^9 \div p^3 t^4$
17. $q^6 u^{12} \div q^2 u^6$
18. $r^{10} v^8 \div r^8 v$
19. $a^7 x^6 \div a^4 x^{10}$
20. $b^8 y^{10} \div b^6 y^{11}$

Rule 3 To raise a power to a power, multiply the powers together.

e.g. $(x^3)^2 = (x \times x \times x)^2$

$= x \times x \times x \times x \times x \times x$

$= x^6$

$\therefore (x^3)^2 = x^{3 \times 2} = x^6$

Example 5

(a) Find the value of $(3^2)^2$

$(3^2)^2 = 3^{2 \times 2} = 3^4 = 81$

(b) Find the value of $(1^5)^6$

$(1^5)^6 = 1^{5 \times 6} = 1^{30} = 1$

Note: 1 to any power is 1.

Example 6

(a) Simplify $(a^4)^3$

$(a^4)^3 = a^{4 \times 3} = a^{12}$

(b) Simplify $(3x^2)^3$

$(3x^2)^3 = (3^3)(x^2)^3$

$= 27 \times x^{2 \times 3} = 27x^6$

Exercise 4.10d

Simplify the following.

1. $(2^3)^2$	2. $(4^2)^2$	3. $(3^3)^2$
4. $(5^2)^2$	5. $(4^3)^2$	6. $(2^5)^2$
7. $(2^2)^3$	8. $(10^2)^3$	9. $(2^3)^3$
10. $(1^3)^3$	11. $(20a)^2$	12. $(3b)^3$
13. $(7c)^3$	14. $(5l^2)^2$	15. $(40m^2)^2$
16. $(2n^2)^3$	17. $(5p^2)^3$	18. $(30q^3)^2$
19. $(4r^3)^3$	20. $(6t^3)^3$	

Because $9^{\frac{1}{2}} \times 9^{\frac{1}{2}} = 9^1 = 9$,

so $9^{\frac{1}{2}} = \sqrt[2]{9} = 3$.

In the same way $8^{\frac{1}{3}} = \sqrt[3]{8} = 2$

Because $2^3 \div 2^4 = \dfrac{2 \times 2 \times 2}{2 \times 2 \times 2 \times 2} = \dfrac{1}{2}$,

and $2^3 \div 2^4 = 2^{3-4} = 2^{-1}$,

so $2^{-1} = \dfrac{1}{2}$.

In the same way, $3^{-2} = \dfrac{1}{3^2} = \dfrac{1}{9}$

Example 7

(a) Find the value of $32^{\frac{1}{5}}$,

$32^{\frac{1}{5}} = \sqrt[5]{32} = 2$

(b) Find the value of 2^{-3},

$2^{-3} = \dfrac{1}{2^3} = \dfrac{1}{8}$

Example 8

Write the following without fractional or negative indices.

(a) $y^{\frac{1}{3}}$ (b) y^{-4}

(a) $y^{\frac{1}{3}} = \sqrt[3]{y}$;   (b) $y^{-4} = \dfrac{1}{y^4}$

Example 9

(a) Simplify $(64a^4)^{\frac{1}{2}}$

$(64a^4)^{\frac{1}{2}} = (64)^{\frac{1}{2}}(a^4)^{\frac{1}{2}}$

$= \sqrt{64} \times a^{4 \times \frac{1}{2}} = 8a^2$

Note: $8a^2 \times 8a^2 = 64a^4$

(b) Simplify $(2a^2)^{-2}$

$(2a^2)^{-2} = 2^{-2} \times (a^2)^{-2}$

$= \dfrac{1}{2^2} \times a^{2 \times -2}$

$= \dfrac{1}{2^2} \times a^{-4}$

$= \dfrac{1}{2^2} \times \dfrac{1}{a^4} = \dfrac{1}{4a^4}$

Exercise 4.10e

Find the value of the following.

1. $81^{\frac{1}{2}}$	2. $10\,000^{\frac{1}{2}}$
3. $64^{\frac{1}{3}}$	4. $16^{\frac{1}{4}}$
5. $100\,000^{\frac{1}{5}}$	6. 5^{-3}
7. 12^{-2}	8. 3^{-4}
9. 15^{-1}	10. 2^{-5}

Write without fractional or negative indices.

11. $x^{\frac{1}{2}}$

12. $(ax)^{\frac{1}{2}}$

13. $(9x)^{\frac{1}{2}}$

14. $(by)^{\frac{1}{3}}$

15. $(8y)^{\frac{1}{3}}$

16. a^{-3}

17. b^{-1}

18. $(3m)^{-2}$

19. $(4n)^{-3}$

20. $(2p)^{-4}$

Simplify the following.

21. $(25a^2)^{\frac{1}{2}}$

22. $(144b^4)^{\frac{1}{2}}$

23. $(49c^8)^{\frac{1}{2}}$

24. $(27p^3)^{\frac{1}{3}}$

25. $(64q^9)^{\frac{1}{3}}$

26. $(3x^4)^{-2}$

27. $(5y^3)^{-2}$

28. $(2z^2)^{-3}$

29. $(5u^5)^{-3}$

30. $(3v^3)^{-4}$

4.11 QUADRATIC EQUATIONS

A quadratic equation is an equation which contains a square as the highest power of the unknown. It is usually written with one side equal to 0 and arranged so that the term containing the square is positive.

e.g. $x^2 + 3x - 4 = 0$ is a typical quadratic equation.

$x^2 - 6x = 7$ is written $x^2 - 6x - 7 = 0$

$6 = 7a - 2a^2$ is written $2a^2 - 7a + 6 = 0$

Exercise 4.11a

Write the following as quadratic equations in the usual way.

1. $x^2 + 8x = -7$
2. $x^2 + 5x = 14$
3. $y^2 - 2y = -1$
4. $z^2 - 3z = 18$
5. $2a^2 + 5a = -2$
6. $3b^2 + 2b = 8$
7. $4c^2 - 8c = -3$
8. $5m^2 - 6m = 8$
9. $x^2 = 15 - 2x$
10. $y^2 = 7y - 12$
11. $2z^2 = 11z + 21$
12. $3t^2 = 5 - 2t$
13. $4 = 5a - a^2$
14. $9 = -15b - 4b^2$
15. $-3 = 2c - c^2$

Some quadratic equations can be solved using the fact that, when $ab = 0$, then either $a = 0$ or $b = 0$.

Example 1

(a) Solve $x(x - 2) = 0$.
Either $x = 0$ or $(x - 2) = 0$
So $x = 0$ or $x = 2$

(b) Solve $(a + 2)(a - 3) = 0$
Either $(a + 2) = 0$ or $(a - 3) = 0$
So $a = -2$ or $a = 3$

(c) Solve $(4y + 5)(3y - 2) = 0$
Either $(4y + 5) = 0$ or $(3y - 2) = 0$
So $4y = -5$ or $3y = 2$
i.e. $y = -1\frac{1}{4}$ or $y = \frac{2}{3}$

Exercise 4.11b

Solve the following quadratic equations

1. $x(x - 4) = 0$
2. $x(x + 3) = 0$
3. $3y(y - 5) = 0$
4. $z(4z + 1) = 0$
5. $5t(2t + 3) = 0$
6. $(x - 3)(x - 5) = 0$
7. $(y - 8)(y + 2) = 0$
8. $(z - 1)(z + 5) = 0$
9. $(u + 2)(u + 4) = 0$
10. $(v - 6)(v - 6) = 0$
11. $(2x - 3)(4x - 5) = 0$
12. $(3y - 4)(5y + 1) = 0$
13. $(2z - 7)(z + 3) = 0$
14. $(p - 5)(3p - 4) = 0$
15. $(3q + 8)(2q + 5) = 0$

Other quadratic equations can be solved by factorisation.

Example 2

(a) Solve $x^2 - 5x = 0$
Factorise the left-hand side to give:

$x(x - 5) = 0$ (x is a common factor)
Either $x = 0$ or $(x - 5) = 0$
So $x = 0$ or $x = 5$

(b) Solve $4x^2 + 18x = 0$
Factorise the left-hand side to give:

$2x(2x + 9) = 0$ ($2x$ is a common factor)
Either $2x = 0$ or $2x + 9 = 0$
So $x = 0$ or $2x = -9$
i.e. $x = 0$ or $x = -4\frac{1}{2}$

Exercise 4.11c

Solve the following quadratic equations

1. $x^2 - 5x = 0$ 2. $y^2 - 9y = 0$
3. $z^2 - z = 0$ 4. $a^2 + 7a = 0$
5. $b^2 + 11b = 0$ 6. $3x^2 - 9x = 0$
7. $4y^2 - 28y = 0$ 8. $5z^2 - 60z = 0$
9. $7m^2 + 35m = 0$ 10. $9n^2 + 54n = 0$
11. $6p^2 - 15p = 0$ 12. $4q^2 - 14q = 0$
13. $15r^2 + 25r = 0$ 14. $12u^2 + 16u = 0$
15. $18v^2 + 21v = 0$ 16. $28l^2 - 35l = 0$
17. $16m^2 - 18m = 0$ 18. $9n^2 + 16n = 0$
19. $8b^2 - 20b = 0$ 20. $10c^2 - 45c = 0$

Example 3

(a) Solve $y^2 - 9 = 0$
    Factorise the left-hand side to give:

$$(y + 3)(y - 3) = 0$$
$y^2 - 9$ is a difference of squares
Either $y + 3 = 0$ or $y - 3 = 0$
So $y = -3$ or $y = +3$

(b) Solve $25a^2 - 49 = 0$
    Factorise the left-hand side to give:

$$(5a + 7)(5a - 7) = 0$$
$25a^2 - 49$ is a difference of squares
Either $5a + 7 = 0$ or $5a - 7 = 0$
So $5a = -7$ or $5a = 7$
i.e. $a = -1\frac{2}{5}$ or $a = 1\frac{2}{5}$

Exercise 4.11d

Solve the following quadratic equations.

1. $x^2 - 36 = 0$ 2. $y^2 - 100 = 0$
3. $z^2 - 49 = 0$ 4. $a^2 - 1 = 0$
5. $b^2 - 64 = 0$ 6. $c^2 - 121 = 0$

7. $l^2 - \frac{1}{4} = 0$ 8. $m^2 - \frac{1}{25} = 0$

9. $n^2 - \frac{1}{9} = 0$ 10. $p^2 - \frac{1}{100} = 0$

11. $4x^2 - 25 = 0$ 12. $9y^2 - 16 = 0$
13. $25z^2 - 64 = 0$ 14. $16u^2 - 81 = 0$
15. $49v^2 - 144 = 0$ 16. $64l^2 - 9 = 0$
17. $81m^2 - 4 = 0$ 18. $36n^2 - 1 = 0$
19. $121q^2 - 100 = 0$ 20. $100r^2 - 49 = 0$

Example 4

(a) Solve $x^2 - x - 6 = 0$
    Factorise the left-hand side to give:

$$(x - 3)(x + 2) = 0$$
$x^2 - x - 6$ is a trinomial
Either $x - 3 = 0$ or $x + 2 = 0$
So $x = 3$ or $x = -2$

(b) Solve $2a^2 - 7a + 6 = 0$
    Factorise the left-hand side to give:

$$(2a - 3)(a - 2) = 0$$
$2a^2 - 7a + 6$ is a trinomial
Either $2a - 3 = 0$ or $a - 2 = 0$
So $2a = 3$ or $a = 2$
i.e. $a = 1\frac{1}{2}$ or $a = 2$

Exercise 4.11e

Solve the following quadratic equations.

1. $x^2 - 9x + 20 = 0$ 2. $x^2 - 12x + 27 = 0$
3. $y^2 - 6y + 5 = 0$ 4. $z^2 - 8z + 16 = 0$
5. $a^2 + 14a + 40 = 0$ 6. $b^2 + 10b + 9 = 0$
7. $c^2 + 18c + 81 = 0$ 8. $l^2 + 2l - 48 = 0$
9. $m^2 + 7m - 60 = 0$ 10. $n^2 + 5n - 36 = 0$
11. $p^2 + p - 56 = 0$ 12. $q^2 - 4q - 21 = 0$
13. $r^2 - 6r - 72 = 0$ 14. $u^2 - 7u - 8 = 0$
15. $v^2 - v - 132 = 0$ 16. $8x^2 - 22x + 15 = 0$
17. $12x^2 - 31x + 7 = 0$ 18. $3y^2 - 19y + 28 = 0$
19. $4z^2 - 20z + 25 = 0$ 20. $9p^2 + 18p + 8 = 0$
21. $5q^2 + 12q + 4 = 0$ 22. $25r^2 + 40r + 16 = 0$
23. $10t^2 + 11t - 6 = 0$ 24. $4u^2 + 12u - 7 = 0$
25. $3v^2 + 14v - 5 = 0$ 26. $6a^2 + a - 12 = 0$
27. $8b^2 - 2b - 15 = 0$ 28. $2c^2 - 5c - 18 = 0$
29. $12m^2 - 8m - 15 = 0$ 30. $2n^2 - n - 28 = 0$

Example 5

(a) Solve $x^2 + 3 = 4x$

    Rearrange to make the right-hand side equal to nought.

    So $x^2 - 4x + 3 = 0$

    Factorise the left-hand side to give:

$$(x - 3)(x - 1) = 0$$
Either $x - 3 = 0$ or $x - 1 = 0$
So $x = 3$ or $x = 1$

(b) Solve $7 - 2x^2 = 5x$

Rearrange to make $2x^2$ positive

$0 = 2x^2 + 5x - 7$

Factorise the right-hand side to give:

$0 = (2x + 7)(x - 1)$
Either $2x + 7 = 0$ or $x - 1 = 0$
So $x = -3\frac{1}{2}$ or $x = 1$

Exercise 4.11f

Solve the following quadratic equations.

1. $x^2 + 2x = 15$
2. $y^2 + 10y = -16$
3. $z^2 - 5z = 24$
4. $m^2 - 3m = -2$
5. $6n^2 + 13n = 5$
6. $3p^2 + 14p = -8$
7. $12q^2 - 7q = 10$
8. $2r^2 - 19r = -42$
9. $t^2 = 11t - 18$
10. $u^2 = 54 - 3u$
11. $10v^2 = 12 - 7v$
12. $2a^2 = 5a + 18$
13. $6 = 5b - b^2$
14. $-28 = 3c - c^2$
15. $10 = -13d - 4d^2$

4.12 SIMPLE PROBLEMS

Solve an algebraic problem as follows.

(i) Let a letter stand for the item which has to be found.
(ii) State clearly the units to be used:
e.g. 'Let the weight be x grams',
or 'Let the distance be y km'.
(iii) From the information given in the question, write statements each containing the unknown quantity.
(iv) Using these statements, make an equation, solve it, and then answer the question.
e.g. 'The weight is 24 g', not '$x = 24$'.
or 'The distance is 5 km', not '$y = 5$'.
(v) Check the answer from the facts given in the question.

Example 1

Find the number which, when multiplied by 5, and 10 is added to this product, makes a total of 25.

Let x be the number required.
Then 5 times the number is $5x$.
So $5x + 10 = 25$
$\qquad 5x = 15$
$\qquad\quad x = 3$
Therefore the number is 3.
Check: $5 \times 3 + 10 = 25$.

Exercise 4.12a

1. A certain number is multiplied by 4, and 9 is then added to the obtained product giving a result of 33. Find the number.
2. I spend 36p on crisps and lemonade. Cans of lemonade cost 15p and packets of crisps cost 7p each. If I only buy one can of lemonade, how many packets of crisps do I buy?
3. A certain number is multiplied by 8 and 11 is then subtracted from the obtained product giving a result of 29. Find the number.
4. Tom has a certain number of marbles. He plays a game and wins: he finds that he has three times as many. He then plays a second game and loses eight of them: he finds that he has 16 left. How many did he have originally?
5. A certain number is divided by 7 and 6 is then added to the obtained quotient, giving a result of 15. Find the number.
6. After a car journey, only a third of the petrol that I had to start with is left in my tank. I call at a garage and have 10 litres put in; I then find that there are 17 litres in the tank altogether. How many litres were in the tank at the start of the journey?
7. A certain number is divided by 5; 4 is then subtracted from the obtained quotient giving a result of 12. Find the number.
8. Mary and three other girls have gathered some nuts. Mary counts the number, divides by 4 and takes her share home. When she gets home she throws 3 nuts away because they are bad and finds that she has 9 left. How many nuts did the girls gather altogether?

Example 2

Find three consecutive odd numbers whose sum is 39.

Let x be the first odd number.
Then $(x + 2)$ is the next one, and
 $(x + 2 + 2)$ is the third.
$\therefore x + (x + 2) + (x + 2 + 2) = 39$
$$3x + 6 = 39$$
$$3x = 33$$
$$x = 11$$
So the numbers are 11, 13 and 15.
Check: $11 + 13 + 15 = 39$.

Exercise 4.12b

1. Find three consecutive numbers whose sum is 48.
2. The figures for the weights of three dogs in kilograms are consecutive. If the total weight of the three dogs is 75 kilograms, find the weight of each dog.
3. Find three consecutive even numbers whose sum is 102.
4. The road from Birmingham to Bromsgrove passes through Selly Oak and Longbridge. The figures for the distances from Birmingham to Selly Oak, Selly Oak to Longbridge and Longbridge to Bromsgrove are consecutive even numbers. If the distance from Birmingham to Bromsgrove is 24 kilometres, find (a) the distance from Birmingham to Selly Oak, (b) the distance from Selly Oak to Longbridge, and (c) the distance from Longbridge to Bromsgrove.
5. Find four consecutive odd numbers whose sum is 120.
6. A train leaves London (King's Cross) and the figures (in minutes) for its time to pass Hatfield, its passing time from Hatfield to Hitchin, and its passing time from Hitchin to Huntingdon are consecutive odd numbers. If it passes Huntingdon 51 minutes after leaving London, find (a) the time taken to pass Hatfield from the start, (b) the passing time from Hatfield to Hitchin, and (c) the passing time from Hitchin to Huntingdon.
7. Find four consecutive numbers which are divisible by 5 and whose sum is 190.
8. In the same innings of a cricket match, four batsmen's scores are consecutive numbers which are divisible by 5. If their total contribution to the innings is 70 runs, how many does each batsman score?

9. Find three consecutive numbers which are divisible by 3 and whose sum is 81.
10. I make a rail journey from London (Victoria) to Sutton Common and I have to change at Clapham Junction and Wimbledon. The times of the three short train journeys from Victoria to Clapham Junction, Clapham Junction to Wimbledon and Wimbledon to Sutton Common are consecutive multiples of 3. Find the time for each stage of the journey if I spend 27 minutes in travelling altogether.

Example 3

A playground is 40 m longer than it is wide. If its perimeter is 272 m, what is its area?

Let the length of the playground be x m.
Then the width is $(x - 40)$ m.
So $x + (x - 40) + x + (x - 40) = 272$
$$4x - 80 = 272$$
$$4x = 352$$
$$x = 88$$

Therefore the length is 88 m and the width is 48 m, giving a perimeter of 272 m.
So the area $= 88 \times 48 = 4224$ m^2.

Exercise 4.12c

1. A tray has a length of 30 cm longer than its width. If its perimeter is 180 cm, find its area.
2. An isosceles triangle has a perimeter of 21 cm, and the unequal side has a length of 3 cm. Find the length of the two equal sides.
3. An isosceles right-angled triangle has a perimeter of 17 cm. If the hypotenuse is 7 cm long, find (a) the length of the two equal sides and (b) the area of the right-angled triangle.
4. The four angles of any quadrilateral total 360°. If a certain quadrilateral has three of its angles equal, find their value when the fourth angle is 135°.
5. A cuboid has a width 1 cm longer than the height and a length 1 cm longer than the width. If the total edge length is 36 cm, find the height, width and length. Find also the volume of the cuboid.

6. The area of the square picture frame illustrated is 92 cm². Find (a) the edge length of the inner square, (b) the edge length of the outer square, (c) the two perimeters of the frame.

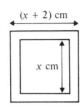

7. A wire of length 32 cm is bent so as to form a square. Find the edge length of the square. If the same wire is now bent so as to form a rectangle with one edge 4 cm longer than the other, find the two edge lengths of the rectangle. Which has the larger area, the square or the rectangle and by how much?

8. Four wooden planks, each having a length 80 cm longer than its width, are arranged as illustrated in the diagram. If the perimeter of the floor space they occupy is 480 cm, find (a) the dimensions of each plank, (b) the area of the floor space that the planks are occupying together.

9. A wire of length 34 cm is bent so as to form a rectangle with one edge 5 cm longer than the other. Find the two edge lengths of the rectangle. If the same wire is now bent into the shape of an isosceles right-angled triangle with a hypotenuse of length 14 cm, find the length of the two equal sides. Which has the larger area, the rectangle or the triangle and by how much?

10. Six paving stones are arranged in a square array as illustrated. If each stone has a length 20 cm greater than its width, find (a) the dimensions of each stone, (b) the area of the ground space that the stones are occupying, (c) the perimeter of the same ground space.

Plot a graph as follows.

(i) Draw the axes at right angles to each other.
(ii) Scale the axes according to the data given.
(iii) Label each axis clearly.
(iv) Plot the points.
(v) Join up these points either by straight lines or by a suitable curve.
(vi) Give the graph a title.

Example 1

On a holiday in France, a boy notes every hour how many litres of petrol are left in the tank of the car. The table lists his results.

Time	Number of litres
10.00	30
11.00	25
12.00	17·5
13.00	12
13.00	40
14.00	40
15.00	40
16.00	32
17.00	25

Using a scale of 2 cm to 1 hour on the horizontal axis and 1 cm to 5 litres on the vertical axis, draw a graph to show the information in the table.

From your graph, find:

(a) during which hour most petrol was used,
(b) between which times the family stopped for lunch,
(c) what was the petrol consumption in kilometres per litre between 12.00 and 13.00 if the car travelled 66 km in this hour.

The graph below is shown smaller than is required to be drawn.

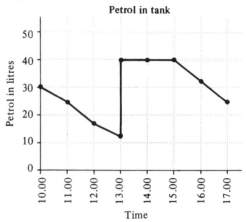

From the graph:

(a) Most petrol was used between 15.00 and 16.00, i.e. 8 litres.
(b) The family stopped for lunch between 13.00 and 15.00 because no petrol was used between 13.00 and 15.00
(c) The petrol used between 12.00 and 13.00 was $5\frac{1}{2}$ litres.
∴ petrol consumption $= 66 \div 5\frac{1}{2}$

$$= 12 \text{ km/l}.$$

Exercise 5.1

1. The following table gives the outside temperature at hourly intervals on a certain day in Winter.

Time	Temp	Time	Temp
6 a.m.	0°C	2 p.m.	14°C
7 a.m.	0°C	3 p.m.	12°C
8 a.m.	1°C	4 p.m.	9°C
9 a.m.	1°C	5 p.m.	5°C
10 a.m.	2°C	6 p.m.	4°C
11 a.m.	4°C	7 p.m.	2°C
12 noon	9°C	8 p.m.	1°C
1 p.m.	10°C		

Draw a graph to show this information.
Use a scale of 2 cm to 1 hour on the horizontal axis, and a scale of 1 cm to 1°C on the vertical axis.

(a) From your graph, estimate the temperature at the following times:

(i) 10.30 a.m.   (ii) 2.30 p.m.
(iii) 4.30 p.m.   (iv) 6.30 p.m.
(v) 1.15 p.m.   (vi) 4.15 p.m.

(b) During which one-hour interval does the temperature rise most quickly, and by how many degrees?

(c) During which one-hour interval does the temperature fall most quickly, and by how many degrees?

2. The following table gives the voltage of the electrical mains supply at hourly intervals for a certain day.

Time	Voltage	Time	Voltage
5 a.m.	250 V	3 p.m.	253 V
6 a.m.	247 V	4 p.m.	255 V
7 a.m.	243 V	5 p.m.	245 V
8 a.m.	235 V	6 p.m.	232 V
9 a.m.	239 V	7 p.m.	235 V
10 a.m.	247 V	8 p.m.	237 V
11 a.m.	250 V	9 p.m.	239 V
12 noon	249 V	10 p.m.	240 V
1 p.m.	245 V	11 p.m.	245 V
2 p.m.	250 V	12 midnight	251 V

Draw a graph to show this information. Use a scale of 1 cm to 1 h on the horizontal axis, and 1 cm to 2 V on the vertical axis starting at 230 V.

(a) From your graph, estimate the voltage at the following times:

(i) 8.30 p.m.   (ii) 12.30 p.m.
(iii) 4.30 p.m.   (iv) 6.15 a.m.
(v) 9.15 a.m.   (vi) 7.45 a.m.

(b) During which one-hour interval does the voltage drop most quickly and by how much?

3. Haweswater in Cumbria is used as a reservoir. The depth of the water in the reservoir on the first day of each month in a certain year is given in the table.

Date	Depth	Date	Depth
1 Jan.	56 m	1 July	42 m
1 Feb.	54 m	1 Aug.	35 m
1 Mar.	58 m	1 Sept.	25 m
1 Apr.	56 m	1 Oct.	33 m
1 May	50 m	1 Nov.	44 m
1 June	45 m	1 Dec.	50 m

Draw a graph to show this information. Use a scale of 2 cm to 1 month on the horizontal axis, and 1 cm to 2 m on the vertical axis starting at 24 m.
From your graph, estimate the depth of water on the following dates:

(i) 16 Jan.   (ii) 15 Apr.
(iii) 16 Aug.   (iv) 15 Nov.
(v) 7 Feb.   (vi) 23 Sept.
(vii) 8 Mar.   (viii) 23 Nov.

4. The atmospheric pressure of 12 midday and 12 midnight for a certain week is given in the table in mm of mercury.

Day	Pressure	Day	Pressure
Mon.	750	Fri.	756
	752		752
Tues.	753	Sat.	755
	754		759
Wed.	762	Sun.	758
	765		764
Thurs.	766		
	762		

Draw a graph to show this information using a scale of 2 cm to 12 h on the horizontal axis, and 1 cm to 1 mm on the vertical axis starting at 750 mm.

(a) From your graph, estimate the atmospheric pressure at the following times:

(i) 6 p.m. Mon.   (ii) 6 a.m. Fri.
(iii) 6 p.m. Sat.   (iv) 6 p.m. Sun.
(v) 3 a.m. Wed.   (vi) 3 p.m. Fri.

(b) During which 12-hour interval does the pressure rise most quickly, and by how many mm?

(c) During which 12-hour interval does the pressure fall most quickly, and by how many mm?

5. A man walks from Romsley to Hagley over the summit of both Walton Hill and Adam's Hill. The table gives his distance from Romsley and his altitude at various points.

	Distance	Altitude
Romsley	0 km	230 m
Whitehall Farm	1·5 km	230 m
Walton Hill Summit	2·0 km	315 m
St. Kenelm's Pass	3·0 km	250 m
Adam's Hill Summit	3·5 km	305 m
Hagley Hall	5·0 km	155 m
Hagley Village	5·5 km	140 m

Draw a graph to show this information using a scale of 4 cm to 1 km on the horizontal axis, and 1 cm to 10 m on the vertical axis starting at 130 m.

(a) From your graph, estimate his altitude at the following points:

(i) 2·5 km from Romsley,
(ii) 4 km from Romsley,
(iii) 1 km from Hagley Village.

(b) Is his first climb or his second climb the steeper?

(c) Find how many metres of altitude he loses for every kilometre he walks down from Adam's Hill to Hagley Hall.

6. The table below gives the details of a day's country walk.
Draw a graph to show the altitude reached at any given time during the day, using a scale of 4 cm to 1 h on the horizontal axis, and 1 cm to 20 m altitude on the vertical axis, starting at 200 m.

(a) From your graph, estimate the altitude at the following times:

(i) 10.15 a.m.   (ii) 12.15 p.m.
(iii) 2.15 p.m.   (iv) 4.15 p.m.
(v) 11.45 a.m.   (vi) 2.45 p.m.

(b) During which interval of half-an-hour is altitude gained most quickly and by how many metres?

(c) During which interval of half-an-hour is altitude lost most quickly and by how many metres?

7. From the table in question 6, draw a graph of the distance walked from Clee Hill Village at any given time. Use a scale of 4 cm to 1 h on the horizontal axis, and 1 cm to 1 km on the vertical axis.

(a) From your graph, estimate the distance walked at the following times:

(i) 12.15 p.m.   (ii) 2.15 p.m.
(iii) 4.15 p.m.   (iv) 11.45 a.m.
(v) 1.45 p.m.   (vi) 2.45 p.m.

(b) During which two half-hour intervals are the slowest speeds achieved? Suggest a reason.

(c) During which two half-hour intervals are the fastest speeds achieved? Suggest a reason.

8. The table shows the distance in km that a certain car can travel on 1 litre of petrol at various speeds.

Speed (km/h)	50	60	70	80	90	100
Fuel (km/l)	21·50	20·25	19·00	17·75	16·50	15·25

Draw a graph to show this information, using a scale of 2 cm to 10 km/h on the horizontal axis, and 2 cm to 1 km/l on the vertical axis.

	Altitude	Distance from Clee Hill Village	Time
Clee Hill Village	380 m	0 km	10 a.m.
Quarry Road Junction	440 m	2·0 km	10.30 a.m.
Titterstone Clee Hill Summit	530 m	3·0 km	11.00 a.m. to 11.30 a.m.
Callowgate Farm	300 m	5·0 km	12.00 noon
Three Horse Shoes Inn	310 m	7·5 km	12.30 p.m.
Banbury Road Junction	320 m	10·0 km	1.00 p.m. to 1.30 p.m.
Boyne Wood	460 m	11·0 km	2.00 p.m.
The Five Springs	460 m	12·5 km	2.30 p.m.
Brown Clee Hill Summit	540 m	14·0 km	3.00 p.m. to 3.30 p.m.
Hillside Farm	300 m	15·5 km	4.00 p.m.
Ditton Priors	230 m	17·5 km	4.30 p.m.

From your graph, find:

(a) The fuel consumption in km/l at

  (i) 54 km/h   (ii) 62 km/h
  (iii) 78 km/h   (iv) 94 km/h

(b) The speed in km/h for the following fuel consumption rates:

  (i) 20·9 km/l
  (ii) 18·4 km/l
  (iii) 17·5 km/l

9. Draw a smooth curve on a graph to show the following information.

Number	Square	Number	Square
0	0	7	49
1	1	8	64
2	4	9	81
3	9	10	100
4	16	11	121
5	25	12	144
6	36		

Use a scale of 1 cm for every 10 on the vertical axis to show the square, and 1 cm for every 1 on the horizontal axis.
Find from your graph the square root of:

(a) 20   (b) 40   (c) 90   (d) 110
(e) 58   (f) 86   (g) 106   (h) 132

10. Draw a smooth curve on a graph to show the length of the perimeter in cm of a square of a given area in cm^2 from the table below. Use a scale of 2 cm to 5 cm on the horizontal axis, and 2 cm to 10 cm^2 on the vertical axis.

Area cm^2	Perimeter cm	Area cm^2	Perimeter cm
1	4	36	24
4	8	49	28
9	12	64	32
16	16	81	36
25	20	100	40

From your graph find:

(a) the area of a square having a perimeter of length: (i) 15 cm (ii) 22 cm
  (iii) 29 cm (iv) 35 cm

(b) the length of perimeter of a square having an area: (i) 20 cm^2 (ii) 45 cm^2
  (iii) 60 cm^2 (iv) 90 cm^2

11. Draw a smooth curve from the information in the table. This shows the distance at which an image is formed by a convex lens of an object at various distances from the lens.

Object distance in cm	Image distance in cm
12	60·0
13	42·0
14	35·0
18	22·5
20	20·0
30	15·0
50	12·5

Show the object distance on the horizontal axis, and the image distance on the vertical axis. Use a scale of 2 cm to 10 cm for each axis.

From your graph find:
(a) the image distance for an object distance of
  (i) 15 cm (ii) 23 cm (iii) 25 cm (iv) 41 cm.
(b) the object distance for which the image distance is:
  (i) twice as long (ii) five times as long
  (iii) half as long.

12. The table shows the temperature of water in a kettle at various times after the electric current has been switched on.

Time	Temp.	Time	Temp.
start	10°C	90s	83°C
15s	36°C	120s	91°C
30s	50°C	150s	96°C
45s	61°C	180s	99°C
60s	70°C	210s	100°C

Draw a graph to show this information. Use a scale of 2 cm to 30 seconds on the horizontal axis, and 2 cm to 10°C on the vertical axis.

(a) Find the temperature of the water after
  (i) 75s (ii) 105s (iii) 2 min 15s
(b) Find the time to heat the water:

  (i) from 10°C to 55°C
  (ii) from 55°C to 100°C

  and suggest a reason for your answers.

A school 'snack bar' sells on one day the following varieties of crisps:

Variety	Plain	Smoky Bacon	Cheese and Onion	Salt 'n' Vinegar	Roast Beef	Roast Chicken
Number sold	42	28	20	16	4	10

Information such as this is often shown graphically as follows:

1. *By a bar chart*

    In this type of graph 'bars' are drawn to represent the numbers sold:

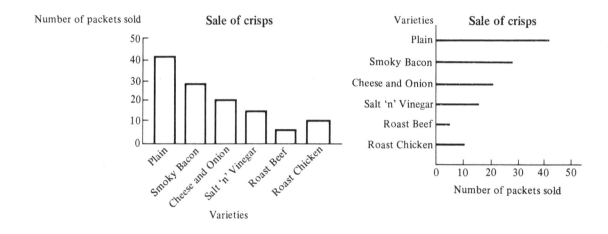

Exercise 5.2a

Draw suitable bar charts to display the following data:

1.
Day of the week	Sunday	Monday	Tuesday	Wednesday	Thursday	Friday	Saturday
Hours of sunshine	5	10	12	8	2	1	7

2.
Favourite lesson	Maths	English	History	Geography	French	Science	Art
Number of pupils	21	19	18	6	2	24	15

3.
Goals scored	0	1	2	3	4	5	6
Number of teams	1	17	12	8	4	3	2

4.
Mark	0	1	2	3	4	5	6	7	8	9	10
Number of pupils	1	4	6	5	8	14	7	3	5	2	1

2. *On a pie chart*

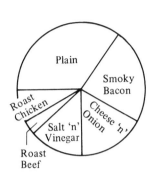

In this type of display a sector of a circle represents the number of packets of crisps that are sold.

The total number of packets sold is 120; this is represented by the angle at the centre of the circle, 360°.

So 1 packet is represented by 360 ÷ 120 = 3° and the sector angle for *Plain Crisps* = 42 × 3 = 126°.

The other angles are:

Smoky bacon = 28 × 3 = 84°
Cheese 'n' onion = 20 × 3 = 60°
Salt 'n' vinegar = 16 × 3 = 48°
Roast beef = 4 × 3 = 12°
Roast chicken = 10 × 3 = 30°

Exercise 5.2b

Draw pie charts to represent the following information.

1.

Sport chosen	Football	Rugby	Hockey	Cross-country	Volley ball
Number	40	32	16	24	8

2.

Transport used	Train	Aeroplane	Bus	Car	Cycle
Number	20	16	8	12	4

3.

Team supported	Spurs	Arsenal	Chelsea	West Ham	Fulham
Number	30	24	18	12	6

4.

Department	Hi fi	Food	Clothes	Furniture	Carpets	Shoes	Stationery
No. of customers	15	90	60	45	40	25	85

5.

Shoe size	1	2	3	4	5	6	7	8	9	10
Number in year	6	9	30	51	60	45	39	18	9	3

6.

Time of bus	8.00 a.m.	10.00 a.m.	12.00 noon	2.00 p.m.	4.00 p.m.	6.00 p.m.	8.00 p.m.
No. of passengers	36	27	18	9	27	45	18

7.

Party	Conservative	Labour	Liberal	Independent	Did not vote
No. of voters	300	240	60	20	100

8.

Day of week	Monday	Tuesday	Wednesday	Thursday	Friday	Saturday	Sunday
No. of cars using ferry	75	60	45	30	90	135	105

3. *By a pictogram*

A pictogram uses a motif to represent a given number of items: in the example below, the sale of 5 packets of crisps.
This type of display is often used in advertising.

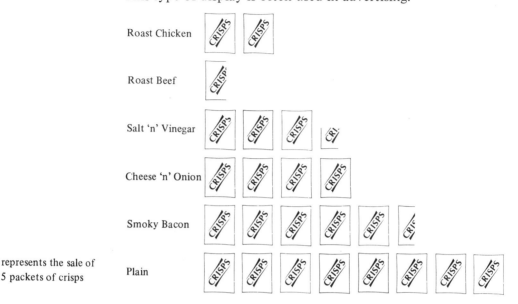

Roast Chicken

Roast Beef

Salt 'n' Vinegar

Cheese 'n' Onion

Smoky Bacon

represents the sale of
5 packets of crisps

Plain

Exercise 5.2c

Draw suitable pictograms to represent the information given in Exercise 5.2b.

Exercise 5.2d

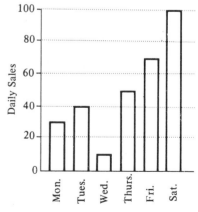

1. The daily sales in Fred's Music Stores of the recording of 'Pie Chart Rock' is shown in the bar chart.

   (a) What was the total number of records sold during the week?
   (b) If each record cost £2·75, how much was taken on Saturday?
   (c) Draw a pie chart to illustrate the same data.
   (d) Illustrate the information on a pictogram.

2. Six hundred pupils were asked which of four subjects they preferred. The pie chart shows how they voted.

   (a) How many preferred Mathematics?
   (b) If 280 voted for 'Art and crafts', what is the sector angle?
   (c) If 15% preferred 'History', what is the sector angle?
   (d) Draw a bar chart to illustrate the data.

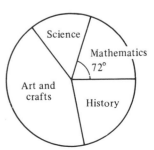

3. The table shows the number of books in a school library.

Subject	Mathematics	History	Geography	Science	Art	Languages	Fiction
Number	240	400	640	320	960	800	1440

Illustrate this data (a) on a bar chart; (b) on a pie chart; (c) on a pictogram.

4. The table gives details of the instruments played in a group of 240 pupils.

Instrument	Clarinet	Guitar	Violin	Trumpet	Drums
Number	36	40	24	58	82

Illustrate this data by means of (a) a pie chart; (b) a bar chart.

5. The table shows the number of children per family from a sample of 100 families.

Number of children	0	1	2	3	4	5	6	7	8	9
Number of families	7	15	21	12	18	11	12	2	0	2

Illustrate this information on a suitable diagram.

6. The pictogram shows the number of ice creams a man sold at Easter

(a) If the man sold 1800 altogether, how many did he sell on each day?
(b) Illustrate this data on a pie chart.

7. The pie chart shows the number of trees cut down in a
large forest for each week in a period of 6 weeks. The total
number of trees cut down was 900.
   (a) How many trees were cut down in the first week?
   (b) If 200 trees were cut down in week 5, what is the
       sector angle for this week?
   (c) One sixth of all the trees were cut down in week 4,
       and one twelfth in week 2. What is the sector angle
       for each of these weeks.
   (d) Illustrate the data on a pictogram.

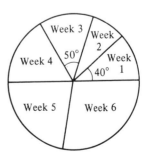

8. The council in Ross-on-Wye spent the money from rates as follows.
   30% on education,   10% on public transport,
   25% on public services,   5% on landscape improvement,
   15% on police,   5% on investments.
   10% on road maintenance

Display this data on a pie chart.

Straight line graphs are the result of plotting quantities which are directly proportional to one another.
Further information can be obtained from points which lie in between those already plotted.

Example 1

For his holiday in France, a boy produced the following table to help him change £'s into francs or francs into £'s.

Number of £'s	1	2	3	4	5	10
Number of francs	9	18	27	36	45	90

Using a scale of 1 cm to £1 on the horizontal axis and 1 cm to 10 francs on the vertical axis, draw the graph.
From your graph determine:
(a)   how many francs he would get for £7·50,
(b)   the value in £'s of 32 francs.

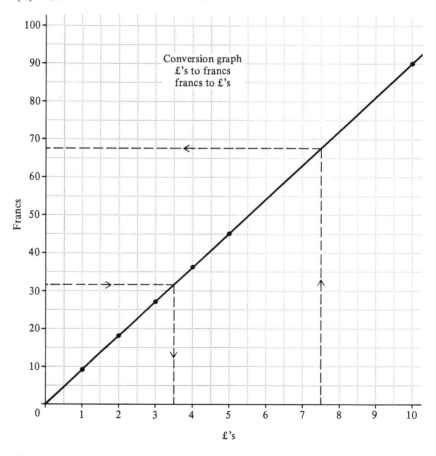

From the above graph:
(a)   he would get approximately 67 francs for £7·50,
(b)   32 francs are worth approximately £3·50.

Exercise 5.3a

1. The table shows the conversion of speed in metres per second to speed in kilometres per hour.

Speed m/s	0	20	40	60
Speed km/h	0	72	144	216

Draw a graph to show this information. Use a scale of 2 cm to 10 m/s on the horizontal axis and 1 cm to 10 km/h on the vertical axis.

(a) From your graph, convert these speeds to km/h:
   (i) 15 m/s; (ii) 35 m/s; (iii) 55 m/s.
(b) From your graph, convert these speeds to m/s:
   (i) 36 km/h; (ii) 90 km/h; (iii) 162 km/h.

2. The speed of a car at various times is shown in the table.

Time s	0	3	6	9	12
Speed km/h	0	27	54	81	108

Show this information on a graph using a scale of 1 cm to 1 second on the horizontal axis and a scale of 2 cm to 10 km/h on the vertical axis.

(a) Find the speed of the car after:
   (i) 5s; (ii) 8s; (iii) 11s.
(b) Find the time when the speed of the car is:
   (i) 18 km/h; (ii) 63 km/h; (iii) 90 km/h.

3. The current in amps in a wire for a given potential difference in volts is shown in the table.

p.d. V	0	3	6	9	12
Current A	0	1·2	2·4	3·6	4·8

Draw a graph of this information, using a scale of 1 cm to 1 V on the horizontal axis and 2 cm to 1 A on the vertical axis.

(a) Find the current for a p.d. of:
   (i) 1 V; (ii) 4 V; (iii) 7 V; (iv) 11 V.
(b) Find the p.d. for a current of:
   (i) 0·8 A; (ii) 2 A; (iii) 3·2 A; (iv) 4·2 A

4. The average train fare for a journey of a given distance is shown in the table.

Distance km	50	100	150
Fare £'s	2	4	6

Show this information on a graph using a scale of 1 cm to 10 km on the horizontal axis and 4 cm to £1 on the vertical axis.

(a) Find the expected fare for a journey of:
   (i) 40 km; (ii) 90 km; (iii) 120 km.
(b) Find the expected journey distance for a fare of:
   (i) £1·20; (ii) £2·80; (iii) £4·40.

5. The real depth of water in a beaker is compared with its apparent depth when viewed from above in the table below.

Real depth cm	3	6	9	12
Apparent depth cm	2·25	4·5	6·75	9

Draw a graph for this, showing the real depth to a scale of 1 cm to 1 cm on the horizontal axis and the apparent depth on the vertical axis to a scale of 2 cm to 1 cm.

(a) Find the apparent depth for a real depth of:
   (i) 1 cm; (ii) 8 cm; (iii) 10 cm.
(b) Find the real depth for an apparent depth of:
   (i) 1·5 cm; (ii) 3·75 cm; (iii) 8·25 cm.

6. A department store is selling everything at a discount dependent on the normal purchase price, as shown in the table.

Price £'s	100	400	600	1000
Discount £'s	8	32	48	80

Draw a graph of this information showing the normal purchase price on the horizontal axis to a scale of 2 cm to £100, and the discount on the vertical axis to a scale of 2 cm to £10.

(a) Find the discount offered on
   (i) a cooker costing £150,
   (ii) a colour television costing £350,
   (iii) a three-piece suite at £800.
(b) Find the normal purchase price of
   (i) a washing machine at a discount of £8,
   (ii) a set of dining chairs at a discount of £36,
   (iii) a bedroom suite at a discount of £60.

Example 2

The table shows the cost of buying gas for cooking and heating.

No. of therms used	5	8	10	12	16	18
Cost (£'s)	£1·75	£2·20	£2.50	£2·80	£3·40	£3·70

The cost consists of a fixed amount (the Standing Charge) plus an amount which is proportional to the number of therms used.
Draw a straight line graph to illustrate the above data. Use a scale of 1 cm to represent 2 therms on the horizontal axis and 1 cm to represent 50p on the vertical axis.

From your graph, estimate:
(a) the cost when 14 therms are used,
(b) the number of therms used when the cost is £3·55,
(c) the Standing Charge,
(d) the average cost per therm excluding the Standing Charge.

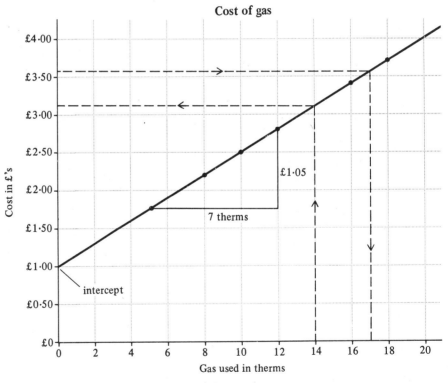

Cost of gas

(a) The cost of 14 therms is £3·10.
(b) The number of therms used is 17.
(c) The Standing Charge is obtained from the *intercept* (the point where the line cuts the vertical axis). This is the cost when no gas is used. The Standing Charge is £1.
(d) Without the Standing Charge, the cost per therm is the *gradient* of the line.

$$\text{Gradient} = \frac{2·80 - 1·75}{12 - 5} = \frac{1·05}{7} = 0·15$$ ∴ Cost per therm = 15p.

Exercise 5.3b

1. The temperature of the water in a hot water tank is recorded at 15 minute intervals after the heater is switched on.

Time min.	15	30	45	60	75	90
Temp. °C	20	30	40	50	60	70

Draw a graph of this information. Show the time from 0 min on the horizontal axis to a scale of 2 cm for 10 min. Show the temperature from 0°C on the vertical axis to a scale of 2 cm for 10°C.

(a) Find the temperature after 27 minutes.
(b) Find the time taken for the temperature to reach 64°C.
(c) Find the temperature when the heater was first switched on.
(d) Find the gradient of the line in °C per min.

2. A spring is stretched by hanging various weights from it, as recorded in the table below.

Weight g	125	250	375	500
Length of spring cm	40	50	60	70

Show this information on a graph using a scale of 4 cm to 100 g from 0 g to 500 g on the horizontal axis, and a scale of 2 cm to 10 cm for the length of spring from 0 cm to 70 cm on the vertical axis.

From your graph, find:
(a) the extended length of the spring when stretched by a 100 g weight.
(b) the weight that will stretch the spring to a length of 58 cm,
(c) the natural length of the spring (from the intercept),
(d) the gradient of the line in cm per g.

3. A gas in an enclosed vessel is heated, and the pressure and temperature are recorded.

Temp. °C	20	40	60	80	100
Pressure mm Hg	800	850	900	950	1000

Draw a graph showing the temperature from 0°C on the horizontal axis to a scale of 2 cm for 10°C and the pressure in millimetres of mercury from 600 mm on the vertical axis to a scale of 4 cm for 100 mm Hg.

From your graph, find:
(a) the pressure at 32°C,

(b) the temperature at 920 mm Hg,
(c) the pressure at 0°C,
(d) the increase in pressure for every °C increase in temperature.

4. The altitude (height above sea level) of a railway incline is given in the table.

Distance from start km	3	6	9	12	15
Altitude m	150	190	230	270	310

Draw a graph of this information. Show the altitude from sea level on the vertical axis to a scale of 5 cm for every 100 m. Show the distance from 0 km to a scale of 1 cm for every 1 km on the horizontal axis.

From your graph, find:
(a) the altitude 4·5 km from the start,
(b) the distance from the start at an altitude of 290 m.
(c) the altitude of the track at the start,
(d) the gradient of the incline.

5. The length of a steel rod is accurately measured at various temperatures. The results are shown in the table.

Temp. °C	Length mm
20	1002·2
40	1002·4
60	1002·6
80	1002·8
100	1003·0

Show this information on a graph. Use a scale from 0°C of 2 cm to 10°C on the horizontal axis, and a scale from 1000 mm of 5 cm to 1 mm on the vertical axis.

From your graph, find:

(a) the length of the rod at 70°C,
(b) the temperature when the length of the rod is 1002·3 mm,
(c) the length of the rod at 0°C,
(d) the increase in length of the rod for every °C rise in temperature.

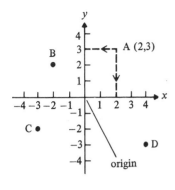

The coordinates of a point are written in the form (x, y).

Thus in the diagram, the point A has coordinates $(2, 3)$.
Its *horizontal* distance along the x-axis from the origin is 2 units.
Its *vertical* distance along the y-axis from the origin is 3 units.

Positive values of x are measured to the right of the origin and negative values to the left.

Positive values of y are measured upwards from the origin and negative values downwards.

Thus point B has coordinates $(-2, 2)$, C is $(-3, -2)$, and D is $(4, -3)$.
The *origin* is the point $(0, 0)$.

Exercise 5.4a

Write down the coordinates of the points in the diagram below.

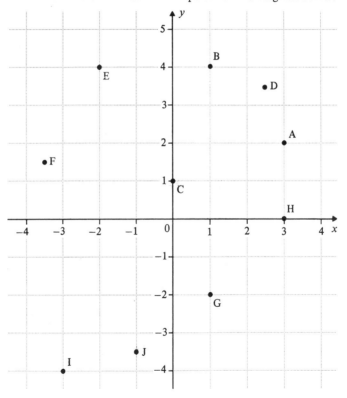

Exercise 5.4b

Plot the following points, using a scale of 2 cm to 1 unit on both axes.

1. $(2, 1)$ **2.** $(1, 3)$ **3.** $(0, 4)$ **4.** $(3, 2\frac{1}{2})$ **5.** $(-1, 3)$
6. $(-2, 0)$ **7.** $(2, -4)$ **8.** $(2\frac{1}{2}, -2)$ **9.** $(-3\frac{1}{2}, -2\frac{1}{2})$ **10.** $(0, -1\frac{1}{2})$

Example 1

Using a scale of 1 cm to 1 unit on both axes, plot the following sets of points and join together with straight lines. Give a name to your picture.

Set A

(2, 10), (3, 8), (3, 4), (1, 2), (3, 0),
(3, −8), (2, −8), (2, −7), (1, −7), (1, −8),
(2, −8), (−1, −8), (−1, −7), (−2, −8),
(−1, −8), (−3, −8), (−3, 0), (−1, 2),
(−3, 4), (−3, 8), (−2, 10), (−1, 8), (1, 8),
(2, 10).

Set B

(−5, 5), (−1, 4), (−5, 3), (−1, 4), (1, 4),
(5, 5), (1, 4), (5, 3).

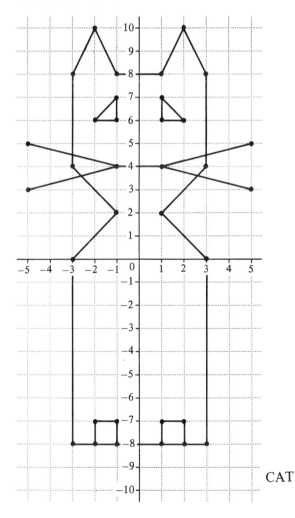

CAT

Set C

(−2, 6), (−1, 7), (−1, 6), (−2, 6).

Set D

(2, 6), (1, 6), (1, 7), (2, 6).

Exercise 5.4c

In all cases plot the given sets of points and join together with straight lines. Use a scale of 1 cm to 1 unit on both axes. Give a name to your picture

1. *Set A*

   (5, 8), (8, 5), (8, −5), (5, −8), (−5, −8), (−8, −5),
   (−8, 5), (−5, 8), (5, 8).

   Set B

   (9, 9), (9, −9), (−9, −9), (−9, 9), (9, 9).

   Set C

   (8, 8), (8, 7), (7, 8), (8, 8).

   Set D

   (8, −8), (8, −7), (7, −8), (8, −8).

   Set E

   (−8, −8), (−8, −7), (−7, −8), (−8, −8).

   Set F

   (−8, 8), (−8, 7), (−7, 8), (−8, 8).

   Set G

   (1, 7), (1, 1), (4, 1), (5, 0), (4, −1), (1, −1), (1, 1),
   (1, −1), (−1, −1), (−1, 1), (1, 1), (−1, 1), (−1, 7),
   (0, 8), (1, 7).

2. *Set A*

$(7, 9), (7, -9), (-2, -9), (-2, 9), (7, 9).$

Set B

$(6, 8), (6, 0), (3, 0), (3, 8), (6, 8).$

Set C

$(6, -3), (6, -8), (3, -8), (3, -3), (6, -3).$

Set D

$(2, -3), (2, -8), (-1, -8), (-1, -3), (2, -3).$

Set E

$(2, 8), (2, 0), (-1, 0), (-1, 8), (2, 8).$

Set F

$(0, -1), (0, -2), (-1, -2), (-1, -1), (0, -1).$

Set G

$(4, -1), (4, -2), (1, -2), (1, -1), (4, -1).$

Set H

$(5, -1), (6, -1), (5, -2).$

3. *Set A*

$(7, 8), (9, 6), (7, 4), (-7, 4), (-7, -6), (-6, -7),$
$(-6, -9), (-9, -9), (-9, -7), (-8, -6), (-8, 9),$
$(-7, 9), (-7, 8), (-7, 4), (-7, 8), (7, 8).$

Set B

$(-5, 7), (-5, 6), (-4, 6), (-4, 7), (-4, 5), (-4, 6),$
$(-5, 6), (-5, 5).$

Set C

$(-3, 7), (-3, 5), (-2, 5), (-2, 7).$

Set D

$(-1, 7), (-1, 5), (0, 5).$

Set E

$(1, 7), (1, 5), (2, 5).$

Set F

$(5, 7), (5, 5).$

Set G

$(6, 7), (7, 7), (7, 5), (6, 5), (6, 7).$

4. *Set A*

$(9, 9), (9, -3), (7, -3), (7, -8), (9, -8), (9, -9),$
$(4, -9), (4, -8), (6, -8), (6, -3), (7, -3),$
$(-7, -3), (-6, -3), (-6, -8), (-4, -8),$
$(-4, -9), (-9, -9), (-9, -8), (-7, -8),$
$(-7, -3), (-9, -3), (-9, 9), (9, 9).$

Set B

$(8, 8), (8, 7), (7, 7), (7, 8), (8, 8).$

Set C

$(8, 6), (8, 5), (7, 5), (7, 6), (8, 6).$

Set D

$(8, 4), (8, 3), (7, 3), (7, 4), (8, 4).$

Set E

$(8, 1), (8, 0), (7, 0), (7, 1), (8, 1).$

Set F

$(8, -1), (8, -2), (7, -2), (7, -1), (8, -1).$

Set G

$(4, 8), (6, 7), (6, -1), (4, -2), (-6, -2), (-8, -1),$
$(-8, 7), (-6, 8), (4, 8).$

5. *Set A*

$(6, 6), (9, 4), (9, 2), (8, 1), (6, 1), (5, 2), (5, 3),$
$(4, 4), (3, 4), (2, 3), (-2, 3), (-3, 4), (-4, 4),$
$(4, 4), (6, -2), (6, -5), (-6, -5), (-6, -2),$
$(-4, 4), (-5, 3), (-5, 2), (-9, 2), (-9, 4), (-6, 6),$
$(6, 6).$

Set B

$(1, 1), (1, -1), (-1, -1), (-1, 1), (1, 1).$

Set C

$(1, 2), (2, 1), (2, -1), (1, -2), (6, -2), (-6, -2),$
$(-1, -2), (-2, -1), (-2, 1), (-1, 2), (1, 2).$

6. *Set A*

$(7, 9), (8, 1), (6, 1), (9, 1), (9, -9), (8, -9),$
$(8, -6), (7, -5), (6, -6), (6, -9), (8, -9), (5, -9),$
$(5, -1), (5, -9), (-8, -9), (-8, -3), (5, -3),$
$(-8, -3), (-7, -1), (5, -1), (5, 1), (6, 1), (7, 9).$

Set B

$(7, 0), (8, -1), (7, -2), (6, -1), (7, 0).$

Set C

$(3, -4), (4, -5), (4, -8), (2, -8), (2, -5), (3, -4).$

Set D

$(0, -4), (1, -5), (1, -8), (-1, -8), (-1, -5),$
$(0, -4).$

Set E

$(-3, -4), (-2, -5), (-2, -8), (-4, -8), (-4, -5),$
$(-3, -4).$

Set F

$(-6, -4), (-5, -5), (-5, -8), (-7, -8), (-7, -5),$
$(-6, -4).$

7. *Set A*

$(4, 9), (6, 9), (7, 8), (8, 5), (6, 6), (5, 6), (4, 7),$
$(2, 6), (2, 4), (6, 0), (6, -3), (4, -5), (3, -5),$
$(3, -7), (4, -7), (4, -8), (5, -9), (4, -9), (3, -8),$
$(3, -9), (2, -9), (2, -8), (1, -9), (0, -9), (1, -8),$
$(1, -7), (2, -7), (2, -5), (0, -5), (-6, -1),$
$(-6, 0), (-7, 1), (-8, 3), (-1, 1), (4, 1), (1, 4),$
$(1, 6), (4, 9).$

Set B

$(6, 8), (6, 7), (5, 8), (6, 8).$

8. $(1, 8), (4, 8), (5, 9), (9, 9), (7, 7), (9, 5), (5, 5),$
$(4, 6), (1, 6), (1, 8), (1, 2), (6, 2), (6, 3), (9, 3),$
$(9, 2), (7, 2), (7, 1), (9, 1), (9, -2), (6, -2),$
$(6, -1), (8, -1), (8, 0), (6, 0), (6, 1), (6, 2), (6, 1),$
$(1, 1), (1, -9), (-1, -9), (-1, 1), (-5, 1), (-5, 2),$
$(-5, -2), (-6, -2), (-8, 1), (-8, -2), (-9, -2),$
$(-9, 3), (-8, 3), (-6, 0), (-6, 3), (-5, 3), (-5, 2),$
$(-1, 2), (-1, 6), (-1, 8), (-1, 6), (-6, 6), (-6, 5),$
$(-9, 7), (-6, 9), (-6, 8), (1, 8).$

9. $(3, 8), (5, 7), (5, 6), (4, 6), (3, 6), (3, 7), (-3, 8),$
$(-3, 7), (-4, 7), (-3, 7), (-3, 8), (3, 7), (3, 6),$
$(4, 6), (6, 1), (7, 0), (8, -5), (8, -7), (8, -5),$
$(9, -5), (9, -7), (8, -7), (7, -8), (6, -8), (6, -9),$
$(5, -9), (5, -8), (6, -8), (-1, -8), (-1, -9),$
$(-2, -9), (-2, -8), (-1, -8), (-5, -8), (-6, -6),$
$(-6, -5), (-5, 0), (-6, -5), (-7, -1), (-9, 1),$
$(-7, 1), (-5, 0), (-4, 1), (-3, 1), (-2, 2), (-1, 2),$
$(0, 3), (1, 3), (1, 2), (-1, 2), (4, 2), (5, 1), (6, 1),$
$(-3, 1), (-4, 3), (-4, 7), (-5, 7), (-5, 9), (-4, 9),$
$(3, 8).$

10. *Set A*

$(2, 9), (3, 7), (1, -1), (2, -5), (4, -5), (4, -9),$
$(1, -9), (1, -5), (0, -2), (-1, -5), (-1, -9),$
$(-4, -9), (-4, -5), (-2, -5), (-1, -1), (-3, 7),$
$(-2, 9), (0, 2), (2, 9).$

Set B

$(0, 1), (1, 0), (0, -1), (-1, 0), (0, 1).$

Set C

$(3, -6), (3, -8), (2, -8), (2, -6), (3, -6).$

Set D

$(-2, -6), (-2, -8), (-3, -8), (-3, -6), (-2, -6).$

Straight lines

All straight lines can be represented by an equation in the general form $y = mx + c$ where m and c are numbers.
To plot a straight line graph:

1. Put the equation in the above form, for example,
$y - 2x = 3$ is written $y = 2x + 3$,
$2y = 3x + 4$ is written $y = \frac{3}{2}x + 2$.
2. Construct a table of values using at least three suitable values of x to find corresponding values of y.
3. Plot the points.

Example 1

Plot the graph of $y = 2x + 3$ for values of x from -2 to $+2$.

(a) Find the intercept
(b) Find the gradient of this line.

Table of values

x	-2	-1	0	1	2
$y = 2x + 3$	-1	1	3	5	7

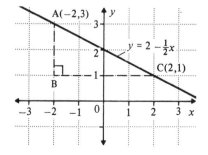

Note:
When $x = -2$, $y = (2 \times -2) + 3 = -1$
$x = -1$, $y = (2 \times -1) + 3 = 1$

(a) The intercept is where the line cuts the y-axis, i.e. $y = 3$
(b) The gradient is

$$\frac{\text{difference between two } y \text{ values}}{\text{difference between two corresponding } x \text{ values}}$$

i.e. gradient $= \dfrac{7 - 3}{2 - 0} = \dfrac{4}{2} = 2$.

Exercise 5.5a

Plot the graph of each equation, then find
(a) the intercept, (b) the gradient of the line.

1. $y = 3x + 5$ for $x = -2$ to $x = +3$
2. $y = 4x + 1$ for $x = -1$ to $x = +3$
3. $y = 2x + 4$ for $x = -3$ to $x = +4$
4. $y = x + 3$　for $x = -5$ to $x = +7$
5. $y = \frac{3}{2}x + 2$ for $x = -4$ to $x = +4$
6. $y = 2x - 7$ for $x = -1$ to $x = +7$
7. $y = 4x - 9$ for $x = -1$ to $x = +4$
8. $y = x - 6$　for $x = -1$ to $x = +10$
9. $y = \frac{5}{2}x - 2$ for $x = -4$ to $x = +4$
10. $y = \frac{1}{2}x - 3$ for $x = -4$ to $x = +12$

Example 2

Plot the graph of $y = 2 - \frac{1}{2}x$ for $x = -2$ to $x = +2$. Then find (a) the intercept, (b) the gradient of this line.

Table of values

x	-2	-1	0	1	2
$y = 2 - \frac{1}{2}x$	3	$2\frac{1}{2}$	2	$1\frac{1}{2}$	1

Note : When $x = -2$,
$y = 2 - (\frac{1}{2} \times -2) = 2 - -1 = 3$
When $x = -1$,
$y = 2 - (\frac{1}{2} \times -1) = 2 - -\frac{1}{2} = 2\frac{1}{2}$

(a) the intercept is where the line cuts the y-axis, i.e. $y = 2$.

(b) The gradient is

$$\frac{\text{difference between two } y \text{ values}}{\text{difference between two corresponding } x \text{ values}}$$

i.e. gradient $= \dfrac{3-1}{-2-2} = \dfrac{2}{-4} = -\dfrac{1}{2}$

If the equation is in the form $y = mx + c$, we can see from examples 1 and 2 that
   m is the value of the gradient
and c is the value of the intercept.

Exercise 5.5b

Plot the graph of each equation, then find
(a) the intercept, (b) the gradient of the line.

1. $y = 5 - 2x$ for $x = -2$ to $x = +6$
2. $y = 7 - 4x$ for $x = -2$ to $x = +3$
3. $y = 3 - 5x$ for $x = -2$ to $x = +2$
4. $y = 8 - 3x$ for $x = -2$ to $x = +4$
5. $y = 6 - x$ for $x = -2$ to $x = +8$
6. $y = -2 - x$ for $x = -4$ to $x = +6$

7. $y = 4 - \frac{3}{2}x$ for $x = -4$ to $x = +8$

8. $y = 2 - \frac{1}{2}x$ for $x = -5$ to $x = +10$

9. $y = 3 - \frac{5}{2}x$ for $x = -4$ to $x = +4$

10. $y = -1 - \frac{1}{2}x$ for $x = -6$ to $x = +4$

The parabola

If an equation contains only y, x, and x^2 as unknowns, then the graph takes the form of a parabola.

Example 3

Find the missing values in the table and then draw the graph of $y = x^2 - 4$.

x	−3	−2	−1	0	1	2	3
x^2	9	4		0		4	
−4	−4	−4	−4	−4	−4	−4	−4
y	5	0	−3	−4	−3	0	5

From your graph estimate

(a) the value of y when $x = +\frac{1}{2}$
(b) the values of x when $y = 0$

The missing values are found thus:

When $x = -1, x^2 = -1 \times -1 = 1$
$x = 1, x^2 = 1 \times 1 = 1$
$x = 3, x^2 = 3 \times 3 = 9$

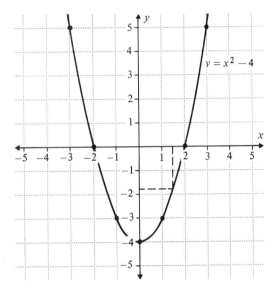

$y = x^2 - 4$

(a) When $x = 1\frac{1}{2}$, we find from the graph that $y = -1.7$
(b) When $y = 0$ (where the curve cuts the x-axis), $x = -2$ or $x = +2$.

Exercise 5.5c

Find the missing values in each table and then draw the graph of the equation to answer the questions.

1. $y = x^2 - 1$

x	−3	−2	−1	0	+1	+2	+3
x^2	+9			0			
−1	−1	−1	−1	−1	−1	−1	−1
y		+3	0	−1			+8

(a) What is the value of y when $x = +2\frac{1}{2}$?
(b) What are the values of x when $y = 0$?

2. $y = x^2 - 9$

x	−4	−3	−2	−1	0	+1	+2	+3	+4
x^2	+16			+1		+1		+9	
−9	−9	−9	−9	−9	−9	−9	−9	−9	−9
y	−9		0	−5	−8	−9	−8	−5	

(a) What is the value of y when $x = +1\frac{1}{2}$?
(b) What are the values of x when $y = 0$?

3. $y = x^2 - 16$

x	-5	-4	-3	-2	-1	0	+1	+2	+3	+4	+5
x^2	+25										
-16	-16	-16	-16	-16	-16	-16	-16	-16	-16	-16	-16
y		0		-12		-16			-7		+9

(a) What is the value of y when $x = -2\frac{1}{2}$?
(b) What are the values of x when $y = 0$?

4. $y = x^2 - \frac{1}{4}$

x	-2	$-1\frac{1}{2}$	-1	$-\frac{1}{2}$	0	$+\frac{1}{2}$	+1	$+1\frac{1}{2}$	+2
x^2	+4	$+2\frac{1}{4}$				$+\frac{1}{4}$	+1		
$-\frac{1}{4}$	$-\frac{1}{4}$	$-\frac{1}{4}$	$-\frac{1}{4}$	$-\frac{1}{4}$	$-\frac{1}{4}$	$-\frac{1}{4}$	$-\frac{1}{4}$	$-\frac{1}{4}$	$-\frac{1}{4}$
y	$+3\frac{3}{4}$					$-\frac{1}{4}$	0	$+\frac{3}{4}$	+2

(a) What is the value of y when $x = +1\frac{3}{4}$?
(b) What are the values of x when $y = 0$?

5. $y = x^2 - 5$

x	-4	-3	-2	-1	0	+1	+2	+3	+4
x^2			+4		0			+9	
-5	-5	-5	-5	-5	-5	-5	-5	-5	-5
y	+11			-4			-1		

(a) What is the value of y when $x = +\frac{1}{2}$?
(b) What are the values of x when $y = 0$?

6. $y = 4x^2 - 9$

x	-2	-1	0	+1	+2
$4x^2$	+16		0	+4	
-9	-9	-9	-9	-9	-9
y	+7		-9	-5	

(a) What is the value of y when $x = 1\frac{1}{4}$?
(b) What are the values of x when $y = 0$?

7. $y = 3x^2 - 2$

x	-2	-1	0	+1	+2
$3x^2$		+3		+12	
-2	-2	-2	-2	-2	-2
y			-2	+10	

(a) What is the value of y when $x = -\frac{1}{2}$?
(b) What are the values of x when $y = 0$?

8. $y = \frac{1}{2}x^2 - 4$

x	-6	-4	-2	0	$+2$	$+4$	$+6$
$\frac{1}{2}x^2$	$+18$		$+2$			$+8$	
-4	-4	-4	-4	-4	-4	-4	-4
y			$+4$	-2	-4		$+14$

(a) What is the value of y when $x = 5$?
(b) What are the values of x when $y = 0$?

9. $y = \frac{1}{2}x^2 - 8$

x	-6	-4	-2	0	$+2$	$+4$	$+6$
$\frac{1}{2}x^2$		$+8$			$+2$		$+18$
-8	-8	-8	-8	-8	-8	-8	-8
y	$+10$		-6				

(a) What is the value of y when $x = 3$?
(b) What are the values of x when $y = 0$?

10. $y = \frac{1}{4}x^2 - 1$

x	-8	-4	0	$+4$	$+8$
$\frac{1}{4}x^2$	$+16$		0		
-1	-1	-1	-1	-1	-1
y		$+3$			$+15$

(a) What is the value of y when $x = 6$?
(b) What are the values of x when $y = 0$?

11. $y = \frac{1}{4}x^2 - 4$

x	-8	-4	0	$+4$	$+8$
$\frac{1}{4}x^2$		$+4$		$+4$	
-4	-4	-4	-4	-4	-4
y			-4		$+12$

(a) What is the value of y when $x = 2$?
(b) What are the values of x when $y = 0$?

12. $y = \frac{1}{3}x^2 - 3$

x	-6	-3	0	$+3$	$+6$
$\frac{1}{3}x^2$	$+12$			$+3$	
-3	-3	-3	-3	-3	-3
y		0			$+9$

(a) What is the value of y when $x = 1\frac{1}{2}$?
(b) What are the values of x when $y = 0$?

Example 4

x	-3	-2	-1	0	1	2	3	4	5
x^2	9	4		0	1	4		16	25
$-2x$	6		2	0	-2	-4	-6	-10	
-8	-8	-8	-8	-8	-8	-8	-8	-8	-8
y			0	-5	-8		-8	-5	0	7

(a) Find the missing values in the above table of the graph $y = x^2 - 2x - 8$.

(b) Using the scale of 2 cm to 2 units on both axes, plot the graph.

(c) From your graph, estimate

    (i) the value of y when $x = -\frac{1}{2}$

    (ii) the values of x when $y = 0$

    (iii) the values of x when $y = 3$

The solution is shown on the opposite page.

(a) The missing values are as follows.

$$x = -1, \quad x^2 = -1 \times -1 = 1;$$
$$x = -2, -2x = -2 \times -2 = 4;$$
$$x = -3, \quad y = 9 + 6 - 8 = 7;$$
$$x = 3, \quad x^2 = 3 \times 3 = 9$$
$$x = 4, \quad -2x = -2 \times 4 = -8$$
$$x = 1, \quad y = 1 - 2 - 8 = -9$$

The completed table is shown below.

x	-3	-2	-1	0	1	2	3	4	5
x^2	9	4	1	0	1	4	9	16	25
$-2x$	6	4	2	0	-2	-4	-6	-8	-10
-8	-8	-8	-8	-8	-8	-8	-8	-8	-8
y	7	0	-5	-8	-9	-8	-5	0	7

(c)   (i) When $x = -\frac{1}{2}$, $y = -6 \cdot 8$

    (ii) When $y = 0$, $x = -2$ or $+4$

    (iii) When $y = 3$, $x = -2 \cdot 5$ or $+4 \cdot 5$

(b) Graph of $y = x^2 - 2x - 8$

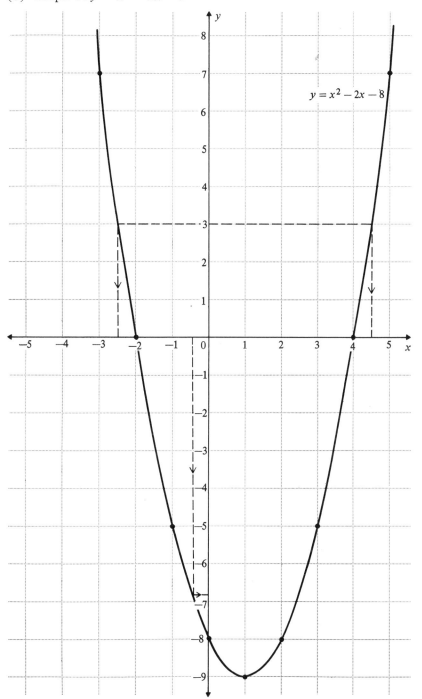

Exercise 5.5d

Find the missing values in each table and then draw the graph of the equation to answer the questions. Use a scale of 2 cm to 1 unit for the x-axis and 1 cm to 1 unit for the y-axis.

1. $y = x^2 - 4x + 3$

x	-2	-1	0	$+1$	$+2$	$+3$	$+4$	$+5$	$+6$
x^2	$+4$		0	$+1$		$+9$		$+25$	
$-4x$		$+4$			-8		-16		-24
$+3$	$+3$	$+3$	$+3$	$+3$	$+3$	$+3$	$+3$	$+3$	$+3$
y	$+15$			$+3$	0	-1		$+8$	

(a) Find the value of y when $x = +3\frac{1}{2}$.
(b) Find the values of x when $y = +1$.
(c) Find the values of x when $y = 0$.

2. $y = x^2 - 5x + 4$

x	-1	0	$+1$	$+2$	$+3$	$+4$	$+5$	$+6$
x^2	$+1$			$+4$		$+16$	$+25$	
$-5x$		0	-5		-15			-30
$+4$	$+4$	$+4$	$+4$	$+4$	$+4$	$+4$	$+4$	$+4$
y		$+4$	0		-2			$+10$

(a) Find the value of y when $x = +5\frac{1}{2}$.
(b) Find the values of x when $y = +3$.
(c) Find the values of x when $y = 0$.

3. $y = x^2 - 6x + 8$

x	-1	0	$+1$	$+2$	$+3$	$+4$	$+5$	$+6$	$+7$
x^2			$+1$		$+9$			$+36$	$+49$
$-6x$	$+6$	0		-12		-24	-30		
$+8$	$+8$	$+8$	$+8$	$+8$	$+8$	$+8$	$+8$	$+8$	$+8$
y			$+8$	$+3$			0	$+3$	$+15$

(a) Find the value of y when $x = +1\frac{1}{2}$.
(b) Find the values of x when $y = +5$.
(c) Find the values of x when $y = 0$.

4. $y = x^2 - 4x$

x	-2	-1	0	$+1$	$+2$	$+3$	$+4$	$+5$	$+6$
x^2		$+1$			$+4$			$+25$	
$-4x$	$+8$		0	-4		-12	-16		-24
y	$+12$		0		-4	-3		$+5$	

(a) Find the value of y when $x = +2\frac{1}{2}$.
(b) Find the values of x when $y = -1$.
(c) Find the values of x when $y = 0$.

5. $y = x^2 - 3x - 4$

x	-3	-2	-1	0	$+1$	$+2$	$+3$	$+4$	$+5$	$+6$
x^2	$+9$		$+1$			$+4$				$+36$
$-3x$		$+6$		0	-3		-9	-12	-15	
-4	-4	-4	-4	-4	-4	-4	-4	-4	-4	-4
y	$+14$		0	-4		-6			$+6$	

(a) Find the value of y when $x = +\frac{1}{2}$.
(b) Find the values of x when $y = +4$.
(c) Find the values of x when $y = 0$.

6. $y = x^2 - x - 6$

x	-4	-3	-2	-1	0	$+1$	$+2$	$+3$	$+4$	$+5$
x^2		$+9$			0				$+16$	$+25$
$-x$	$+4$		$+2$	$+1$		-1	-2	-3		
-6	-6	-6	-6	-6	-6	-6	-6	-6	-6	-6
y	$+14$		0		-6		-4		$+6$	

(a) Find the value of y when $x = -1\frac{1}{2}$.
(b) Find the values of x when $y = +9$.
(c) Find the values of x when $y = 0$.

7. $y = x^2 + 2x - 3$

x	-5	-4	-3	-2	-1	0	$+1$	$+2$	$+3$
x^2				$+4$			$+1$		$+9$
$+2x$	-10	-8	-6		-2	0		$+4$	
-3	-3	-3	-3	-3	-3	-3	-3	-3	-3
y	$+12$			-3			0	$+5$	

(a) Find the value of y when $x = -2\frac{1}{2}$.
(b) Find the values of x when $y = -2$.
(c) Find the values of x when $y = 0$.

8. $y = x^2 + x - 12$

x	-5	-4	-3	-2	-1	0	$+1$	$+2$	$+3$	$+4$
x^2	$+25$		$+9$		$+1$	0		$+4$		$+16$
$+x$		-4		-2			$+1$		$+3$	
-12	-12	-12	-12	-12	-12	-12	-12	-12	-12	-12
y	$+8$	0			-12			-6		

(a) Find the value of y when $x = -4\frac{1}{2}$.
(b) Find the values of x when $y = -5$.
(c) Find the values of x when $y = 0$.

Simultaneous equations

1. Put each equation in the form
   $y = mx + c$.
2. Plot the graphs of the two straight lines.
3. The point of intersection of the straight lines is the point at which x and y satisfy both equations.

Example 1

Solve graphically the simultaneous equations $x + y = 3$
$$3x + 2y = 7$$

1. $x + y = 3$ becomes $y = 3 - x$
   $3x + 2y = 7$ becomes $y = 3\frac{1}{2} - 1\frac{1}{2}x$.
2. Draw up tables of values, using suitable values of x. Remember, only three points are found when plotting straight lines.

x	0	1	2
$y = 3 - x$	3	2	1

x	0	1	2
$y = 3\frac{1}{2} - 1\frac{1}{2}x$	$3\frac{1}{2}$	2	$\frac{1}{2}$

3. Plot each graph using the same axes.
4. The point of intersection of the straight lines gives the solution.
   i.e. $x = 1, y = 2$

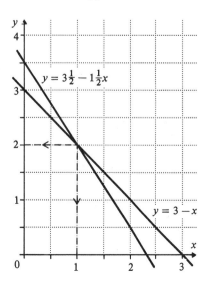

Exercise 5.6a

Solve graphically the following pairs of simultaneous equations. For each straight line plot values of y for values of x from 0 to 6.

1. $x + y = 5$
   $3x + y = 9$
2. $x + y = 6$
   $2x + y = 10$
3. $3x + y = 10$
   $4x + y = 13$
4. $2x + y = 9$
   $3x + 2y = 14$
5. $3x + y = 11$
   $x + 2y = 7$
6. $y - x = 7$
   $y - 4x = 1$
7. $y - 2x = 5$
   $y - 3x = 2$
8. $y - 3x = 1$
   $2y - x = 12$
9. $x - y = 2$
   $2x - y = 6$
10. $3x - y = 8$
    $x - 2y = 1$

Quadratic equations

This type of equation is usually solved graphically by drawing a parabola and finding the values of x at the points where the curve is cut by a straight line.

Example 2

Draw the graph of $y = x^2 - 3x$ from the following table of values, using a scale of 1 cm to 1 unit on both axes.

x	-2	-1	0	$+1$	$+2$	$+3$	$+4$	$+5$
x^2	$+4$	$+1$	0	$+1$	$+4$	$+9$	$+16$	$+25$
$-3x$	$+6$	$+3$	0	-3	-6	-9	-12	-15
y	$+10$	$+4$	0	-2	-2	0	$+4$	$+10$

Use your graph to find:

(a) the solution of the equation
$$x^2 - 3x = 0$$
(b) the solution of the equation
$$x^2 - 3x = 4$$

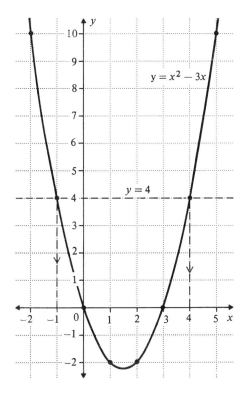

The graph shows $y = x^2 - 3x$ and the line $y = 4$.

1. Draw the graph.

2. (a) The solution of the equation
  $x^2 - 3x = 0$ is solved at the points
  where the line $y = 0$ cuts the
  parabola.
  Note: The line $y = 0$ is the x-axis.
  i.e. when $x = 0$ or $x = 3$.

  (b) The solution of the equation
  $x^2 - 3x = 4$ is solved at the points
  where the line $y = 4$ cuts the
  parabola.
  i.e. when $x = 4$ or $x = -1$.

Exercise 5.6b

Find the missing values in each table and then
draw the graph of the equation to answer the
questions. Use a scale of 2 cm to 1 unit for
the x-axis and 1 cm to 1 unit for the y-axis.

1. $y = x^2 - 2x$

x	-3	-2	-1	0	$+1$	$+2$	$+3$	$+4$	$+5$
x^2	$+9$		$+1$	0		$+4$		$+16$	$+25$
$-2x$	$+6$	$+4$	$+2$	0	-2		-6	-8	-10
y		$+8$			-1	0	$+3$		

Find the solution of:
(a) the equation $x^2 - 2x = 0$.
(b) the equation $x^2 - 2x = 8$.

2. $y = x^2 - 5x$

x	-2	-1	0	$+1$	$+2$	$+3$	$+4$	$+5$	$+6$	$+7$
x^2		$+1$		$+1$	$+4$			$+25$	$+36$	$+49$
$-5x$	$+10$	$+5$	0			-15	-20	-25		-35
y	$+14$		0	-4	-6	-6	-4		$+6$	

Find the solution of:
(a) the equation $x^2 - 5x = 0$
(b) the equation $x^2 - 5x = 6$
(c) the equation $x^2 - 5x = -4$
(d) the equation $x^2 - 5x = -6$

3. $y = x^2 - x$

x	-3	-2	-1	0	$+1$	$+2$	$+3$	$+4$
x^2	$+9$	$+4$			$+1$			$+16$
$-x$	$+3$		$+1$	0	-1	-2	-3	
y			$+2$	0		$+2$	$+6$	$+12$

Find the solution of:
(a) the equation $x^2 - x = 0$
(b) the equation $x^2 - x = 2$

4. $y = x^2 - 6x$

x	-1	0	$+1$	$+2$	$+3$	$+4$	$+5$	$+6$	$+7$
x^2	$+1$			$+4$		$+16$	$+25$	$+36$	
$-6x$			-6	-12	-18		-30		-42
y	$+7$		-5		-9	-8		0	$+7$

Find the solution of:
(a) the equation $x^2 - 6x = 0$
(b) the equation $x^2 - 6x = -5$

5. $y = x^2$

x	-4	-3	-2	-1	0	$+1$	$+2$	$+3$	$+4$
y	$+16$			$+1$			$+4$		

Find the solution of:
(a) the equation $x^2 = 9$
(b) the equation $x^2 = 4$
(c) the equation $x^2 = 2$
(d) the equation $x^2 - 3 = 0$

Draw up tables as above, then solve graphically
the equations for each question below.

6. (a) $x^2 + 3x = 0$
   (b) $x^2 + 3x = 4$

7. (a) $x^2 + 2x = 0$
   (b) $x^2 + 2x = 3$

8. (a) $x^2 + 4x = 0$
   (b) $x^2 + 4x = -3$

9. (a) $x^2 + x = 0$
   (b) $x^2 + x = 6$

Travel graphs are usually straight lines. The horizontal axis is always used for the time taken and the vertical axis for the distance travelled.

Note:

(a) Units must correspond.
   If the speed is given in km/h, then the distance is measured in km and the time in h.
(b) distance travelled = speed × time.

Example 1

At 13.00 h, a boy starts to cycle to visit a friend who lives 25 km away. If he can cycle at an average speed of 10 km/h, at what time will he arrive?

1. Draw suitable axes, with the time on the horizontal axis scaled from 13.00 h and the distance on the vertical axis scaled to 25 km.
2. The cyclist travels at 10 km/h. So after 1 hour (i.e. at 14.00 h), he will be 10 km away from his starting point. After 2 hours (i.e. at 15.00 h), he will be 20 km away from his starting point.
   Using the information, plot the straight line to show his journey.
3. Read off his time of arrival from the graph, i.e. 15.30 h.

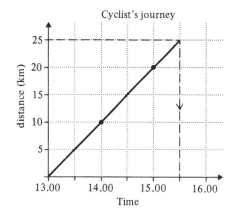

Exercise 5.7a

1. A man travels from Brighton to Portsmouth on his motor cycle at an average speed of 40 km/h. If he starts at 5.00 p.m., find from a travel graph the time when he reaches:

   (a) Worthing, 20 km from Brighton
   (b) Bognor Regis, 50 km from Brighton
   (c) Chichester, 60 km from Brighton
   (d) Havant, 80 km from Brighton
   (e) Portsmouth, 90 km from Brighton.

2. A train to Leeds leaves London at 8.00 p.m. Its average speed is 100 km/h. Draw a travel graph to show this journey and find when the train reaches:

   (a) Hatfield, 25 km from London
   (b) Hitchin, 50 km from London
   (c) Peterborough, 125 km from London
   (d) Retford, 225 km from London
   (e) Doncaster, 250 km from London
   (f) Leeds, 300 km from London

3. A train to Plymouth leaves London at 4.0 p.m. Its average speed is 120 km/h. Draw a travel graph to show this journey and find when the train reaches:

   (a) Reading, 60 km from London
   (b) Newbury, 90 km from London
   (c) Westbury, 150 km from London
   (d) Castle Cary, 180 km from London
   (e) Exeter, 270 km from London
   (f) Plymouth, 360 km from London

4. A bus from Darlington to Hawes leaves at 2.00 p.m. and travels at an average speed of 28 km/h. Draw a suitable graph to find when it reaches:

   (a) Scotch Corner, 14 km from Darlington
   (b) Richmond, 21 km from Darlington
   (c) Leyburn, 42 km from Darlington
   (d) Aysgarth, 56 km from Darlington
   (e) Bainbridge, 63 km from Darlington
   (f) Hawes, 70 km from Darlington

5. A man leaves Maidstone at 6.00 a.m. and cycles to Guildford at an average speed of 16 km/h. Draw a suitable graph to find when he reaches:

   (a) Westerham, 32 km from Maidstone
   (b) Oxted, 40 km from Maidstone
   (c) Godstone, 44 km from Maidstone
   (d) Redhill, 52 km from Maidstone
   (e) Dorking, 64 km from Maidstone
   (f) Albury, 76 km from Maidstone
   (g) Guildford, 84 km from Maidstone

6. An athlete runs at an average speed of 12 km/h from Kidderminster to Birmingham. If he leaves at 3.00 p.m., find from a suitable graph the time he arrives at:

   (a) Blakedown, 6 km from Kidderminster
   (b) Halesowen, 15 km from Kidderminster
   (c) Quinton, 18 km from Kidderminster
   (d) Bearwood, 21 km from Kidderminster
   (e) Birmingham, 27 km from Kidderminster.

7. A party of walkers leave Buxton at 11.00 a.m. If their average speed is 4 km/h, find from a suitable graph their time of arrival at:

   (a) Blackwell, 7 km from Buxton
   (b) Miller's Dale, 9 km from Buxton
   (c) Monsal Dale, 16 km from Buxton
   (d) Ashford-in-the-Water, 18 km from Buxton
   (e) Bakewell, 20 km from Buxton

8. A man leaves Birmingham by car at 2.00 p.m. and travels along the M6 motorway towards Preston at an average speed of 96 km/h. Find from a travel graph the time at which he passes the service areas:

   (a) 24 km from Birmingham
   (b) 72 km from Birmingham
   (c) 144 km from Birmingham
   (d) 168 km from Birmingham

Example 2

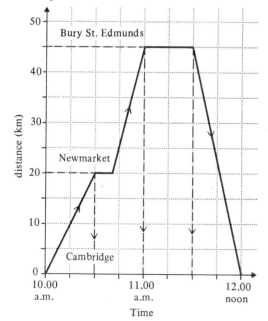

The graph shows a motorist's journey from Cambridge to Bury St. Edmunds and back.

From this graph, find:

(a) the time at which the motorist reached Newmarket,
(b) the distance from Newmarket to Bury St. Edmunds,
(c) how long the motorist stayed in Bury St. Edmunds,
(d) his average speed for the journey from Bury St. Edmunds back to Cambridge.

From the graph:

(a) he reaches Newmarket at 10.30 h

(b) distance from Cambridge to Newmarket = 20 km
distance from Cambridge to Bury St. Edmunds = 45 km.
so distance from Newmarket to Bury St. Edmunds = 45 − 20 = 25 km

(c) He reached Bury St. Edmunds at 11.00h
He left Bury St. Edmunds at 11.30 h
So he stayed 30 minutes.

(d) Time taken from Bury St. Edmunds to Cambridge = 30 minutes
Distance from Bury St. Edmunds to Cambridge = 45 km

So speed $= \dfrac{\text{distance in km}}{\text{time taken in hours}}$

$= \dfrac{45}{30/60} = \dfrac{45}{\frac{1}{2}}$

$= 90 \text{ km/h}$

Exercise 5.7b

1. The graph opposite shows a return journey from Hull to Beverley that a man made by bus. He stopped in Dunswell on the way and caught the next bus.

   (a) At what time did he reach Beverley?
   (b) Find the distance from Dunswell to Beverley?
   (c) How long did he stop in Dunswell?
   (d) How long did he stop in Beverley?
   (e) Find the average speed of his journey
       (i) from Hull to Dunswell,
       (ii) from Dunswell to Beverley,
       (iii) from Beverley back to Hull.

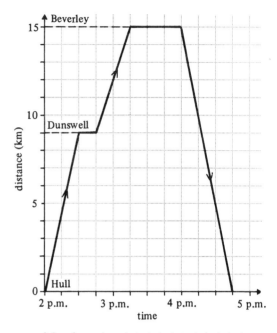

2. The graph opposite shows a return journey from Sevenoaks to London that a man made by train. He changed trains at Orpington on the way in order to meet a friend. He returned from London to Sevenoaks direct.

   (a) At what time did he reach Orpington?
   (b) At what time did he reach London?
   (c) How far is it from Sevenoaks to Orpington?
   (d) Find the distance from Orpington to London.
   (e) Find how long he waited at Orpington.
   (f) How long did he stay in London?
   (g) Find the average speed of his train
       (i) from Sevenoaks to Orpington,
       (ii) from Orpington to London,
       (iii) from London back to Sevenoaks.

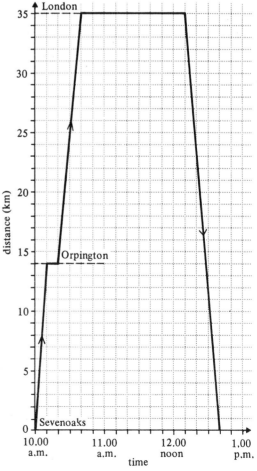

3. The graph opposite shows the return journey of
a delivery van from Birmingham to Wolverhampton.
The journey to Wolverhampton is direct, but the
van stops in West Bromwich on the way back.

   (a) When did the van reach Wolverhampton?
   (b) When did the van leave West Bromwich?
   (c) Find the distance from Wolverhampton to
   West Bromwich.
   (d) How long did the van stop in West Bromwich?
   (e) Find the average speed of the van:
       (i) from Birmingham to Wolverhampton,
       (ii) from Wolverhampton to West Bromwich,
       (iii) from West Bromwich to Birmingham.

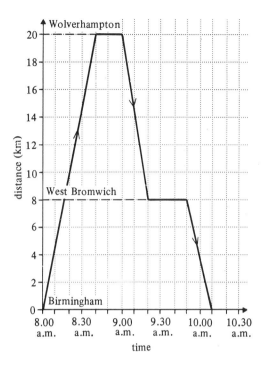

4. The graph opposite shows the return journey a
man made by car from Newcastle to Carlisle. On
the way out, he stopped in Hexham, but returned
from Carlisle direct.

   (a) When did he reach Carlisle?
   (b) How far is Carlisle from Hexham?
   (c) How long did he stay in Carlisle?
   (d) Find the average speed:
       (i) from Carlisle back to Newcastle,
       (ii) from Newcastle to Hexham,
       (iii) from Hexham to Carlisle.

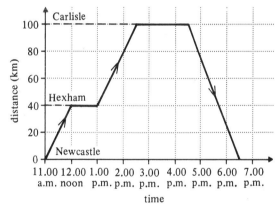

5. The graph opposite shows a return train journey
a man made from Manchester to Carlisle. He
caught a through train on the outward journey,
but on the way back he had to change at Preston.

   (a) Estimate the time he left Preston.
   (b) How long did he stay in Carlisle?
   (c) Estimate the distance from Carlisle to
   Preston.
   (d) How long do you think he waited in Preston?
   (e) Estimate the average speed of his train:
       (i) from Manchester to Carlisle,
       (ii) from Carlisle to Preston,
       (iii) from Preston back to Manchester.

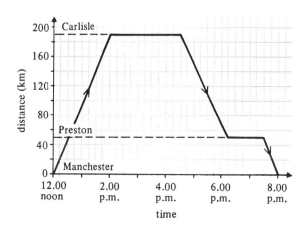

Example 3

At noon a man starts cycling from Northampton to Rugby 30 km away at a steady speed of 10 km/h. He stops for lunch at 1.00 p.m. for 30 minutes, and then resumes his journey at the same speed.
A car leaves Rugby at 12.30 p.m. and drives to Northampton at a steady speed of 50 km/h.
When does the cyclist meet the car and how far from Northampton?

1. Draw suitable axes with the time on the horizontal axis scaled from 12 noon and the distance on the vertical axis scaled to 30 km.
2. The cyclist travels at 10 km/h. After 1 hour (i.e. at 1 p.m.) he will be 10 km from Northampton. He then stops for lunch. From this information, plot the two straight lines to show his journey.
3. The car leaves Rugby at 12.30 p.m. and travels 25 km in $\frac{1}{2}$ hour, i.e. it is 5 km from Northampton at 1.00 p.m. Plot the straight line for the car from this information.
4. The point of intersection of the two lines gives the required information.
   They meet at 12.55 p.m. 9 miles from Northampton.

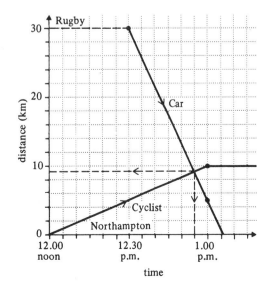

Exercise 5.7c

1. At 2.00 p.m. Jane leaves her house and walks along a main road at 6 km/h. At 2.30 p.m. her sister Anne leaves the house and cycles along the same main road at 15 km/h. Find from a graph:
   (a) the time at which they meet,
   (b) the distance that they both are from their house at that time.

2. At 10.00 a.m. Peter leaves his house and cycles along a main road at 20 km/h. At 10.15 a.m. his father leaves the house and drives his car along the same main road at 45 km/h. Find from a graph:
   (a) the time at which they meet and,
   (b) the distance that they both are from their house at that time.

3. At 9.00 a.m. a lorry leaves a warehouse in Manchester and sets off for Liverpool Docks at an average speed of 50 km/h. At 9.10 a.m. the haulage firm's manager starts the same journey in his car and his average speed is 75 km/h. Find from a graph:
   (a) the time at which the manager catches up with the lorry driver,
   (b) the distance from Manchester that they have both covered.

4. At 8.00 a.m. a local train leaves London (St. Pancras) towards Kettering at an average speed of 75 km/h. At 8.10 a.m. an express departs from St. Pancras and travels at an average speed of 100 km/h on a parallel track to the one used by the local train. Find from a graph:
   (a) the time at which the express overtakes the local train,
   (b) the distance from London that both trains have then reached.

5. At 9.00 a.m. a boy starts to cycle from Worcester to Malvern at an average speed of 18 km/h. At exactly the same time another boy starts to walk from Malvern to Worcester at an average speed of 6 km/h. If the distance between the two places is 12 km, find from a graph
   (a) the time when they meet,
   (b) how far they both are from Worcester at that time.

6. At 3.00 p.m. a man begins to drive his car from Oxford to Banbury at an average speed of 50 km/h. At exactly the same time another man begins to cycle from Banbury to Oxford at an average speed of 10 km/h. If the distance between the two places is 36 km, find from a graph:
   (a) the time when they meet
   (b) how far they both are from Oxford at that time.

7. At 7.00 p.m. a man begins to drive his car from Birmingham to Luton. He travels via the motorway at an average speed of 100 km/h. At exactly the same time another man begins to drive a lorry from Luton to Birmingham by the same route at an average speed of 60 km/h. If the distance between the two places is 144 km, find from a graph:
   (a) the time when they met,
   (b) how far they both are from Birmingham at that time.

8. At 4.00 p.m. John begins to drive his moped from Evesham to Alcester at an average speed of 24 km/h. At 4.30 p.m. his younger brother David starts to walk from Alcester to Evesham at an average speed of 6 km/h. If the distance between the two places is 17 km, find from a graph:
   (a) the time when they meet,
   (b) how far they both are from Evesham at that time.

9. At 5.00 p.m. a bus leaves Norwich and heads for Thetford travelling at an average speed of 30 km/h. At 5.30 p.m. a man starts to drive his car from Thetford to Norwich at an average speed of 45 km/h. If the distance between the two places is 50 km, find from a graph:
   (a) the time when they meet,
   (b) how far they both are from Norwich at that time.

10. At 6.00 p.m. a train leaves Liverpool for Leeds and travels at an average speed of 84 km/h. At 6.45 p.m. a train leaves Leeds and heads the opposite way to Liverpool at an average speed of 87 km/h. If the distance between the two cities is 120 km find from a graph:
    (a) the time at which they pass each other,
    (b) the distance from Liverpool that they have both reached at that time.

11. At 5.00 p.m. a man begins to drive a van from Carlisle to Glasgow. At 5.30 p.m. a lady begins to make the same journey in her car. The car overtakes the van before reaching Glasgow. Details for both vehicles are given below.

distance	location	time (van)	time (car)
—	Carlisle	5.00 p.m.	5.30 p.m.
40 km	Lockerbie	5.40 p.m.	6.00 p.m.
96 km	Abington	6.36 p.m.	6.42 p.m.
152 km	Glasgow	7.32 p.m.	7.24 p.m.

Find from a graph:

(a) the time when they meet,
(b) how far they both are from Carlisle at that time
(c) the average speed of both vehicles.

12. At 2.00 p.m. a train leaves Bristol for London (Paddington). At 2.15 p.m. another train begins to make the same journey. The second train overtakes the first when they are travelling on parallel tracks. Details for both trains are given below.

distance	location	time (1st train)	time (2nd train)
—	Bristol	2.00 p.m.	2.15 p.m.
70 km	Swindon	2.35 p.m.	2.43 p.m.
130 km	Reading	3.05 p.m.	3.07 p.m.
190 km	London (Paddington)	3.35 p.m.	3.31 p.m.

Find from a graph:

(a) the time when they meet,
(b) how far they both are from Bristol at that time
(c) the average speed of both trains.

You should use only a pencil, a ruler and a pair of compasses for constructions. The following sketches are to remind you of the basic constructions which you should know.

1. The bisector of a line

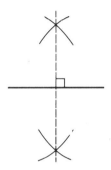

2. The bisector of an angle

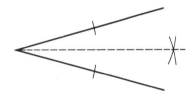

3. The perpendicular from a point on a line.

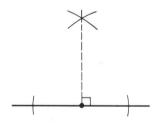

4. An angle of 60°

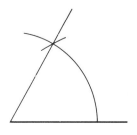

5. A right angle

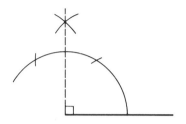

Example 1

Construct a triangle ABC in which AB = 4 cm, AC = 6 cm, and BC = 3 cm.

(a) Draw a straight line AB = 4 cm.
(b) With centre A and radius 6 cm, draw an arc.
(c) With centre B and radius 3 cm, draw an arc to cut the first arc at C.
(d) Join AC and BC.

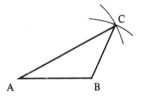

Exercise 6.1a

Construct the following triangles.

1. Triangle ABC in which AB = 5 cm, BC = 6 cm, AC = 7 cm.
2. Triangle LMN in which LM = 8 cm, MN = 9 cm, LN = 6 cm.
3. Triangle PQR in which PQ = 4 cm, QR = 5 cm, PR = 8 cm.
4. Triangle UVW in which UV = 60 mm, VW = 65 mm, UW = 110 mm.
5. Triangle XYZ in which XY = 40 mm, YZ = 70 mm, XZ = 80 mm.
6. Triangle BCD in which BC = 45 mm, CD = 75 mm, BD = 60 mm.
7. Triangle KLM in which KL = 6 cm, LM = 7 cm, KM = 7 cm.
8. Triangle QRS in which QR = 9 cm, RS = 6 cm, QS = 6 cm.
9. Triangle TUV in which TU = 105 mm, UV = 75 mm, TV = 75 mm.
10. Triangle DEF in which DE = 65 mm, EF = 65 mm, DF = 65 mm.

Example 2

Construct a triangle ABC in which
AB = 8 cm, $C\hat{A}B = 30°$, $C\hat{B}A = 45°$.

(a) Draw a straight line AB = 8 cm.
(b) At A construct an angle of 30° (i.e.
    construct an angle of 60° and bisect
    this angle).
(c) At B construct an angle of 45° (i.e.
    construct an angle of 90° and bisect
    this angle).
(d) Complete the triangle ABC.

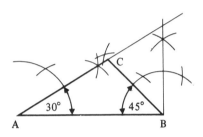

Exercise 6.1b

Construct the following figures.

 1. Triangle ABC in which AB = 6 cm, $C\hat{A}B = 45°$,
    $C\hat{B}A = 60°$.
 2. Triangle LMN in which LM = 9 cm, $N\hat{L}M = 30°$,
    $N\hat{M}L = 60°$.
 3. Triangle PQR in which PQ = 55 mm, $R\hat{P}Q = 90°$,
    $R\hat{Q}P = 45°$.
 4. Triangle UVW in which UV = 70 mm, $W\hat{U}V = 30°$,
    $W\hat{V}U = 120°$.
 5. Triangle XYZ in which XY = 50 mm, $Z\hat{X}Y = 45°$,
    $Z\hat{Y}X = 105°$.
 6. The quadrilateral ABCD in which AB = 6 cm,
    BC = 4 cm, $D\hat{A}B = 90°$, $A\hat{B}C = 120°$, $B\hat{C}D = 60°$.
 7. The parallelogram KLMN in which KL = 8 cm,
    LM = 5 cm, $N\hat{K}L = 60°$, $K\hat{L}M = 120°$, $L\hat{M}N = 60°$.
 8. The quadrilateral PQRS in which PQ = 5 cm,
    QR = 7 cm, $S\hat{P}Q = 90°$, $P\hat{Q}R = 120°$, $Q\hat{R}S = 45°$.
 9. The cyclic quadrilateral in which $W\hat{X} = 6$ cm,
    XY = 3 cm, $Z\hat{W}X = 60°$, $W\hat{X}Y = 90°$, $X\hat{Y}Z = 120°$
10. The quadrilateral BCDE in which BC = 7 cm,
    CD = 5 cm, $E\hat{B}C = 90°$, $B\hat{C}D = 105°$, $C\hat{D}E = 60°$.

Example 3

Construct a right-angled triangle ABC, in
which AB = BC = 6 cm and $A\hat{B}C = 90°$.
Bisect angle A internally and mark a point
Z on this bisector so that AZ = 4 cm.
Find all the points that are 4 cm from Z
and also 4 cm from the line AB.

(a) Draw the triangle line AB = 6 cm.
(b) At B, construct a right angle.
(c) Draw BC = 6 cm.
(d) Complete the triangle ABC.
(e) Bisect angle A and mark point Z.
(f) Centre Z, draw a circle radius 4 cm to
    find all the points which are 4 cm
    from Z.
(g) Draw the two lines which are parallel
    to AB and 4 cm away from AB.
(h) Mark clearly all the points which
    satisfy both conditions.

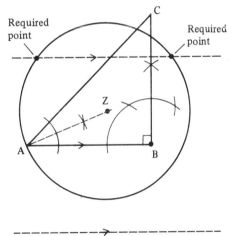

Exercise 6.1c

 1. Construct a right-angled triangle ABC in which
    AB = 4 cm, BC = 7 cm, $A\hat{B}C = 90°$. Bisect $\hat{A}$
    internally and mark a point P on this bisector
    such that AP = 5 cm. Find the points that are
    5 cm from P and also 5 cm from the line AB.
 2. Construct a right-angled triangle LMN in which
    LM = 105 mm, MN = 60 mm, $L\hat{M}N = 90°$.
    Bisect $\hat{L}$ internally and mark a point Q on this
    bisector such that LQ = 70 mm. Find the points
    that are 70 mm from Q and also 70 mm from the
    line LM.

3. Construct a right-angled triangle XYZ in which XY = 80 mm, YZ = 80 mm, $X\hat{Y}Z = 90°$. Bisect $\hat{X}$ internally and mark a point R on this bisector such that XR = 60 mm. Find the points that are 60 mm from R and also 60 mm from the line YZ.

4. Construct a right-angled triangle BCD in which BC = 5 cm, CD = 9 cm, $B\hat{C}D = 90°$. Bisect $\hat{B}$ internally and mark a point U on this bisector such that BU = 4 cm. Find the points that are 4 cm from U and also 4 cm from the line CD.

5. Construct a right-angled triangle KLM such that KL = 70 mm, LM = 40 mm, $K\hat{L}M = 90°$. Bisect $\hat{K}$ internally and mark a point X on this bisector such that KX = 45 mm. Find the points that are 45 mm from X and also 45 mm from the line LM.

6. Construct a right-angled triangle QRS such that QR = 60 mm, RS = 60 mm, $Q\hat{R}S = 90°$. Bisect the line RS internally and mark a point M on this bisector which is 50 mm from the line RS. Find the points that are 50 mm from M and also 50 mm from the line QR.

7. Construct a right-angled triangle UVW such that UV = 7 cm, VW = 4 cm and $U\hat{V}W = 90°$. Bisect the line VW internally and mark a point P on this bisector which is 5 cm from the line VW. Find the points that are 5 cm from P and also 5 cm from the line UV.

8. Construct an isosceles triangle ABC such that AB = 7 cm and BC = AC = 5 cm. Bisect $\hat{C}$ internally and mark a point N on this bisector such that CN = 6 cm. Find the points which are 6 cm from N and also 6 cm from the line AB.

9. Construct an equilateral triangle PQR such that PQ = QR = RP = 60 mm. Bisect $\hat{R}$ internally and mark a point T on this bisector such that RT = 55 mm. Find the points which are 55 mm from T and also 55 mm from the line PQ.

10. Construct an isosceles triangle XYZ such that XY = 105 mm and YZ = XZ = 60 mm. Bisect $\hat{Z}$ internally and mark a point L on this bisector such that ZL = 50 mm. Find the points that are 50 mm from L and also 50 mm from the line XY.

6.2 ANGLES AND BEARINGS

Adjacent angles

$\hat{a} + \hat{b} = 180°$

$\hat{a} + \hat{b} + \hat{c} + \hat{d} = 180°$

The sum of adjacent angles on a straight line is 180°.

Example 1

(a) Find $\hat{a}$, giving reasons.

$\hat{a} = 180° - 44° - 90°$
    $= 46°$

(b) Find $\hat{x}$ if $\hat{y} = 3x$.

$\hat{x} + \hat{y} = 180°$
$\therefore \hat{x} + 3\hat{x} = 180°$
because $\hat{y} = 3\hat{x}$
Hence $\hat{x} = 45°$

Exercise 6.2a

Find the unknown angles, giving reasons.

1.

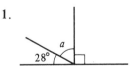

2.

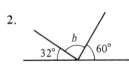

3.

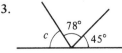

4.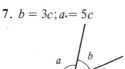

5. $y = 5x$

6. $t = \frac{1}{4}z$

7. $b = 3c; a = 5c$

8. $m = 2n; l = 9n$

9. $p = 2q; q = 3r$

10. $x = 6y; y = 2z$

Angles at a point

$$\hat{a} + \hat{b} + \hat{c} = 360°$$

The sum of all angles at a point is 360°.

Example 2

(a) Find $\hat{a}$, giving reasons.

$$\hat{a} = 360° - 110° - 30° - 86°$$
$$= 134°.$$

(b) Find $\hat{y}$, giving reasons

$$4\hat{y} + 4\hat{y} + \hat{y} = 360°$$
$$9\hat{y} = 360°$$
$$\hat{y} = 40°$$

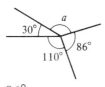

Exercise 6.2b

Find the unknown angles, giving reasons.

1.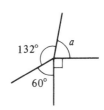

2.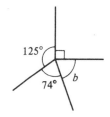

3.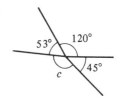

4.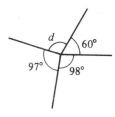

5. $y = 2x,$
$z = 5x,$
$t = 4x$

6. $q = 3p;$
$r = 4p;$
$s = 7p.$

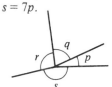

7. $b = 2a;$
$c = 3a.$

8. $l = 5k; n = 3k;$
$m = 4k; p = 7k$

9. $v = 2u;$
$x = 4u;$
$y = 5u;$
$z = 6u.$

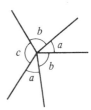

10. $b = 2a;$
$c = 4a;$
$d = 9a;$
$e = 8a.$

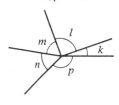

Vertically opposite angles

$\hat{a} = \hat{b}; \hat{x} = \hat{y}$

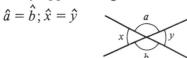

When two straight lines cross, the vertically opposite angles are equal.

Example 3

Find $\hat{a}$, $\hat{b}$, and $\hat{c}$.

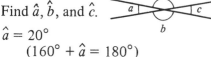

$\hat{a} = 20°$
    $(160° + \hat{a} = 180°)$

$\hat{b} = 160°$
    (vertically opposite to 160°)

$\hat{c} = 20°$
    (vertically opposite to $\hat{a}$)

Exercise 6.2c

Find the unknown angles.

1.   140°

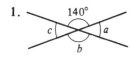

2.

3.

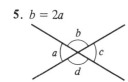

4.

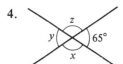

5. $b = 2a$

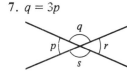

6. $m = 5l$

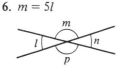

7. $q = 3p$

8. $y = 8x$

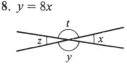

9. $c = 4b$

10. $l = 9k$

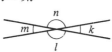

When two angles have a sum of 90°, they are called *complementary* angles.
e.g. 40° is the complement of 50°.

When two angles have a sum of 180°, they are called *supplementary* angles.
e.g. 120° is the supplement of 60°.

Exercise 6.2d

1. Find the complement of the following angles.
    (a) 40° (b) 72° (c) $62\frac{1}{2}°$ (d) 79·4°
    (e) 5° 30′
2. Find the supplement of the following angles.
    (a) 140°   (b) 18°   (c) $102\frac{1}{2}°$
    (d) 26·2°   (e) 80° 12′

Bearings

Bearings are always measured in a clockwise direction from North. They are written down using three figures.

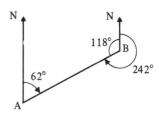

The bearing of B from A is 062°.
The bearing of A from B is 242°; this is the *reciprocal bearing* of B from A.

Example 4

Give the three-figure bearings of the following compass directions.

(a) E   (b) SE   (c) NNW

(a) Bearing is 090°
(b) Bearing is 135°
(c) Bearing is $337\frac{1}{2}°$

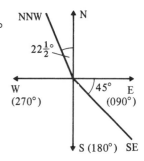

Example 5

Find the reciprocal bearings of the following.

(a) Bearing of A from B is 078°.

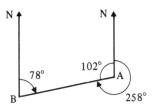

Bearing of B from A is 258°.
    (78° + 180° = 258°).

(b) Bearing of X from Y is 320°.

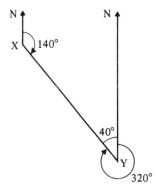

Bearing of Y from X is 140°.

because 320° + 180° = 500°
and   500° − 360° = 140°

Exercise 6.2e

1. Give the three-figure bearings of each compass direction.

    (a) NE   (b) ENE   (c) ESE   (d) SW
    (e) NW   (f) NNE   (g) SSE   (h) SSW
    (i) WSW   (j) WNW

In questions 2 to 10, find the reciprocal bearing and illustrate your answer with a diagram.

2. Bearing of A from B is 060°.
3. Bearing of P from Q is 075°.
4. Bearing of U from V is 034°.
5. Bearing of X from Y is 100°.
6. Bearing B from C is 125°.
7. Bearing of Q from R is 250°.

8. Bearing of Y from Z is 225°.
9. Bearing of A from B is 236°.
10. Bearing of P from Q is 310°.

A *course* is plotted from the bearing and distance given by drawing to a suitable scale.

Example 6

A ship travels 50 km due North and then 70 km on a bearing of 030°. Finally it sails 130 km on a bearing of 280°.
Using a scale of 1 cm to 20 km, draw an accurate plot of this journey.
From your diagram find the distance and heading the ship must now sail to return direct to its starting point.

(a) Draw a vertical line AB, 2·5 cm long to represent a course of 50 km due North.
(b) Draw BC, 3·5 cm long at an angle of 30° from North to represent a course of 70 km on a bearing of 030°.
(c) Draw CD, 6·5 cm long at an angle of 280° from North to represent a course of 130 km on a bearing of 280°.
(d) Measure the angle DA makes with North.
    Bearing = 146° and distance is 162 km.

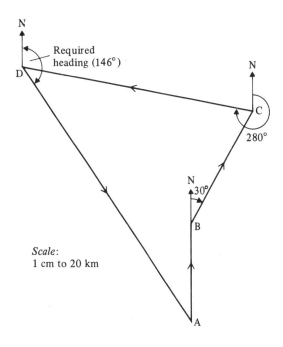

Scale:
1 cm to 20 km

Exercise 6.2f

1. A ship travels 40 km due North and then 15 km on a heading of 060°. Using a scale of 1 cm to 5 km, draw an accurate diagram of this journey. From your diagram measure and give the heading on which the ship must now travel in order to return direct to its starting point.

2. A ship travels 55 km due North and then 40 km on a heading of 120°. Using a scale of 1 cm to 5 km, draw an accurate diagram of this journey. From your diagram measure and give the heading on which the ship must now travel in order to return direct to its starting point.

3. A ship travels 28 km due North and then 56 km on a heading of 135°. Using a scale of 1 cm to 4 km, draw an accurate diagram of this journey. From your diagram measure and give the heading on which the ship must now travel in order to return direct to its starting point.

4. A helicopter flies 48 km due North to where it lands. It then flies 66 km on a heading of 240° to where it lands again. Using a scale of 1 cm to 6 km, draw an accurate diagram of this two-stage flight. From your diagram measure and give the heading on which the helicopter must now fly in order to return direct to its starting point.

5. An aeroplane flies 150 km due North to where it lands. It then flies 390 km on a heading of 210° to where it lands again. Using a scale of 1 cm to 30 km, draw an accurate diagram of this two-stage flight. Find from your diagram the heading on which the aeroplane must now fly in order to return direct to its starting point.

6. An aeroplane flies 180 km due North to where it lands. It then flies 480 km on a heading of 300° to where it lands again. Using a scale of 1 cm to 30 km, draw an accurate diagram of this two-stage flight. Find from your diagram the heading on which the aeroplane must now fly in order to return direct to its starting point.

7. A ship travels 16 km due North, then 16 km on a heading of 045°, followed by 32 km on a heading of 285°. Draw an accurate diagram of this journey by using a scale of 1 cm to 4 km. Find from your diagram the heading on which the ship must now travel in order to return direct to its starting point.

8. A ship travels 28 km due North, then 40 km on a heading of 045°, followed by 56 km on a heading of 255°. Draw an accurate diagram of this journey by using a scale of 1 cm to 4 km. Find from your diagram the heading on which the ship must now travel in order to return direct to its starting point.

9. An aeroplane flies 160 km due North to where it lands and then on a heading of 330° for 400 km to where it lands again. It then flies on a heading of 105° for 560 km to where it lands for the third time. Using a scale of 1 cm to 40 km, draw an accurate diagram of this three-stage flight. Find from your diagram the heading that the aeroplane must now fly to return direct to its starting point.

10. A helicopter flies due North for 80 km to where it lands and then on a heading of 285° for 56 km to where it lands again. It then flies on a heading of 090° for 80 km to where it lands for the third time. Using a scale of 1 cm to 8 km, draw an accurate diagram of this three-stage flight. Find from your diagram the heading that the helicopter must now fly to return direct to its starting point.

6.3 PARALLEL LINES

A pair of parallel lines is always the same distance apart.

Alternate angles

Alternate angles are equal.

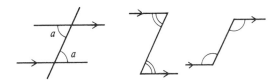

Example 1

Find $\hat{a}$ and $\hat{b}$, giving reasons.

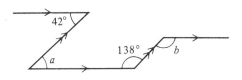

$\hat{a} = 42°$ (alternate with 42°)

$\hat{b} = 138°$ (alternate with 138°)

Exercise 6.3a

Find each one of the lettered angles giving reasons.

1.
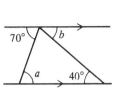

2.
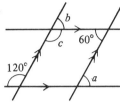

3.
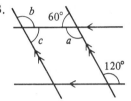

4.
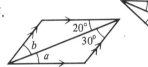

5.
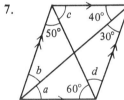

6.
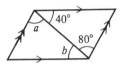

7.

8.

9.
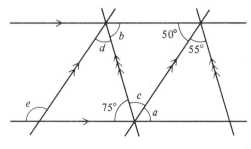

10.
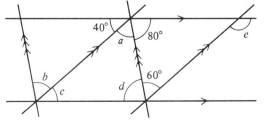

Corresponding angles

Corresponding angles are equal.

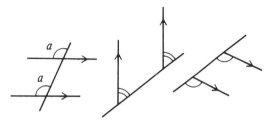

Example 2

Find $\hat{x}$ and $\hat{y}$, giving reasons.

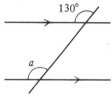

$\hat{x} = 119°$ (corresponding with 119°)
$\hat{y} = 119°$ (corresponding with 119°)

Exercise 6.3b

Find each of the lettered angles giving reasons.

1.

2.
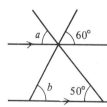

3.
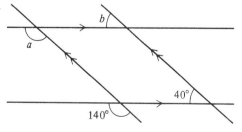

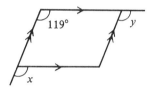

4.

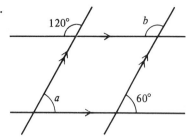

9.

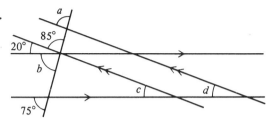

5.

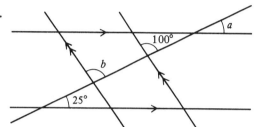

10.

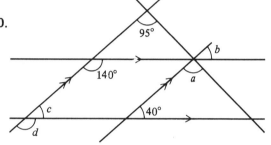

6.

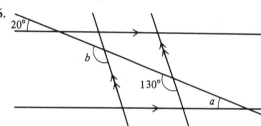

Allied angles

Allied angles are supplementary.

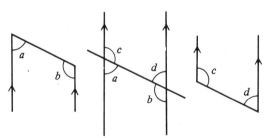

In the above diagrams

$\hat{a} + \hat{b} = 180°$
$\hat{c} + \hat{d} = 180°$

7.

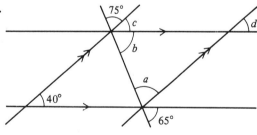

Example 3

Find $\hat{a}$, $\hat{b}$, and $\hat{c}$.

$\hat{a} = 120°$
  (allied to 60°)
$\hat{b} = 60°$
  (allied to a)
$\hat{c} = 120°$
  (allied to b)

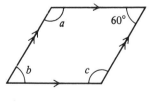

8.

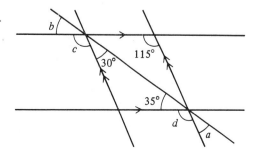

Exercise 6.3c

Find each of the lettered angles giving reasons.

1.

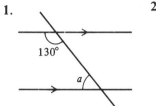

2.

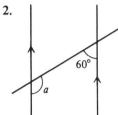

7.

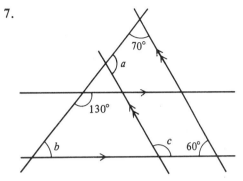

3.

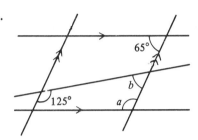

8.

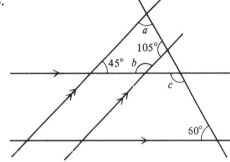

4.

5.

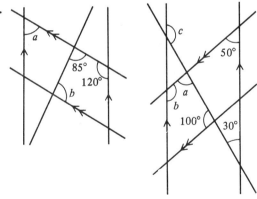

9.

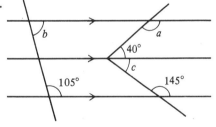

6.

10.

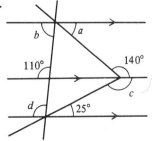

Some questions on parallel lines may use other angle properties. Extra lines may also need drawing.

Example 4

Find $\hat{a}$, $\hat{b}$, and $\hat{c}$, giving reasons.

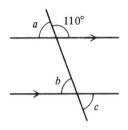

$\hat{a} = 70°$ (adjacent to 110°)

$\hat{b} = 70°$ (corresponding with $\hat{a}$)

$\hat{c} = 70°$ (vertically opposite to $\hat{b}$)

Example 5

Find $\hat{x}$, giving reasons.

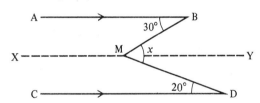

Draw a straight line XMY, parallel to AB.

$B\hat{M}Y = 30°$ (alternate with $A\hat{B}M$)

$D\hat{M}Y = 20°$ (alternate with $M\hat{D}C$)

$\therefore \hat{x} = 50°$ ($B\hat{M}Y + D\hat{M}Y$)

Exercise 6.3d

Find each of the lettered angles giving reasons.

1.

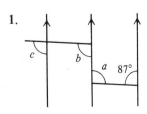

2.

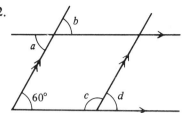

3.

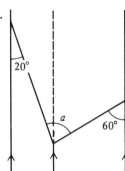

4.   **5.**

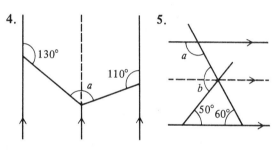

6.

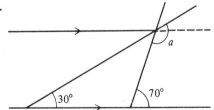

7.   **8.**

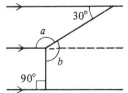   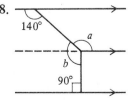

9.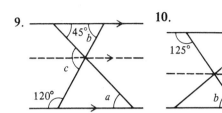

10.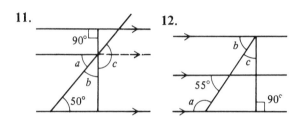

13.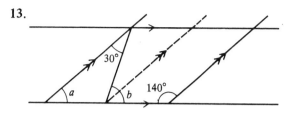

11.

12.

14.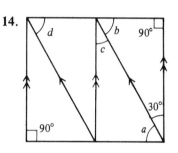

6.4 TRIANGLES

Types of triangles

Right-angled

One angle must
equal 90°.

Equilateral

Three sides equal.
All angles = 60°

Isosceles

Two sides equal.
Angles opposite the
equal sides are
equal.

Obtuse-angled

One angle is greater
than 90° and less
than 180°.

Acute-angled

All three angles
must be less
than 90°.

Scalene

No sides equal.
No angles equal.

Exercise 6.4a

Choose the correct answer from those listed A to G below for each question. The same answer may be used more than once.

A. acute-angled and scalene
B. obtuse-angled and scalene
C. acute-angled and isosceles
D. obtuse-angled and isosceles
E. right-angled
F. right-angled and isosceles
G. equilateral

1.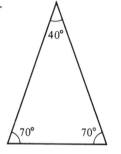

2.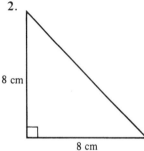

3.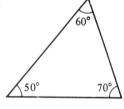

4.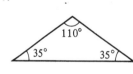

5.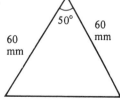

6.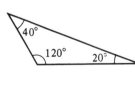

7.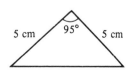

8.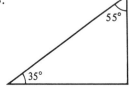

The sum of the interior angles of a triangle is 180°.

$$\hat{a} + \hat{b} + \hat{c} = 180°$$

The exterior angle of a triangle is equal to the sum of the two opposite interior angles.

$$\hat{x} = \hat{a} + \hat{b}$$

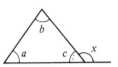

Example 1

Find the size of $\hat{a}$, $\hat{b}$, and $\hat{c}$, giving reasons.

$\hat{a} = 70°$
    (isosceles triangle)

$\hat{b} = 40°$
    (sum of angles in a triangle is 180°)

$\hat{c} = 110°$
    (adjacent angles on a straight line.)

Example 2

Find the size of $\hat{x}$ and $\hat{y}$, giving reasons.

$\hat{x} = 180° - 136°$
     $= 44°$

$\hat{x} + \hat{y} = 90°$ (the exterior angle)

so $\hat{y} = 46°$ i.e. $90° - 44°$

Exercise 6.4b

Find each lettered angle, giving reasons.

1.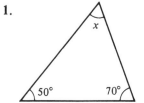

2.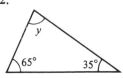

3.

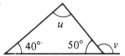

4.

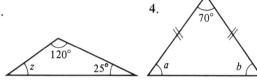

Questions using the angle properties of triangles may also require the use of the angle properties of straight lines and of parallel lines.

Example 3

Find the size of AB̂C, giving reasons.

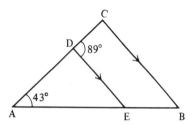

5.

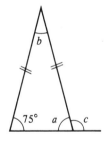

6.

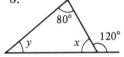

7.

8.

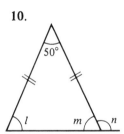

ED̂C is an exterior angle of triangle ADE,

so 89° = 43° + AÊD

AÊD = 46°

∴ AB̂C = 46° (corresponding angles).

9.

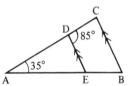

10.

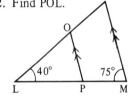

Exercise 6.4c

1. Find AB̂C.

2. Find PÔL.

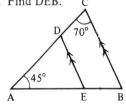

11.

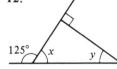

12.

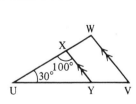

3. Find UV̂W.

4. Find DÊB.

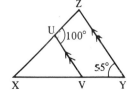

13.

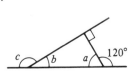

14.

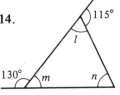

5. Find SP̂T.

6. Find UX̂V.

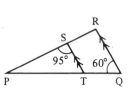

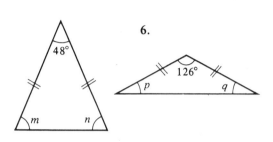

7. Find (a) DB̂C
      (b) BD̂C
      (c) DB̂A
      (d) Â

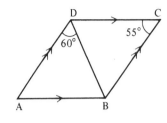

8. Find (a) KM̂N
      (b) NK̂M
      (c) KM̂L
      (d) L̂

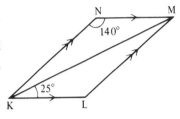

9. Find (a) QŜR
      (b) RQ̂S
      (c) PŜQ
      (d) P̂

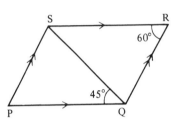

10. Find (a) ZŴY
       (b) WŶZ
       (c) YŴX
       (d) X̂

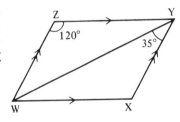

11. Find (a) LM̂O
       (b) NM̂O
       (c) LÔM
       (d) ML̂O

12. Find (a) DÊB
       (b) AÊD
       (c) AĈB
       (d) EĈB
       (e) DB̂E
       (f) AB̂C
       (g) Â

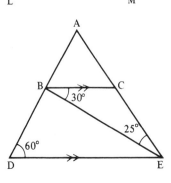

13. Find (a) Ŝ
       (b) RQ̂T
       (c) PQ̂R
       (d) PR̂Q
       (e) QR̂T
       (f) QT̂R
       (g) PT̂S

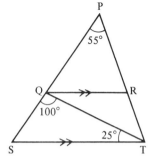

14. Find (a) FB̂C
       (b) FÊB
       (c) CÊD
       (d) FD̂C
       (e) BĈD
       (f) FD̂A
       (g) AB̂C

15. Find (a) DÊC
       (b) CÊF
       (c) AF̂E
       (d) BF̂C
       (e) BĈF
       (f) FĈE

16. Find (a) FÊC
       (b) ED̂F
       (c) BD̂C
       (d) CB̂D
       (e) CB̂A
       (f) BÂD
       (g) AD̂B
       (h) AB̂D
       (i) ED̂B

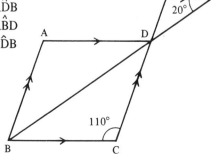

A polygon is a closed figure bounded by straight lines.
A regular polygon has all its sides and all its angles equal.

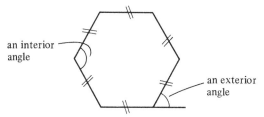

In a regular polygon:

the exterior angle $= \dfrac{360°}{\text{number of sides}}$

the interior angle $= 180° - (\text{exterior angle})$

Example 1

Find (a) the exterior angle
       (b) the interior angle of a regular 12-sided polygon.

(a) exterior angle $= \dfrac{360}{12} = 30°$

(b) interior angle $= 180 - 30 = 150°.$

Example 2

The interior angle of a regular polygon is 162°. How many sides has the polygon?

exterior angle $= 180 - 162 = 18°$

number of sides $= \dfrac{360}{18} = 20.$

Exercise 6.5a

In questions **1** to **10**, find:

(a) the exterior angle
(b) the interior angle of the regular polygon from its number of sides.

1. 5 sides	2. 3 sides	3. 10 sides
4. 15 sides	5. 8 sides	6. 30 sides
7. 24 sides	8. 36 sides	9. 45 sides
10. 60 sides		

In questions **11** to **20**, find the number of sides that the regular polygon has from the size of the interior angle.

11. 120°	12. 90°	13. 140°
14. 160°	15. 171°	16. 157° 30'
17. 172° 30'	18. 168° 45'	19. 165° 36'
20. 172° 48'		

The sum of the interior angles of any polygon is found from the formula

sum $= (2n - 4)$ right angles where n is the number of sides.

Example 3

Find the sum of the interior angles in a 22-sided polygon.

sum $= (2 \times 22 - 4) \times 90$
     $= (44 - 4) \times 90$
     $= 40 \times 90 = 3600°.$

Exercise 6.5b

Find the sum of the interior angles of the following polygons.

1. 5 sides	2. 7 sides	3. 8 sides
4. 10 sides	5. 14 sides	6. 20 sides
7. 24 sides	8. 52 sides	9. 62 sides
10. 77 sides		

Example 4

If the sum of the interior angles of a polygon is 3600°, how many sides has it?

$3600 = (2n - 4) \times 90$
$\quad 40 = 2n - 4$
$\quad 44 = 2n$
$\quad 22 = n$

∴ the polygon has 22 sides.

Exercise 6.5c

Find the number of sides of the following polygons if the sum of the interior angles is:

1. 720°	2. 360°	3. 180°
4. 1260°	5. 1800°	6. 1620°
7. 2700°	8. 4500°	9. 5040°
10. 10 440°		

A regular tessellation is a pattern of regular polygons all of the same kind filling the whole plane.

Only squares, equilateral triangles, and hexagons can be used for these patterns.

Example 5

Triangles

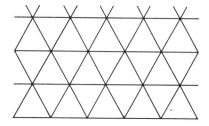

Hexagons

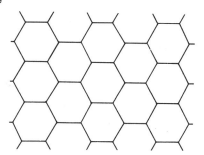

A semi-regular tessellation is a pattern using more than one regular polygon.

Example 6

Hexagon and triangle

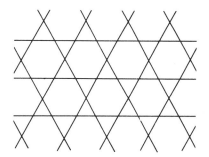

Octagon and square

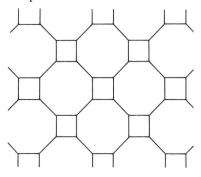

Exercise 6.5d

Draw tessellations using the shapes in questions **1** to **7**.
First make paper models of each polygon using a side length of 3 cm.

1. Regular 12-sided polygons, and equilateral triangles.
2. Regular 12-sided polygons, squares, and equilateral triangles.
3. Regular 12-sided polygons, regular hexagons, and squares.
4. Regular hexagons, squares, and equilateral triangles.
5. Regular 12-sided polygons, regular hexagons, squares, and equilateral triangles.
6. Equilateral triangles and regular hexagons.
7. Squares and equilateral triangles.

8. Name the three regular plane figures, each of which will produce a regular tessellation. Draw a sketch each time to illustrate your answer.
9. Draw diagrams to illustrate how six equilateral triangles may be placed with at least one side of each equilateral triangle touching one side of another triangle to form
   (a) a hexagon   (b) a parallelogram
10. Draw diagrams to illustrate the different tessellations that can be made with
    (a) a 45° set square
    (b) a 60° set square

Any figures which are exactly the same size and shape are said to be *congruent*.

Example 1

Which of the following shapes is 'the one that is different'? i.e. not congruent to the other two.

(a)   (b)   (c)

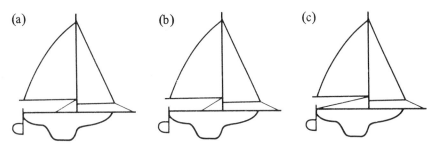

The one that is different is (c).

Exercise 6.6a

Which of the following drawings is the 'One that is different.' (Not congruent to the other two.)

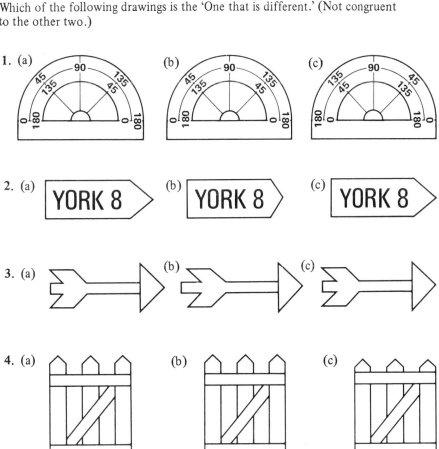

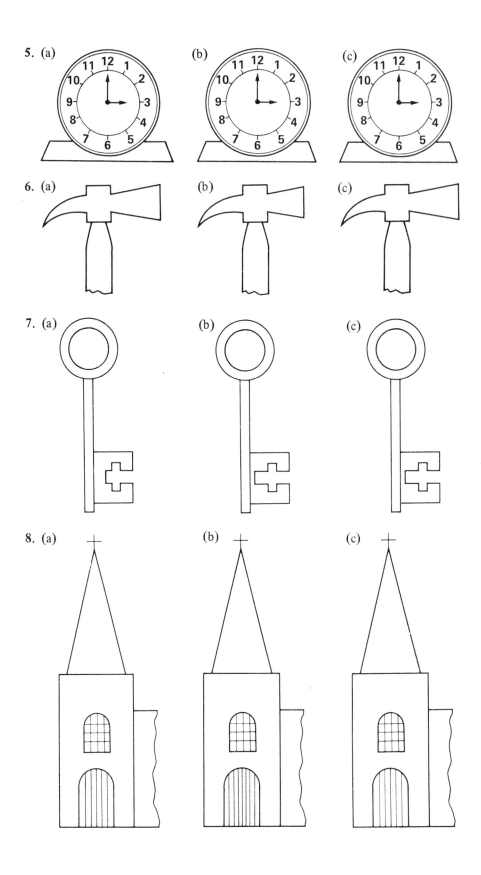

5. (a)   (b)   (c)

6. (a)   (b)   (c)

7. (a)   (b)   (c)

8. (a)   (b)   (c)

Congruent triangles

Triangles are congruent if they have:

1. Three sides equal (SSS)

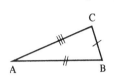 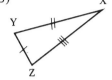

Triangle ABC = Triangle XYZ

2. Two sides and the included angle equal (SAS)

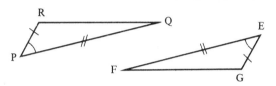

Triangle PQR = Triangle EFG

3. Two angles and a corresponding side equal (AAS)

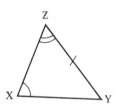 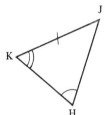

Triangle XYZ = Triangle HJK

4. Right angle, hypotenuse, and a corresponding side equal (RHS)

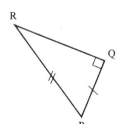 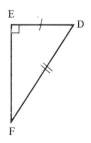

Triangle PQR = Triangle DEF

Exercise 6.6b

State if the following pairs of triangles are congruent, and if so, give the reason.

1.

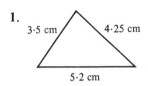 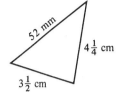

3·5 cm   4·25 cm   52 mm $4\frac{1}{4}$ cm

5·2 cm $3\frac{1}{2}$ cm

2.

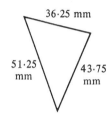

$4\frac{3}{8}$ cm $3\frac{5}{8}$ cm   36·25 mm

$5\frac{1}{8}$ cm   51·25 mm   43·75 mm

3.

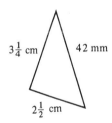

3·4 cm   2·5 cm $3\frac{1}{4}$ cm   42 mm

4·2 cm $2\frac{1}{2}$ cm

4.

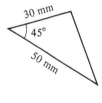

5 cm   30 mm   45°

45°   50 mm

3 cm

5.

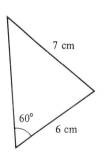

6 cm   7 cm

60°   60°

7 cm   6 cm

6.

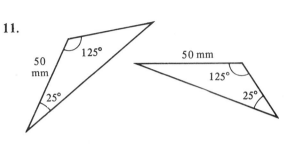

45 mm
30°
70 mm

70 mm
30°
45 mm

11.

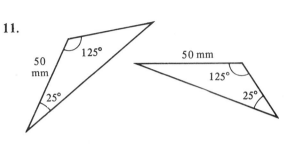

50 mm
125°
25°

50 mm
125°
25°

7.

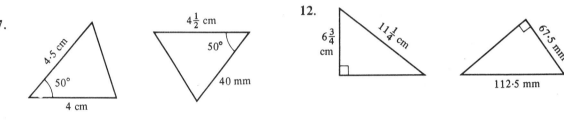

4·5 cm
50°
4 cm

4½ cm
50°
40 mm

12.

$6\frac{3}{4}$ cm
$11\frac{1}{4}$ cm

67·5 mm
112·5 mm

8.

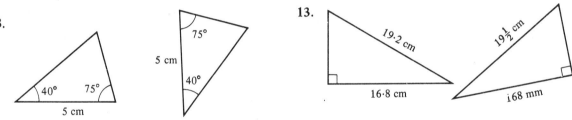

40°   75°
5 cm

75°
5 cm
40°

13.

19·2 cm
16·8 cm

$19\frac{1}{2}$ cm
168 mm

9.

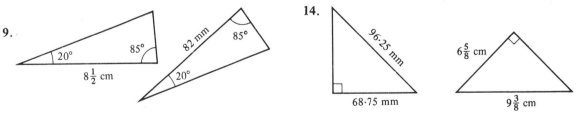

20°   85°
8½ cm

82 mm   85°
20°

14.

96·25 mm
68·75 mm

$6\frac{5}{8}$ cm
$9\frac{3}{8}$ cm

10.

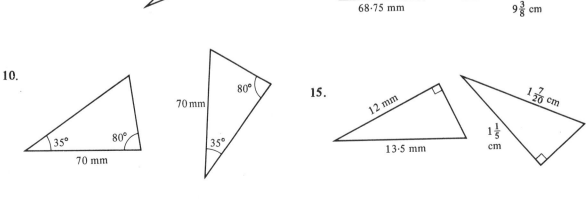

35°   80°
70 mm

70 mm   80°
35°

15.

12 mm
13·5 mm

$1\frac{7}{20}$ cm
$1\frac{1}{5}$ cm

Similar figures are figures which are the same shape. One similar figure can be said to be a scale model of another.

Example 2

Which of the following shapes is 'the one that is different'? i.e. not similar to the other two.

(a)   (b) 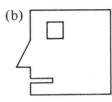   (c)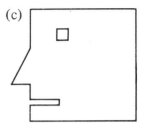

The one that is different is (c).

Exerc   .6c

In each question, of the three drawings name 'the one that is different', i.e. not similar to the other two.

1. (a) 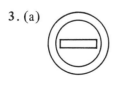   (b) 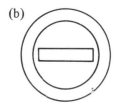   (c)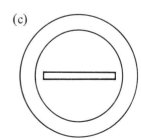

2. (a) 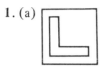   (b) 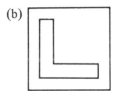   (c)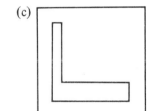

3. (a) 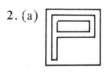   (b) 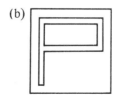   (c)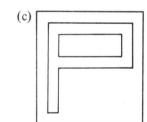

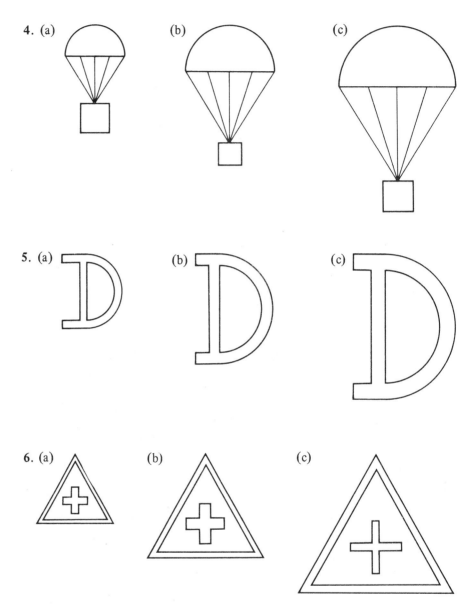

Two triangles are *similar* if all the angles in one are equal to the corresponding angles in the other.

Example 3

Which of the following triangles is 'the one that is different'? i.e. not similar to the other two.

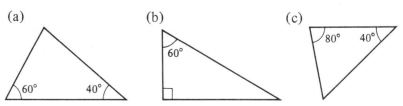

(a)   (b)   (c)

The one that is different is (b) because its angles are 90°, 60°, 30°.
The other two have angles of 80°, 60°, 40°.

Exercise 6.6d

In each question, of the three triangles name 'the one that is different', i.e. not similar to the other two.

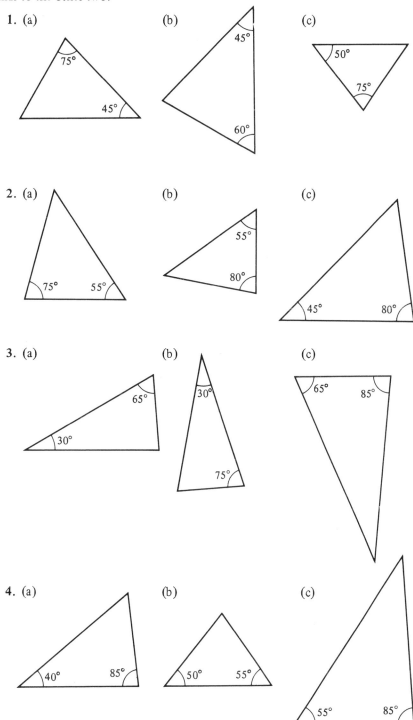

1. (a) 75° 45° (b) 45° 60° (c) 50° 75°

2. (a) 75° 55° (b) 55° 80° (c) 45° 80°

3. (a) 65° 30° (b) 30° 75° (c) 65° 85°

4. (a) 40° 85° (b) 50° 55° (c) 55° 85°

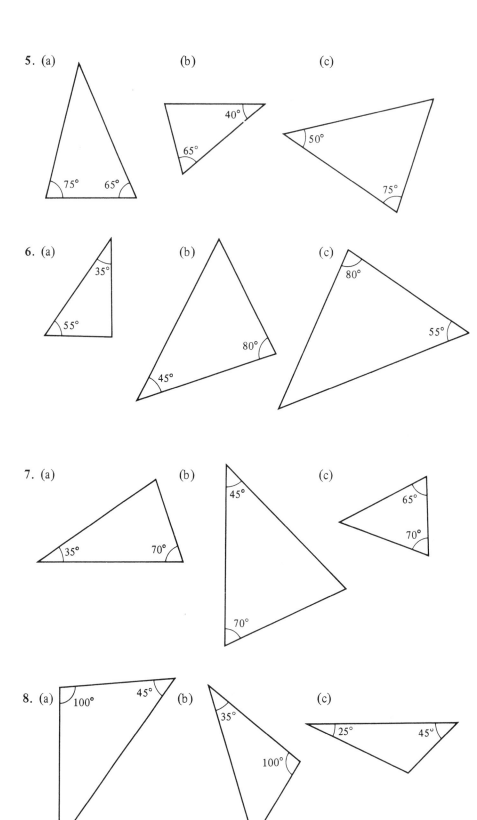

5. (a)

(b)

(c)

75°   65°

40°

65°

50°

75°

6. (a)

(b)

(c)

35°

55°

80°

45°

80°

55°

7. (a)

(b)

(c)

35°   70°

45°

70°

65°

70°

8. (a)

(b)

(c)

100°   45°

35°

100°

25°   45°

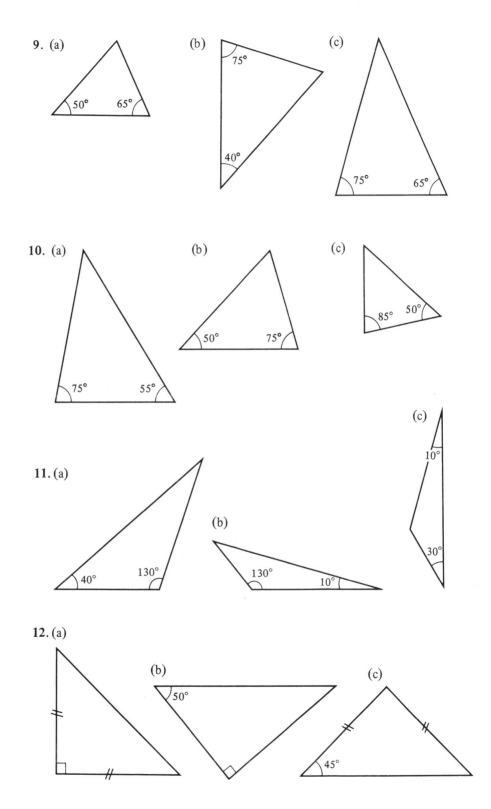

9. (a) 50° 65°
   (b) 75° 40°
   (c) 75° 65°

10. (a) 75° 55°
    (b) 50° 75°
    (c) 85° 50°

11. (a) 40° 130°
    (b) 130° 10°
    (c) 10° 30°

12. (a)
    (b) 50°
    (c) 45°

If two triangles are similar, the sides opposite the equal angles are in the same ratio.

Example 4

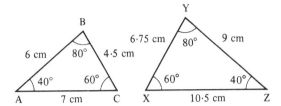

Triangle ABC is similar to triangle ZYX. Write down the ratio of the correspond-ing sides and its numerical value.

The equal ratios are:

(a) $\dfrac{BC}{XY}$ (which are sides opposite the 40° angle)

$$= \dfrac{4\cdot5}{6\cdot75} = \dfrac{2}{3}$$

(b) $\dfrac{AB}{YZ}$ (which are sides opposite the 60° angle)

$$= \dfrac{6}{9} = \dfrac{2}{3}$$

(c) $\dfrac{AC}{XZ}$ (which are sides opposite the 80° angle)

$$= \dfrac{7}{10\cdot5} = \dfrac{2}{3}$$

Exercise 6.6e

Write down the ratio of the corresponding sides for each pair of triangles and check that it is the same.

1.

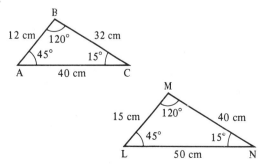

2.

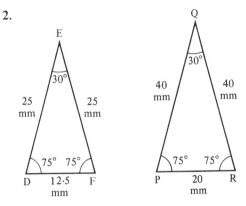

3.

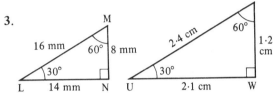

4.

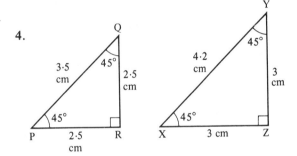

5.

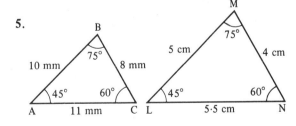

6.

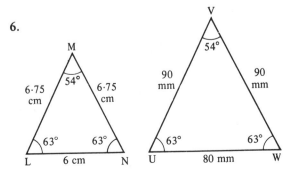

Example 5

Triangles ABC and XYZ are similar. If AB = 5 cm, AC = 4 cm, BC = 3 cm, XY = 4 cm, and $\hat{C} = \hat{Z}$, find the length of YZ.

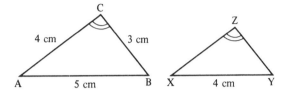

From the triangles:

$$\frac{AB}{XY} = \frac{CB}{YZ}; \quad \therefore \frac{5}{4} = \frac{3}{YZ}$$

$$5\,YZ = 12$$

$$\therefore YZ = 2\cdot4 \text{ cm}$$

Exercise 6.6f

In questions **1** to **6**, find the length of all unmarked sides of the pair of similar triangles.

1.

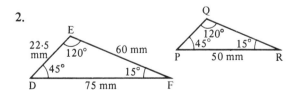

2.

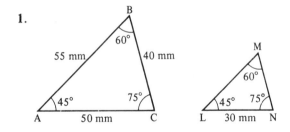

3.

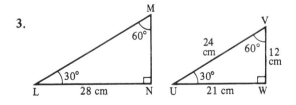

4.

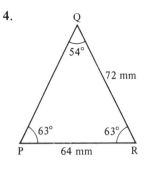

5.

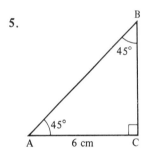

6.

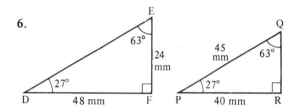

7. A sheet of cardboard measuring 20 cm by 15 cm is placed near a lamp and its shadow is cast onto a nearby wall. If the longer edge of the shadow measures 36 cm, find the length of the shorter edge.
8. A wooden stake of length 49 cm and width 28 mm is held horizontally between a lamp and the floor. If the length of the shadow is 63 cm, find its width.
9. A projector slide has a height of 30 mm and a width of 36 mm. If the height of the image on the screen is 1·25 m, find its width.
10. The front cover of a book measures 32 cm by 20 cm. When photographed for use in a magazine, the longer edge of the book measures 60 mm. Find the length of the shorter edge of the book in the photograph.

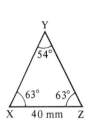

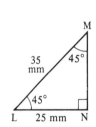

Line symmetry

If a shape can be divided by a straight line so that each side is the mirror image of the other, the shape is said to be *symmetrical* about the line. This line is called the axis of symmetry.

Examples 1

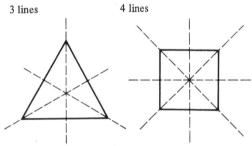

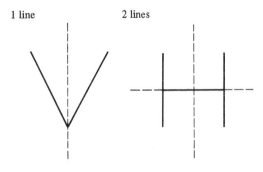

Exercise 6.7a

State the number of axes of line symmetry for each figure.

1. An isosceles triangle.
2. A rectangle.
3. A semicircle.
4. A kite.
5. A rhombus.
6. An isosceles trapezium.
7. A regular pentagon.
8. A regular hexagon.
9. A regular octagon.
10. A television screen.
11. A 45° set square.
12. A 50p piece.
13. The figure eight.
14. The letter M.
15. The letter N.

Exercise 6.7b

Copy and complete the following diagrams in which the dotted lines are axes of symmetry.

1.
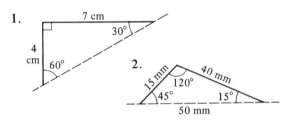

2.

3.
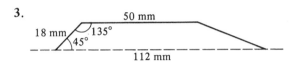

4.   5.
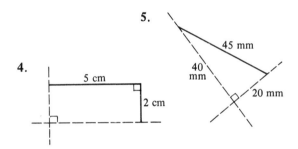

6.
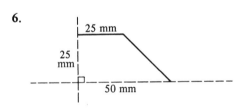

7.   8.
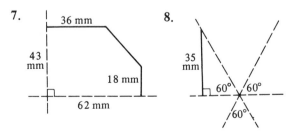

9.

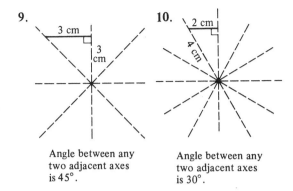

Angle between any
two adjacent axes
is 45°.

10.

Angle between any
two adjacent axes
is 30°.

3.

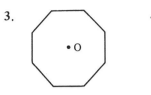

4.

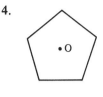

5.

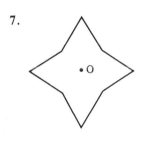

6.

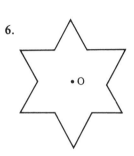

Rotational symmetry

If a figure is unchanged after a rotation
through an angle about a centre O, it is
said to have *rotational* symmetry. The
number of times this is possible in a
rotation of 360° is called the order
of rotational symmetry.

7.

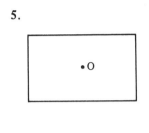

8.

Example 2

An equilateral triangle has rotational
symmetry, order 3 when rotated
about 0.

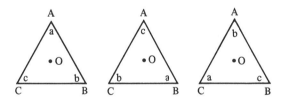

9.

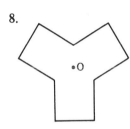

10.

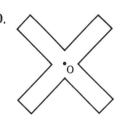

Exercise 6.7c

State the order of rotational symmetry of each of
the following.

1.

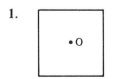

2.

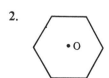

11.

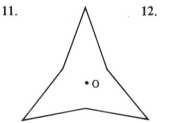

12.

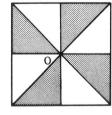

Example 3

Copy a nine-square grid and shade in:

(a) 3 squares so that the resultant figure has two lines of symmetry and rotational symmetry, order 2.
(b) 5 squares so that the resultant figure has four lines of symmetry and rotational symmetry, order 4.

(a)   (b)

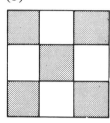

Exercise 6.7d

Copy this grid for questions **1** to **3**, then answer the questions.

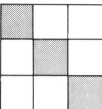

1. Shade in 4 squares so that the resultant pattern has 4 lines of symmetry and rotational symmetry of order 4. Produce two different patterns which fulfil the conditions.
2. Shade in 8 squares so that the resultant pattern has 4 lines of symmetry and rotational symmetry of order 4. Produce two different patterns which fulfil the conditions.
3. Shade in 12 squares so that the resultant pattern has 4 lines of symmetry and rotational symmetry of order 4. Produce two different patterns which fulfil the conditions.

For questions **4** to **6**, copy the illustrated triangle which contains sixteen small triangles.

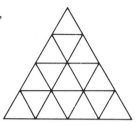

4. Shade in 4 triangles so that the resultant pattern has 3 lines of symmetry and rotational symmetry of order 3. Produce three different patterns which fulfil the conditions.
5. Shade in 7 triangles so that the resultant pattern has 3 lines of symmetry and rotational symmetry of order 3. Produce two different patterns which fulfil the conditions.
6. Shade in 10 triangles so that the resultant pattern has 3 lines of symmetry and rotational symmetry of order 3. Produce two different patterns which fulfil the conditions.

For questions **7** and **8** copy the illustrated square which contains twenty-five small squares.

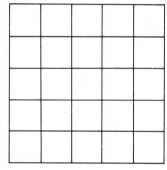

7. Shade in 8 squares such that the resultant pattern has 4 lines of symmetry and rotational symmetry of order 4. Produce three different patterns which fulfil the conditions.
8. Shade in 13 squares such that the resultant pattern has 4 lines of symmetry and rotational symmetry of order 4. Produce three different patterns which fulfil the conditions.

A quadrilateral is a four-sided figure; the sum of its interior angles is 360°.

Types of quadrilateral

1. The parallelogram

(a) Opposite sides are equal and parallel.
(b) Opposite angles are equal.
(c) Diagonals bisect each other.
(d) There are no axes of symmetry.
(e) Rotational symmetry, order 2.

2. The rectangle

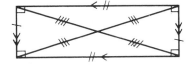

(a) Opposite sides are equal and parallel.
(b) All angles are right angles.
(c) Diagonals are equal and bisect each other.
(d) There are 2 axes of symmetry.
(e) Rotational symmetry, order 2.

3. The rhombus

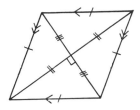

(a) All sides are equal.
(b) Opposite sides are parallel.
(c) Opposite angles are equal.
(d) Diagonals bisect each other at right angles.
(e) There are 2 axes of symmetry.
(f) Rotational symmetry, order 2.

4. The square

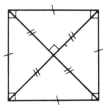

(a) All sides are equal.
(b) Opposite sides are parallel.
(c) All angles are right angles.
(d) Diagonals are equal and bisect each other at right angles.
(e) There are 4 axes of symmetry.
(f) Rotational symmetry, order 4.

5. The kite

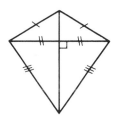

(a) There are two pairs of equal sides with the opposite sides not equal.
(b) Diagonals cut each other at right angles.
(c) One diagonal is bisected by the other.
(d) There is 1 axis of symmetry.

6. The trapezium

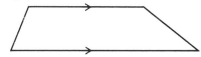

One pair of opposite sides is parallel.

7. The isosceles trapezium

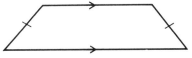

(a) One pair of opposite sides is parallel.
(b) The non-parallel sides are equal.
(c) There is 1 axis of symmetry.

Example 1

ABCD is a parallelogram. What kind of quadrilateral is it if:

(a) AC = BD and AC is perpendicular to BD?
(b) AC is perpendicular to BD but is not equal to it?
(c) AC = BD but AC is not perpendicular to BD?

(a) The shape is a square
(b) The shape is a rhombus
(c) The shape is a rectangle

Exercise 6.8a

1. ABCD is a quadrilateral. What kind of quadrilateral is it if:
   (a) The diagonals cut each other perpendicularly at X such that AX = XC?
   (b) The diagonals and the sides form four congruent right-angled triangles?
2. PQRS is a quadrilateral. What kind of quadrilateral is it if:
   (a) PQ is parallel to RS and the diagonals are equal in length?
   (b) PQ and RS are parallel and $\hat{P} = \hat{R}$?

3. KLMN is a quadrilateral with $\hat{K} = \hat{M} = 90°$. What kind of quadrilateral is it if:
   (a) KL = LM?
   (b) KM is a symmetry axis?
   (c) a symmetry axis perpendicularly bisects KL?
4. ABCD is a parallelogram. Its shape is altered by interchanging the lengths of AB and BC. What kind of quadrilateral is ABCD after this change?
5. UVWX is an isosceles trapezium with UV parallel to WX. Its shape is altered by interchanging the sizes of angles $\hat{U}$ and $\hat{X}$. What kind of quadrilateral is UVWX after this change?
6. PQRS is a kite with $\hat{Q} = \hat{S} = 90°$. Its shape is altered by interchanging the sizes of $Q\hat{P}R$ and $Q\hat{R}P$. What kind of quadrilateral is PQRS after this change?
7. What kinds of quadrilaterals are formed when the mid-points of the sides of the following are joined?
   (a) a rectangle,
   (b) a rhombus,
   (c) a kite,
   (d) an isosceles trapezium.

8. What special kinds of quadrilaterals can always be made from the following combinations of triangles?
   (a) Two congruent right-angled triangles.
   (b) Four congruent right-angled triangles.
   (c) Two congruent scalene triangles.
   (d) Two congruent isosceles triangles.

Example 2

ABCD is a rectangle whose diagonals intersect at M.

If $A\hat{M}D = 38°$, find $M\hat{C}B$.

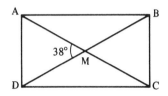

$B\hat{M}C = 38°$ (vertically opposite angles)

∴ $M\hat{C}B = 71°$ because triangle BMC is isosceles (diagonals on a rectangle are equal and bisect each other).

Example 3

PQRS is a rhombus. If $R\hat{S}P = 110°$,

find: (a) $P\hat{Q}R$ (b) $Q\hat{R}S$

(c) $S\hat{X}P$ where X is the point of intersection of the diagonals.

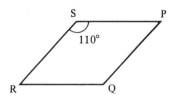

(a) $P\hat{Q}R = 110°$ (opposite angles in a rhombus are equal).

(b) $Q\hat{R}S = 70°$ (allied angles are supplementary).

(c) $S\hat{X}P = 90°$ (diagonals of a rhombus intersect at right angles).

Exercise 6.8b

1.

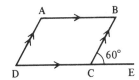

ABCD is a parallelogram. If DC is produced to E and $B\hat{C}E = 60°$, find
(a) $A\hat{B}C$ (b) $B\hat{A}D$

2.

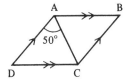

ABCD is a rhombus. If $D\hat{A}C = 50°$, find
(a) $A\hat{C}D$ (b) $C\hat{A}B$ (c) $A\hat{B}C$

3.

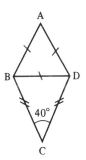

ABCD is a kite. If $B\hat{C}D = 40°$, find
(a) $B\hat{D}C$ (b) $A\hat{B}C$

4.

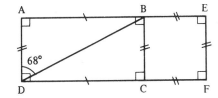

ABCD is a rectangle and BEFC is a square.
If $A\hat{D}B = 68°$, find
(a) $D\hat{B}C$ (b) $D\hat{B}F$ (c) $B\hat{D}C$

5.

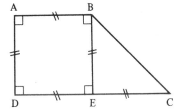

ABCD is a trapezium and ABED is a square.
If BE = EC, find
(a) $B\hat{A}E$ (b) $A\hat{B}C$
(c) What shape is the figure ABCE?

6.

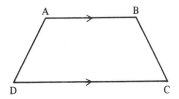

ABCD is a trapezium. What is the relationship between
(a) $B\hat{A}D$ and $A\hat{D}C$,
(b) $A\hat{B}C$ and $B\hat{C}D$?
If AD = BC, what is the relationship between
(c) $A\hat{B}C$ and $B\hat{A}D$,
(d) $A\hat{D}C$ and $B\hat{C}D$?

7.

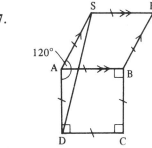

ABCD is a square and ABRS is a rhombus.
If $S\hat{A}D = 120°$, find
(a) $A\hat{S}D$ (b) $S\hat{R}B$

8. KLMN is a rectangle whose diagonals cut at X.
If $K\hat{N}L = 35°$, find $L\hat{X}M$.

9. PQRS is a rhombus. If $P\hat{Q}S = 25°$, find $Q\hat{R}S$.

10. WXYZ is a parallelogram whose diagonals bisect at Q. If $W\hat{Q}Z = 50°$, $X\hat{Z}Y = 20°$, $Y\hat{X}Z = 35°$, find:
(a) $Y\hat{W}Z$ (b) $W\hat{Y}X$
(c) $Y\hat{W}X$ (d) $W\hat{Y}Z$

11. ABCD is a kite and $\hat{A} = \hat{C}$. If $C\hat{A}D = 70°$, $C\hat{B}D = 65°$, find:
(a) $B\hat{C}D$ (b) $A\hat{D}C$

12. KLMN is an isosceles trapezium whose diagonals cut at X and KL is parallel to NM. If $K\hat{N}L = 25°$, $K\hat{M}N = 30°$, find
(a) $K\hat{X}N$ (b) $M\hat{L}N$

In any right-angled triangle, the side opposite the right angle is called the *hypotenuse*.

Pythagoras' Theorem

The square on the hypotenuse is equal to the sum of the squares on the other two sides.

Example 1

In the figure

$BC^2 = AC^2 + AB^2$

Find BC when
$AB = 5$ cm, $AC = 12$ cm.

$$BC^2 = 5^2 + 12^2$$
$$= 25 + 144 = 169$$
$$BC = \sqrt{169} = 13 \text{ cm.}$$

Example 2

In the figure

$YZ^2 = XY^2 + XZ^2$

Find YZ when
$XY = 21$ cm, $XZ = 36$ cm

$$YZ^2 = 21^2 + 36^2$$
$$= 441 + 1296 = 1737$$
$$YZ = \sqrt{1737} = 41\cdot67 \text{ (from tables)}$$
$$\therefore YZ = 41\cdot7 \text{ (correct to 3 s.f.)}$$

Exercise 6.9a

Triangle ABC is right-angled at $\hat{A}$.
Find BC in each question.

1. $AB = 3$ cm, $AC = 4$ cm.
2. $AB = 7$ cm, $AC = 24$ cm.
3. $AB = 12$ cm, $AC = 9$ cm.
4. $AB = 8$ cm, $AC = 15$ cm.
5. $AB = 24$ cm, $AC = 10$ cm.
6. $AB = 32$ mm, $AC = 24$ mm.
7. $AB = 40$ mm, $AC = 9$ mm.
8. $AB = 42$ mm, $AC = 40$ mm.
9. $AB = 60$ mm, $AC = 11$ mm.
10. $AB = 42$ mm, $AC = 144$ mm.
11. $AB = 0\cdot8$ cm, $AC = 0\cdot6$ cm.
12. $AB = 1\cdot8$ cm, $AC = 2\cdot4$ cm.
13. $AB = 0\cdot28$ cm, $AC = 0\cdot96$ cm.
14. $AB = 19\cdot2$ cm, $AC = 5\cdot6$ cm.
15. $AB = 1\cdot2$ cm, $AC = 0\cdot9$ cm.

Exercise 6.9b

Triangle XYZ is right-angled at X.
Find YZ in each question.

1. $XY = 4$ cm, $XZ = 2$ cm.
2. $XY = 6$ cm, $XZ = 3$ cm.
3. $XY = 5$ cm, $XZ = 5$ cm.
4. $XY = 1$ cm, $XZ = 3$ cm.
5. $XY = 2$ cm, $XZ = 11$ cm.
6. $XY = 12$ mm, $XZ = 14$ mm.
7. $XY = 18$ mm, $XZ = 16$ mm.
8. $XY = 16$ mm, $XZ = 8$ mm.
9. $XY = 13$ mm, $XZ = 19$ mm.
10. $XY = 17$ mm, $XZ = 9$ mm.
11. $XY = 1\cdot1$ cm, $XZ = 0\cdot7$ cm.
12. $XY = 1\cdot7$ cm, $XZ = 0\cdot9$ cm.
13. $XY = 0\cdot8$ cm, $XZ = 0\cdot4$ cm.
14. $XY = 0\cdot6$ cm, $XZ = 0\cdot2$ cm.
15. $XY = 1\cdot9$ cm, $XZ = 0\cdot3$ cm.

The Theorem of Pythagoras can be used to find any side of a right-angled triangle, given the length of the other two sides.

Example 3

Find the length of
PR.

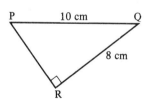

In triangle PQR

$$PQ^2 = PR^2 + QR^2$$
$$PR^2 = PQ^2 -- QR^2$$
$$= 10^2 - 8^2$$
$$= 100 - 64 = 36$$
$$PR = \sqrt{36} = 6 \text{ cm}$$

Example 4

Find the length of AC.

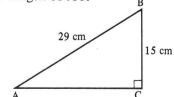

In triangle ABC

$AB^2 = AC^2 + BC^2$
$AC^2 = AB^2 - BC^2$
$\quad\ \ = 29^2 - 15^2$
$\quad\ \ = 841 - 225 = 616$
$AC\ \ = \sqrt{616} = 24 \cdot 82$
$\therefore AC\ \ = 24 \cdot 8$ (correct to 3 s.f.)

Exercise 6.9c

Triangle PQR is right-angled at $\hat{R}$.
In questions **1** to **8** find QR.

1. PQ = 20 mm,	PR = 16 mm.
2. PQ = 25 mm,	PR = 15 mm.
3. PQ = 29 mm,	PR = 21 mm.
4. PQ = 65 mm,	PR = 63 mm.
5. PQ = 2·6 cm,	PR = 2·4 cm.
6. PQ = 3·7 cm,	PR = 3·5 cm.
7. PQ = 3·4 cm,	PR = 3·0 cm.
8. PQ = 5·2 cm,	PR = 4·8 cm.

In questions **9** to **15** find PR.

9. PQ = 85 mm,	QR = 84 mm.
10. PQ = 50 mm,	QR = 48 mm.
11. PQ = 85 mm,	QR = 77 mm.
12. PQ = 53 mm,	QR = 28 mm.
13. PQ = 6·5 cm,	QR = 6·0 cm.
14. PQ = 7·8 cm,	QR = 7·2 cm.
15. PQ = 11·3 cm,	QR = 11·2 cm.

Exercise 6.9d

Triangle ABC is right-angled at $\hat{C}$.
In questions **1** to **8** find BC.

1. AB = 7 cm,	AC = 3 cm.
2. AB = 6 cm,	AC = 4 cm.
3. AB = 9 cm,	AC = 1 cm.
4. AB = 17 mm,	AC = 13 mm.
5. AB = 18 mm,	AC = 12 mm.
6. AB = 1·3 cm,	AC = 0·7 cm.
7. AB = 1·4 cm,	AC = 0·4 cm.
8. AB = 1·9 cm,	AC = 1·1 cm.

In questions **9** to **15** find AC.

9. AB = 8 cm,	BC = 2 cm.
10. AB = 11 cm,	BC = 9 cm.
11. AB = 22 mm,	BC = 18 mm.
12. AB = 19 mm,	BC = 9 mm.
13. AB = 21 mm,	BC = 19 mm.
14. AB = 2·1 cm,	BC = 0·9 cm.
15. AB = 1·6 cm,	BC = 1·4 cm.

Example 5

Bristol is 168 km due West of London;
Liverpool is 224 km due North of Bristol.
Find the direct distance from London to
Liverpool.

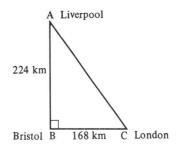

The three cities form a right-angled triangle
in which:

$AC^2 = AB^2 + BC^2$
$AC^2 = 168^2 + 224^2$
$\quad\ \ = 28\ 220 + 56\ 180$
$\quad\ \ = 78\ 400$
$AC\ \ = \sqrt{78\ 400} = 280$

So the distance from London to Liverpool
is 280 km.

Exercise 6.9e

1. Maidenhead is 40 km due West of London and
   Tring is 30 km due North of Maidenhead. Find
   the direct distance from London to Tring.
2. Leeds is 80 km due West of Hull and Blyth is
   150 km due North of Leeds. Find the direct
   distance from Hull to Blyth.
3. Crianlarich is 70 km due West of Perth and
   Durness is 240 km due North of Crianlarich.
   Find the direct distance from Perth to Durness.
4. Banbury is 90 km due East of Hereford and
   Mansfield is 120 km due North of Banbury.
   Find the direct distance from Hereford to
   Mansfield.
5. Brighton is 15 km due East of Worthing and
   Cuckfield is 20 km due North of Brighton.
   Find the direct distance from Worthing to
   Cuckfield.
6. Leeds is 50 km due East of Burnley and Durham
   is 120 km due North of Leeds. Find the direct
   distance from Burnley to Durham.
7. Cannock is 45 km due East of Shrewsbury and
   Worcester is 60 km due South of Cannock. Find
   the direct distance from Shrewsbury to Worcester.

Example 6

The supporting wire to the top of a vertical pole is 17 m long and it is fastened to the ground 8 m from the foot of the pole. How high is the pole?

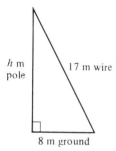

Let the height of the pole be h m. The pole, the wire and the ground form a right-angled triangle,

$$17^2 = h^2 + 8^2$$
$$17^2 - 8^2 = h^2$$
$$289 - 64 = h^2$$
$$\therefore h = \sqrt{225} = 15$$

So the height of the pole is 15 m.

Exercise 6.9f

1. The supporting wire from the top of a vertical pole to the ground is 13 m long and is fastened to the ground 5 m from the foot of the pole. What is the height of the pole?
2. A ladder 10 m long rests with its foot 6 m from the base of a vertical wall. How high up the wall will the ladder reach?
3. A ladder 15 m long rests against a window sill. If the window sill is 12 m above the ground, how far is the foot of the ladder from the wall?
4. The length of the diagonal of a rectangle is 25 cm. If the length of one side of the rectangle is 7 cm, what is
   (a) the length of the longer side,
   (b) the area of the rectangle?

5.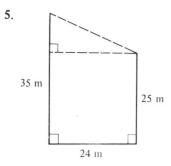

Two vertical posts are 24 m apart, as shown in the diagram. If their tops are 35 m and 25 m above the ground, find the distance between the tops.

6. The diagonals of a rhombus are 40 cm and 42 cm long. Calculate the length of one of its sides.
7. The heights of two vertical masts are 66 m and 55 m. If the shortest distance between their tops is 61 m, find how far they are apart.

8.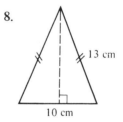

An isosceles triangle has a base of 10 cm and its equal sides are 13 cm long. Calculate
   (a) its height,
   (b) its area.
9. The slanting edge of a square-based pyramid is 41 cm long. If the vertical height of the pyramid is 40 cm, find
   (a) the length of the diagonal of the base of the pyramid,
   (b) the area of the base,
   (c) the volume of the pyramid.

The angle at the centre is twice the angle at the circumference.

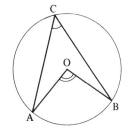

$$A\hat{O}B = 2 A\hat{C}B$$

Example 1

If $A\hat{C}B = 40°$, find (a) $A\hat{O}B$ (b) $O\hat{B}A$

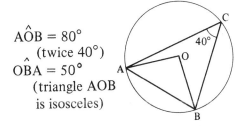

$A\hat{O}B = 80°$
  (twice 40°)
$O\hat{B}A = 50°$
  (triangle AOB
  is isosceles)

Exercise 6.10a

If O is the centre of each circle, find the size of each lettered angle.

1.
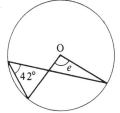

2.
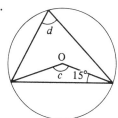

3.
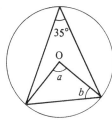

4.
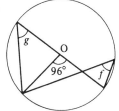

5.
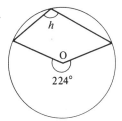

6.
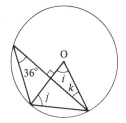

7.
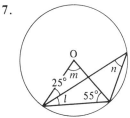

8.
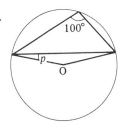

9.
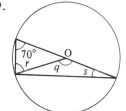

10.
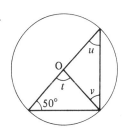

Angles in the same segment of a circle are equal.

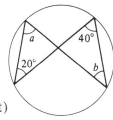

$$P\hat{T}Q = P\hat{S}Q = P\hat{R}Q$$

Example 2

Find: (a) $\hat{a}$ (b) $\hat{b}$

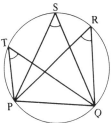

$\hat{a} = 40°$ (same segment)
$\hat{b} = 20°$ (same segment)

Exercise 6.10b

Find the size of each lettered angle in questions **1** to **4**.

1.
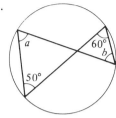

2.
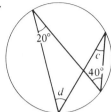

3.
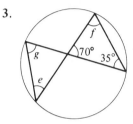

4.
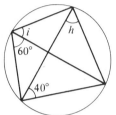

5. If O is the centre of the circle, find:

(a) O$\hat{A}$B

(b) O$\hat{B}$A

(c) O$\hat{C}$D

(d) O$\hat{D}$C

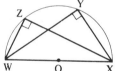

6. If O is the centre of the circle, find:

(a) O$\hat{P}$Q

(b) O$\hat{Q}$P

(c) Q$\hat{O}$R

(d) O$\hat{R}$Q

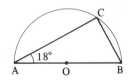

7. Find the following:

(a) A$\hat{C}$D

(b) B$\hat{D}$C

(c) C$\hat{A}$B

(d) B$\hat{D}$A

(e) C$\hat{A}$D

(f) D$\hat{B}$C

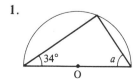

8. Find the following:

(a) E$\hat{F}$H

(b) F$\hat{H}$G

(c) F$\hat{E}$G

(d) G$\hat{F}$H

(e) E$\hat{H}$F

(f) E$\hat{G}$F

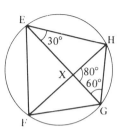

9. If AB = BC, find:

(a) A$\hat{D}$B

(b) C$\hat{A}$D

(c) D$\hat{B}$C

(d) B$\hat{A}$C

(e) B$\hat{D}$C

(f) A$\hat{C}$D

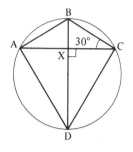

The angle in a semi-circle is a right angle.
If O is the centre of the circle:

$$W\hat{Z}X = W\hat{Y}X = 90°$$

Example 3

Find A$\hat{B}$C if O is the centre of the circle:

$$A\hat{C}B = 90°$$
(angle is a
semi-circle)
$$\therefore A\hat{B}C = 72°$$
(sum of angles in triangle ABC = 180°).

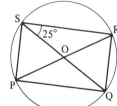

Exercise 6.10c

The centre of the circle is O in each question.
Find the size of each lettered angle in questions **1** to **6**.

1.
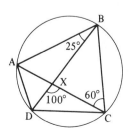

2.
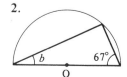

3.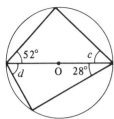

4.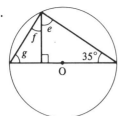

The opposite angles in a cyclic quadrilateral are supplementary.

$\hat{A} + \hat{C} = 180°$

$\hat{B} + \hat{D} = 180°$

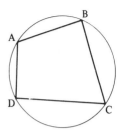

5.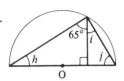

6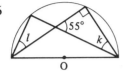

Example 4

If QR̂X = 95°, find SP̂Q.

SR̂Q = 85°
(on a straight line)

∴ SP̂Q = 95°
(opposite angles in a cyclic quadrilateral)

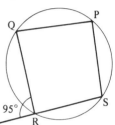

7. Find the following:

(a) AX̂D

(b) CB̂D

(c) BÂC

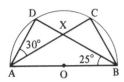

8. If PT = QT, find:

(a) SP̂T

(b) RQ̂T

(c) RP̂Q

(d) SQ̂P

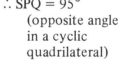

Exercise 6.10d

Find the size of each lettered angle in questions **1** to **6**.

1.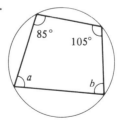

2.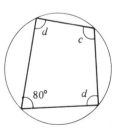

9. Find the following:

(a) CÂD

(b) CÂB

(c) CB̂A

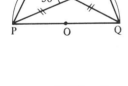

10. If QZ = QY and YZ is parallel to WX, find:

(a) YX̂Z

(b) YŴZ

(c) XẐY

(d) WŶZ

(e) XŴY

(f) WX̂Y

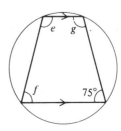

3.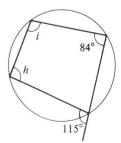

4.

5.

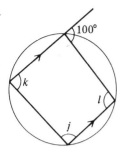

6.

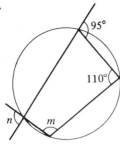

7. If AT is parallel to
DC and $B\hat{C}D = 78°$
find:

(a) $B\hat{A}D$

(b) $C\hat{B}T$

(c) $A\hat{D}C$

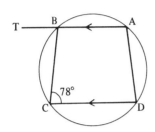

8. If XY = YZ and
$Y\hat{X}Z = 35°$, find:

(a) $X\hat{Y}Z$

(b) $X\hat{W}Z$

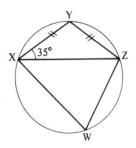

9. If BAD = 60° and
$A\hat{C}B = 2\ A\hat{C}D$, find:

(a) $A\hat{C}D$

(b) $A\hat{C}B$

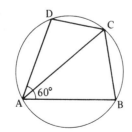

10. If $P\hat{Q}R = 90°$ and
$P\hat{S}Q = 5\ Q\hat{S}R$
find:

(a) $Q\hat{S}R$

(b) $P\hat{S}Q$

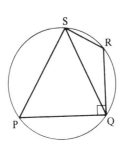

Remember that many questions on the circle will require the use of some or all of the above properties.

Example 5

PQRS is a cyclic quadrilateral in which
$P\hat{Q}R = 110°$, $Q\hat{R}S = 60°$ and $Q\hat{S}P = 20°$.

Find: (a) $P\hat{S}R$

(b) $P\hat{R}Q$

(c) $S\hat{Q}R$

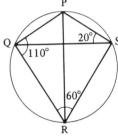

$P\hat{S}R = 70°$ (opposite angles in the cyclic quadrilateral PQRS)

$P\hat{R}Q = 20°$ (in the same segment as $Q\hat{S}P$)

$Q\hat{S}R = 50°$ $(70° - 20°)$

$S\hat{Q}R = 70°$ (sum of the angles in triangle QSR is 180°).

Exercise 6.10e

The centre of the circle is O in each question. Find the size of each lettered angle in questions **1** to **4**.

1.   **2.**

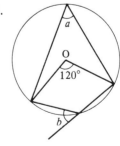

3.   **4.**

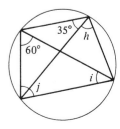

5. If $\hat{CBD} = 30°$,
   $\hat{ABD} = 60°$,
   and $\hat{BAD} = 80°$,
   find:

   (a) $\hat{ACD}$

   (b) $\hat{ACB}$

   (c) $\hat{ADB}$

   (d) $\hat{BDC}$

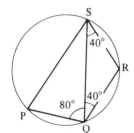

6. If $\hat{PQS} = 80°$
   $\hat{QSR} = 40°$,
   and $\hat{RQS} = 40°$
   find:

   (a) $\hat{QRS}$

   (b) $\hat{SPQ}$

   (c) $\hat{PSQ}$

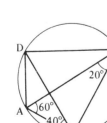

7. If $\hat{KOL} = 80°$,
   and $\hat{MKN} = 70°$,
   find:

   (a) $\hat{KML}$

   (b) $\hat{KLO}$

   (c) $\hat{KMN}$

   (d) $\hat{NKT}$

8. If $\hat{BAC} = 60°$,
   $\hat{ACB} = 20°$,
   and $\hat{ABD} = 40°$,
   find:

   (a) $\hat{ADB}$

   (b) $\hat{ACD}$

   (c) $\hat{BDC}$

   (d) $\hat{CBD}$

9. If $\hat{WYZ} = 35°$,
   $\hat{XZY} = 45°$,
   and $\hat{XWZ} = 120°$,
   find:

   (a) $\hat{WXZ}$

   (b) $\hat{WZX}$

   (c) $\hat{WYX}$

   (d) $\hat{YXZ}$

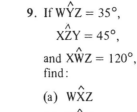

10. If $\hat{PXS} = 90°$,
    and $\hat{SPR} = 25°$,
    find:

    (a) $\hat{PSQ}$

    (b) $\hat{QSR}$

    (c) $\hat{QRS}$

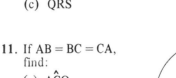

11. If $AB = BC = CA$,
    find:

    (a) $\hat{ACQ}$

    (b) $\hat{CAQ}$

    (c) $\hat{BCQ}$

    (d) $\hat{COB}$

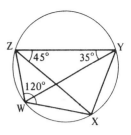

12. If $\hat{ADC} = 102°$,
    and $\hat{ABO} = 22°$,
    find:

    (a) $\hat{BAO}$

    (b) $\hat{AOB}$

    (c) $\hat{ACB}$

    (d) $\hat{ABC}$

    (e) $\hat{BAC}$

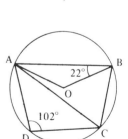

A tangent is at right angles to a radius at the point of contact with the circle.

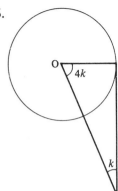

$O\hat{T}A = 90°$

5.

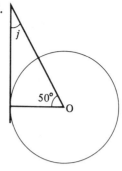

6.

7.

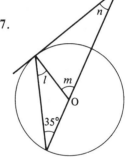

8.

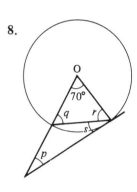

Example 1

If AT is a tangent and $B\hat{A}T = 40°$, find:

(a) $O\hat{A}B$ (b) $A\hat{B}O$ (c) $A\hat{O}B$

(a) $O\hat{A}B = 50°$

    $(O\hat{A}T = 90°)$

(b) $A\hat{B}O = 50°$

    $(OA = OB)$

(c) $A\hat{O}B = 80°$

    $(180° - 50° - 50°)$

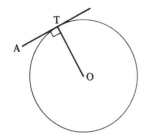

Exercise 6.11a

If O is the centre of each circle, find the size of each lettered angle.

1.

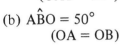
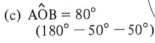

2.

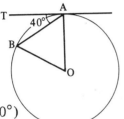

9.

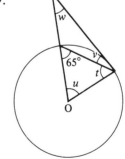

10.

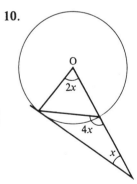

3.

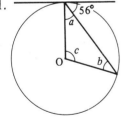

4.

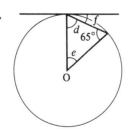

Two tangents only can be drawn to a circle from a point. These tangents are equal in length.

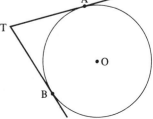

$TA = TB$

Example 2

If AX and AY are tangents to a circle and
$X\hat{A}Y = 70°$, find $X\hat{O}Y$.

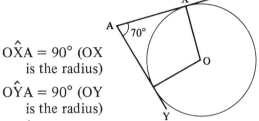

$O\hat{X}A = 90°$ (OX
   is the radius)
$O\hat{Y}A = 90°$ (OY
   is the radius)
$\therefore X\hat{O}Y = 110°$
$(360° - 90° - 90° - 70°)$

Exercise 6.11b

The centre of the circle is O in each question.
Find the size of each lettered angle in questions
1 to **5**.

1.
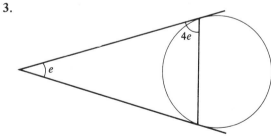

2.

3.

4.

5.
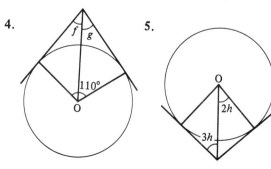

6. If $A\hat{B}O = 40°$,
  and $A\hat{X}O = 90°$,
  find:

(a) $B\hat{A}C$,

(b) $B\hat{C}A$,

(c) $O\hat{A}C$,

(d) $A\hat{C}O$,

(e) $A\hat{O}C$

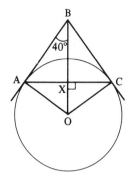

7. If $R\hat{O}S = 65°$,
  and $P\hat{X}S = 90°$,
  find:

(a) $P\hat{R}S$

(b) $P\hat{S}R$

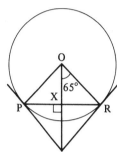

8. If $Y\hat{X}Z = 63°$,
  and $X\hat{Q}Y = 90°$,
  find:

(a) $X\hat{O}Y$

(b) $X\hat{Y}Z$

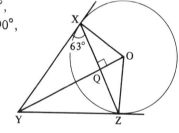

9. B

If $A\hat{B}O = 18°$, find:

(a) $A\hat{O}B$,

(b) $D\hat{A}O$,

(c) $A\hat{D}C$

10. If $O\hat{S}R = 40°$,
  and $P\hat{O}S = 50°$,
  find:

(a) $O\hat{R}Q$

(b) $Q\hat{P}S$

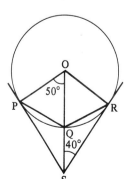

The angle between a chord and a tangent is equal to the angle subtended by the chord in the alternate segment.

$$A\hat{T}Y = T\hat{B}A$$

$$X\hat{T}B = T\hat{A}B$$

Example 3

If XTY is a tangent, $X\hat{T}B = 80°$, and $T\hat{B}A = 60°$, find (a) $Y\hat{T}A$,   (b) $T\hat{A}B$.

$Y\hat{T}A = 60°$
(alternate segment)

$T\hat{A}B = 80°$
(alternate segment)

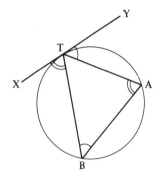

Exercise 6.11c

Find the size of each lettered angle in questions **1** to **4**.

1.

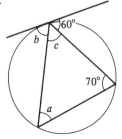

2.

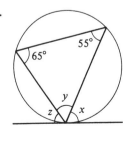

3.

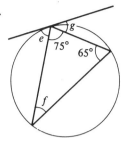

4.

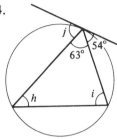

5. If $Y\hat{X}Z = 75°$ and XY = YZ find:

(a) $X\hat{Z}Y$

(b) $T\hat{Y}Z$

(c) $P\hat{Y}X$

(d) $X\hat{Y}Z$

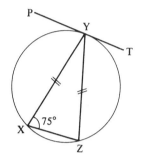

6. If $X\hat{Y}Z = 50°$ and XY = XZ, find:

(a) $X\hat{Z}Y$

(b) $Y\hat{X}Z$

(c) $B\hat{Y}Z$

(d) $A\hat{Y}X$

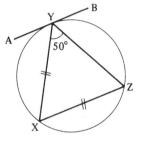

Find the size of each lettered angle in questions **7** and **8**.

7.

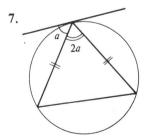

8.

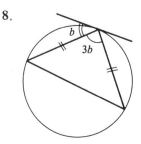

Remember that questions on the angle properties of the circle usually require knowledge of more than one angle property.

Example 4

O is the centre of the circle and CT is a tangent. If $T\hat{C}A = 32°$, find:

(a) $O\hat{T}C$ (b) $D\hat{A}T$ (c) $T\hat{O}C$

(d) $T\hat{A}B$ (e) $B\hat{T}C$

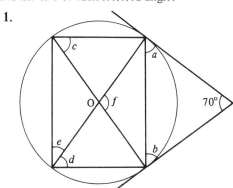

(a) $O\hat{T}C = 90°$ (OT is a radius, TC is a tangent)

(b) $D\hat{A}T = 90°$ (angle in a semi-circle)

(c) $T\hat{O}C = 58°$ (sum of angles in triangle TOC is 180°)

(d) $T\hat{A}B = 29°$ (angle at centre equals twice angle at circumference).

(e) $B\hat{T}C = 29°$ (alternate segment)

Exercise 6.11d

The centre of the circle is O in each question. Find the size of each lettered angle.

1.

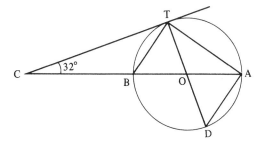

2.

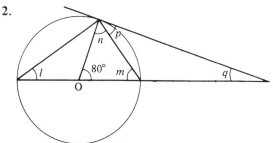

3.

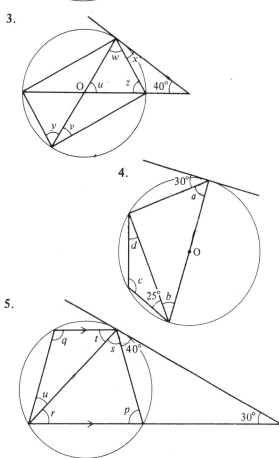

4.

5.

6.

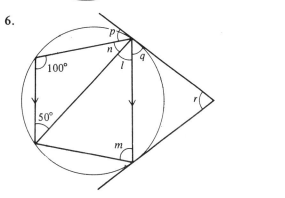

The locus of a point (plural: loci) is the path traced out by the point moving under given conditions. All points on the locus satisfy these conditions.
Some common loci are given below.

1. The locus of a point which moves in a plane at a given distance from a fixed point O, is a circle with centre O.

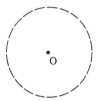

2. The locus of a point which moves in a plane, equidistant from two fixed points A and B is the perpendicular bisector of the straight line joining A to B.

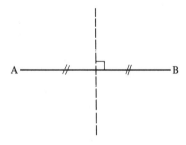

3. The locus of a point which moves in a plane, equidistant from two fixed lines AB and CD is the bisector of the angle formed where the two lines meet.

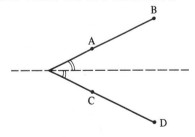

4. The locus of a point which moves in a plane, equidistant from a line AB produces parallel lines on each side of the line AB.

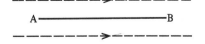

Example 1

Draw the locus of a point 1 cm from the end A of a straight line AB 2.5 cm long. On the same figure, draw the locus of a point 2 cm from B. Letter the points X and Y which are both 1 cm from A and 2 cm from B.

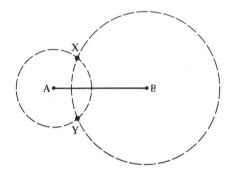

Example 2

Draw and describe the locus of a point 3 cm from a straight line AB 5 cm long. On the same figure, draw and describe the locus of a point equidistant from A and B. Letter the points X and Y which are both 3 cm from AB and equidistant from A and B. Join AX, BX, AY and BY, and describe the figure AXBY.

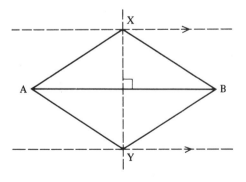

The locus from the line AB is a line parallel to AB on both sides of it and 3 cm from it. The locus equidistant from A and B is the perpendicular bisector of AB.
AXBY is a rhombus.

Exercise 6.12

1. A straight line AB is 7 cm long. Draw and describe the locus of the point which is:

   (a) 5 cm from A   (b) 5 cm from B

   Mark the two points X and Y which are 5 cm from both A and B. Join AX, AY, BX, and BY, and describe the figure AXBY.

2. A$\hat{B}$C = 90°, and AB = BC = 4 cm. Draw and describe the locus of a point which is:

   (a) 4 cm from A   (b) 4 cm from C

   Mark the point D which is 4 cm from both A and C. Join AD and CD, and describe the figure ABCD.

3. A straight line PQ is 7 cm long. Draw and describe the locus of a point which is:

   (a) 4 cm from P
   (b) equidistant from P and Q

   Mark the points U and V which are 4 cm from both P and Q. Join PU and PV, and describe the figure PUV.

4. A straight line LN is 14 cm long and M is the mid-point. Draw and describe the locus of a point which is:

   (a) 5 cm from L   (b) 5 cm from N
   (c) equidistant from L and M
   (d) equidistant from M and N

   Mark the points P and Q which are 5 cm from both L and M, and the points R and S which are 5 cm from N and M, P and R being on the same side of LN. Join PQ, QS, SR, and RP, and describe the figure PQSR.

5. A straight line AB is 10 cm long. Draw and describe the locus of a point which is:

   (a) always 5 cm from the line AB
   (b) equidistant from A and B. Mark the two points X and Y which are 5 cm from AB and equidistant from A and B. Join AX, AY, BX, and BY and describe the figure AXBY.

6. A straight line PR is 18 cm long. Draw and describe the locus of a point which is:

   (a) 4 cm from the line PR
   (b) 8 cm from Q, the mid-point of PR

   Mark A, B, C, and D, the four points which are

4 cm from the line PR and 8 cm from Q. Join AB, BC, CD, and DA, and describe the figure ABCD. Join AQ and BQ and describe the figure AQD.

7. A straight line UW is 16 cm long and V is its mid-point. Draw and describe the locus of a point which is:

   (a) 7 cm from the line UW
   (b) 8 cm from U
   (c) 8 cm from W

   Mark the two points A and B which are 7 cm from UW and 8 cm from U. Also mark the two points C and D which are 7 cm from UW and 8 cm from W, C being on the same side of UW as A. Join AB, BD, DC, and CA, and describe the figure ABDC. Also join AV and BV and describe the figure AVB.

8. A straight line AB is 80 mm long. Draw and describe the locus of a point which is:

   (a) 35 mm from AB
   (b) 40 mm from A
   (c) 40 mm from B

   Mark the two points P and Q which are 35 mm from AB and 40 mm from A, and also the two points R and S which are 35 mm from AB and 40 mm from B, R being on the same side of AB as P. Join AP, PR, RB, BS, SQ, and QA, and describe the figure APRBSQ.

9. A$\hat{B}$C = 60° and BA = BC = 8 cm. The mid-points of BA, BC are M and N respectively. Draw and describe the locus of a point which is:

   (a) equidistant from BA and BC
   (b) 4 cm from M
   (c) 4 cm from N

   Mark the point P which is 4 cm from both M and N and equidistant from BA and BC. Join MP and NP, and describe the figure BMPN.

10. X$\hat{Y}$Z = 120° and YZ = YX = 4 cm. Draw and describe the locus of a point which is:

    (a) equidistant from the two lines YZ and YX
    (b) 4 cm from Y

    Mark the two points P and Q which are equidistant from YZ and YX and 4 cm from Y. Join XP, PZ, ZQ, and QX, and describe the figure XPZQ. Join XZ and describe the figures XZP and XZQ.

Table of natural sines

A	0'	6'	12'	18'	24'	30'	36'	42'	48'	54'	1'	2'	3'	4'	5'
(b)→ 31°	.5150	5165	5180	5195	5210	5225	5240	[5255]	5270	5284	2	5	7	10	12
32°	.5299	5314	5329	5344	5358	5373	5388	5402	5417	5432	2	5	7	10	12
33°	.5446	5461	5476	5490	5505	5519	5534	5548	5563	5577	2	5	7	10	12
34°	.5592	5606	5621	5635	5650	5664	5678	5693	5707	5721	2	5	7	10	12
35°	.5736	5750	5764	5779	5793	5807	5821	5835	5850	5864	2	5	7	9	12
(a)→ 36°	.5878	5892	5906	5920	5934	5948	5962	5976	5990	6004	2	5	7	9	12
37°	.6018	6032	6046	6060	6074	6088	6101	6115	6129	6143	2	5	7	9	12
(c)--→ 38°	.6157	[6170]	6184	6198	6211	6225	6239	6252	6266	6280	2	5	7	[9]	11

An angle of 1° is divided into 60 smaller parts called minutes.

$$1° = 60'$$

Trigonometric tables, as shown above, consist of three main columns, A, B, and C. Column A lists the angles from 0° to 90°. Column B shows the division of each degree at 6 minute intervals (0·1°) from 0' to 54'. Column C is called the mean *difference* *column*.

Example 1

From the table above find the sine of an angle of (a) 36° (b) 31° 42' (c) 38° 10'

(a) Find 36° in column **A** of the tables and on the same row find the number under 0' in column **B**.
   ∴ sin 36° = 0·5878
(b) Find 31° in column **A** of the tables and on the same row find the number under 42' in column **B**.
   ∴ sin 31° 42' = 0·5255
(c) Find 38° in column **A** of the tables. On the same row, find the number under 6' in column **B**, i.e. 0·6170.

Table of natural sines

Again using the same row, find the number under 4' in column **C**, i.e. 0·0009.
   ∴ sin 38° 10' = sin [38° (6 + 4)']
       = 0·6170 + 0·0009
       = 0·6179

Exercise 7.1a

Use tables to find the sine of each angle.

1. 15°	2. 30°	3. 70°
4. 45°	5. 25°	6. 50°
7. 72°	8. 66°	9. 48°
10. 8°	11. 40° 42'	12. 54° 48'
13. 15° 12'	14. 61° 30'	15. 73° 6'
16. 12° 42'	17. 37° 54'	18. 7° 36'
19. 29° 18'	20. 89° 24'	21. 34° 15'
22. 62° 22'	23. 52° 26'	24. 13° 2'
25. 18° 38'	26. 16° 50'	27. 7° 43'
28. 68° 35'	29. 42° 8'	30. 33° 25'
31. 47° 14'	32. 56° 40'	33. 70° 16'
34. 44° 34'	35. 25° 51'	36. 13° 57'
37. 10° 34'	38. 23° 59'	39. 8° 53'
40. 25° 35'		

Example 2

Find from the table below, the angle whose sine is (a) 0·9511 (b) 0·9641 (c) 0·9757

A	0'	6'	12'	18'	24'	30'	36'	42'	48'	54'	1'	2'	3'	4'	5'
71°	.9455	9461	9466	9472	9478	9483	9489	9494	9500	9505	1	2	3	4	5
(a)→ 72°	.9511	9516	9521	9527	9532	9537	9542	9548	9553	9558	1	2	3	4	4
73°	.9563	9568	9573	9578	9583	9588	9593	9598	9603	9608	1	2	2	3	4
(b)→ 74°	.9613	9617	9622	9627	9632	9636	[9641]	9646	9650	9655	1	2	2	3	4
75°	.9659	9664	9668	9673	9677	9681	9686	9690	9694	9699	1	1	2	3	4
76°	.9703	9707	9711	9715	9720	9724	9728	9732	9736	9740	1	1	2	3	3
(c)--→ 77°	.9744	9748	9751	[9755]	9759	9763	9767	9770	9774	9778	1	1	[2]	3	3
78°	.9781	9785	9789	9792	9796	9799	9803	9806	9810	9813	1	1	2	2	3

To find an angle from its sine

1. If the exact value can be found in column B, read off the required angle.
2. If the exact value cannot be found in column **B**,
   (i) find the nearest value *below* it and note its angle.
   (ii) find the difference in column **C** and note its angle.
   (iii) add (i) and (ii) to give the required angle.

Hence:
(a) the angle whose sine is 0·9511 is 72° 0′
(b) the angle whose sine is 0·9641 is 74° 36′
(c) the angle whose sine is 0·9757 is 77° 18′ + 3′ = 77° 21′

Exercise 7.1b

Use tables to find the angle of each sine.

1. 0·9848	2. 0·1736	3. 0·3420
4. 0·5736	5. 0·9659	6. 0·9063
7. 0·9945	8. 0·8290	9. 0·2079
10. 0·5446	11. 0·6388	12. 0·9668
13. 0·1633	14. 0·7826	15. 0·6468
16. 0·9898	17. 0·0976	18. 0·8202
19. 0·9198	20. 0·9992	21. 0·3327
22. 0·9148	23. 0·8219	24. 0·6788
25. 0·5657	26. 0·5093	27. 0·6408
28. 0·9067	29. 0·5232	30. 0·9183
31. 0·6314	32. 0·8710	33. 0·2328
34. 0·4488	35. 0·5719	36. 0·4723
37. 0·3557	38. 0·6228	39. 0·4359
40. 0·2979		

Tangent tables are used in exactly the same way as sine tables. But remember that angles greater than 45° have ratios greater than 1.

Exercise 7.1c

Use tables to find the tangent of each angle.

1. 10°	2. 40°	3. 25°
4. 36°	5. 14°	6. 50°
7. 65°	8. 75°	9. 63°
10. 81°	11. 12° 30′	12. 33° 42′
13. 24° 12′	14. 32° 6′	15. 40° 54′
16. 51° 18′	17. 62° 36′	18. 48° 30′
19. 57° 48′	20. 70° 24′	21. 14° 44′
22. 39° 25′	23. 47° 31′	24. 50° 49′
25. 16° 13′	26. 36° 32′	27. 35° 1′
28. 32° 44′	29. 11° 15′	30. 47° 43′
31. 13° 38′	32. 53° 49′	33. 22° 50′
34. 18° 20′	35. 25° 44′	36. 51° 31′
37. 10° 28′	38. 48° 26′	39. 11° 58′
40. 28° 34′		

Exercise 7.1d

Use tables to find the angle of each tangent.

1. 0·7002	2. 0·3640	3. 0·2867
4. 0·1405	5. 0·9004	6. 1·4281
7. 1·7321	8. 2·7475	9. 1·5399
10. 2·0503	11. 0·8214	12. 0·2830
13. 0·8785	14. 0·4791	15. 0·0892
16. 1·3127	17. 1·8418	18. 1·1463
19. 2·9375	20. 4·1976	21. 0·4129
22. 0·6826	23. 1·2138	24. 1·1799
25. 0·6954	26. 0·3647	27. 0·1408
28. 1·0507	29. 0·3333	30. 0·2428
31. 0·2914	32. 0·6350	33. 0·3574
34. 0·8026	35. 1·2587	36. 1·6720
37. 0·1552	38. 1·1969	39. 0·5057
40. 1·4341		

Example 3

Find from the table below, the cosine of an angle of
(a) 45°   (b) 46° 54′   (c) 48° 23′

Table of natural cosines

A	0′	6′	12′	18′	24′	30′	36′	42′	48′	54′	1′	2′	3′	4′	5′
(a) → 45°	.7071	7059	7046	7034	7022	7009	6997	6984	6972	6959	2	4	6	8	10
(b) → 46°	.6947	6934	6921	6909	6896	6884	6871	6858	6845	6833	2	4	6	8	11
47°	.6820	6807	6794	6782	6769	6756	6743	6730	6717	6704	2	4	6	9	11
(c) → 48°	.6691	6678	6665	6652	6639	6626	6613	6600	6587	6574	2	4	7	9	11

(B spans 0′ to 54′; C — Subtract — spans 1′ to 5′)

The cosine ratio gets smaller as the angle gets larger, so the numbers in column **C** are *subtracted*. Apart from this, the tables are used in the same way as the other tables.

Hence:
(a) cos 45° = 0·7071
(b) cos 46° 54′ = 0·6833
(c) cos 48° 23′ = 0·6652 − 0·0011
= 0·6641

Exercise 7.1e

Use tables to find the cosine of each angle.

1. 40°	**2.** 30°	**3.** 70°
4. 55°	**5.** 85°	**6.** 15°
7. 24°	**8.** 72°	**9.** 9°
10. 84°	**11.** 25° 12′	**12.** 70° 30′
13. 32° 54′	**14.** 40° 6′	**15.** 50° 18′
16. 61° 36′	**17.** 28° 42′	**18.** 12° 24′
19. 54° 48′	**20.** 85° 12′	**21** 33° 10′
22. 19° 35′	**23.** 55° 27′	**24.** 67° 13′
25. 58° 2′	**26.** 75° 3′	**27.** 60° 38′
28. 34° 55′	**29.** 24° 14′	**30.** 67° 39′
31. 79° 32′	**32.** 64° 21′	**33.** 30° 51′
34. 72° 38′	**35.** 23° 53′	**36.** 28° 22′
37. 63° 41′	**38.** 86° 5′	**39.** 77° 28′
40. 80° 46′		

Example 4

Find from the table below the angle whose cosine is
(a) 0·5000 (b) 0·4571 (c) 0·4070

To find an angle from its cosine.

1. If the exact value can be found in column **B**, read off the required angle.
2. If the exact value cannot be found in column **B**:
   (i) find the nearest value *above* it and note its angle.
   (ii) find the difference in column **C** and note the number of minutes,
   (iii) add (i) and (ii) to give the required angle.

Hence:
(a) the angle whose cosine is 0·5000 is 60°
(b) the angle whose cosine is 0·4571 is 62° 48′
(c) the angle whose cosine is 0·4070 is 65° 54′ + 5′ = 65° 59′.

Exercise 7.1f

Use tables to find the angle of each cosine.

1. 0·6428	**2.** 0·9848	**3.** 0·9397
4. 0·8192	**5.** 0·2588	**6.** 0·9962
7. 0·8090	**8.** 0·5592	**9.** 0·9511
10. 0·4384	**11.** 0·7478	**12.** 0·9449
13. 0·3206	**14.** 0·7965	**15.** 0·8957
16. 0·4305	**17.** 0·9627	**18.** 0·5329
19. 0·8396	**20.** 0·1461	**21.** 0·9753
22. 0·7971	**23.** 0·5383	**24.** 0·7001
25. 0·2411	**26.** 0·3250	**27.** 0·8391
28. 0·7800	**29.** 0·8649	**30.** 0·7688
31. 0·1936	**32.** 0·9439	**33.** 0·8587
34. 0·2782	**35.** 0·8305	**36.** 0·1524
37. 0·4263	**38.** 0·0860	**39.** 0·2150
40. 0·1103		

Table of natural cosines

	A	B										C (Subtract)					
		0′	6′	12′	18′	24′	30′	36′	42′	48′	54′	1′	2′	3′	4′	5′	
	59°	.5150	5135	5120	5105	5090	5075	5060	5045	5030	5015	3	5	8	10	13	
(a) →	60°	.5000	4985	4970	4955	4939	4924	4909	4894	4879	4863	3	5	8	10	13	
	61°	.4848	4833	4818	4802	4787	4772	4756	4741	4726	4710	3	5	8	10	13	
(b) →	62°	.4695	4679	4664	4648	4633	4617	4602	4586	4571	4555	3	5	8	10	13	←
	63°	.4540	4524	4509	4493	4478	4462	4446	4431	4415	4399	3	5	8	10	13	
	64°	.4384	4368	4352	4337	4321	4305	4289	4274	4258	4242	3	5	8	11	13	
(c) --→	65°	.4226	4210	4195	4179	4163	4147	4131	4115	4099	4083	3	5	8	11	13	←--

In a right-angled triangle, the sides are named as follows:
The *hypotenuse* is the side opposite the right angle.
The *opposite* side is always opposite to a marked or given angle.
The *adjacent* side is always next to the given angle.

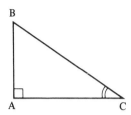

BC is the hypotenuse;
AB is the opposite side;
AC is the adjacent side.

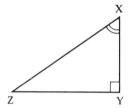

XZ is the hypotenuse;
YZ is the opposite side;
YX is the adjacent side.

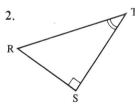

QR is the hypotenuse;
PR is the opposite side;
PQ is the adjacent side.

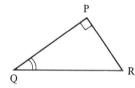

Exercise 7.2a

For each triangle, state which is the hypotenuse, the opposite side, and the adjacent side with respect to the marked angle.

1.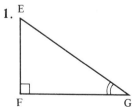

2.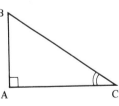

3.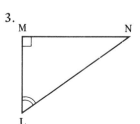

4.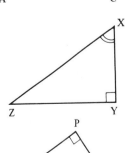

5.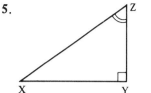

6.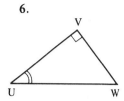

7.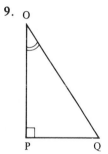

8.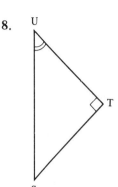

9.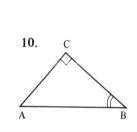

10.

The ratio $\dfrac{\text{opposite}}{\text{adjacent}}$ is called the tangent (or tan) of the angle.

$$\tan \hat{C} = \frac{AB}{AC}$$

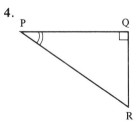

$$\tan \hat{X} = \frac{YZ}{YX}$$

$$\tan \hat{Q} = \frac{PR}{PQ}$$

Exercise 7.2b

Write down the tangent ratio of each angle marked with a small letter.

1.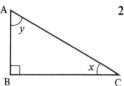

2.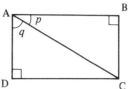

3.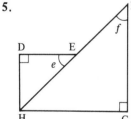

4.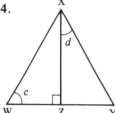

5.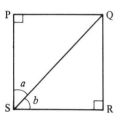

6.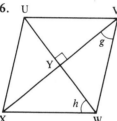

7.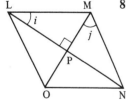

8.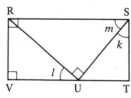

9.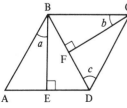

10. Give *two* answers for each angle.

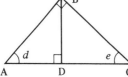

Follow the steps below in using the tangent ratio to find a missing side or a missing angle in a right-angled triangle.

1. Name the sides from the given angle (see Exercise 7.2a).
2. Write down the ratio (see Exercise 7.2b).
3. Solve the equation as shown in the following examples, using tables of natural tangents.

(a) Finding a missing side.

Example 1a

Find the length of AB in the diagram.

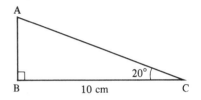

1. AB is the opposite side; BC is the adjacent side.

2. $\dfrac{AB}{BC} = \tan \hat{C}$

3. $\dfrac{AB}{10} = \tan 20°$

$AB = 10 \times \tan 20° = 10 \times 0\cdot3640$
$\qquad\qquad\qquad\quad = 3\cdot64 \text{ cm}$

Example 1b

Find the length of BC in the diagram.

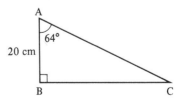

1. BC is the opposite side; AB is the adjacent side.

2. $\dfrac{BC}{AB} = \tan \hat{A} = \dfrac{BC}{20}$

3. $BC = 20 \times \tan 64° = 20 \times 2\cdot050$
$\qquad\qquad\qquad\qquad\quad = 41 \text{ cm}$

Exercise 7.2c

In this exercise, triangle ABC is right-angled at B.

Find BC in questions **1** to **10**.

1. AB = 5 cm, $\hat{A} = 35°$
2. AB = 8 cm, $\hat{A} = 31°$
3. AB = 16 cm, $\hat{A} = 42°$
4. AB = 25 mm, $\hat{A} = 27°$
5. AB = 4 m, $\hat{A} = 53°$
6. AB = 15 mm, $\hat{A} = 61°$
7. AB = 18 mm, $\hat{A} = 17°$
8. AB = 9 cm, $\hat{C} = 28°$
9. AB = 12 cm, $\hat{C} = 42°$
10. AB = 7 m, $\hat{C} = 68°$

Find AB in questions **11** to **18**.

11. BC = 16 cm, $\hat{C} = 13°$
12. BC = 12 cm, $\hat{C} = 22°$
13. BC = 5 cm, $\hat{C} = 39°$
14. BC = 24 mm, $\hat{C} = 58°$
15. BC = 11 mm, $\hat{C} = 66°$
16. BC = 3 m, $\hat{A} = 24°$
17. BC = 4 m, $\hat{A} = 29°$
18. BC = 6 cm, $\hat{A} = 63°$

In questions **19** to **22**, find the height h metres of the tower, given its angle of $e°$ from a point x metres from the base of the tower.

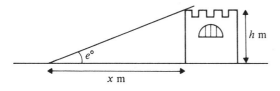

19. $x = 120$ m, $e = 12°$
20. $x = 90$ m, $e = 8°$
21. $x = 50$ m, $e = 20°$
22. $x = 80$ m, $e = 10°$

(b) Finding a missing angle.

Example 2a

Find the size of angle BAC in the diagram.

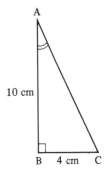

1. BC is the opposite side; AB is the adjacent side.

2. $\dfrac{BC}{AB} = \tan B\hat{A}C$

3. $\dfrac{4}{10} = \tan B\hat{A}C = 0·4$

   So B$\hat{A}$C is the angle whose tangent is 0·4. From tables of natural tangents, B$\hat{A}$C = 21° 48′

Example 2b

Find the size of $\hat{Z}$ in the diagram.

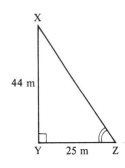

1. XY is the opposite side; YZ is the adjacent side.

2. $\dfrac{XY}{YZ} = \tan \hat{Z}$

3. $\dfrac{44}{25} = \tan \hat{Z} = 1·76$

   So $\hat{Z}$ is the angle whose tangent is 1·76. From tables $\hat{Z} = 60°$ 24′

Exercise 7.2d

In this exercise, triangle PQR is right-angled at $\hat{Q}$.

Find the size of $\hat{P}$ in questions **1** to **7**.

1. QR = 12 cm, PQ = 30 cm
2. QR = 24 cm, PQ = 80 cm
3. QR = 1·8 m, PQ = 5 m
4. QR = 4·2 m, PQ = 4·8 m
5. QR = 33 mm, PQ = 75 mm
6. QR = 85 mm, PQ = 50 mm
7. QR = 132 mm, PQ = 80 mm

Find the size of $\hat{R}$ in questions **8** to **12**.

8. PQ = 85 cm, QR = 125 cm

9. PQ = 138 cm, QR = 150 cm
10. PQ = 141 mm, QR = 120 mm
11. PQ = 7·05 mm, QR = 3 m
12. PQ = 2·01 m, QR = 1·5 m
13. A cliff is 36 m high. What is the angle of elevation of the top of the cliff at a point 120 m from its base?
14. A church steeple is 27 m high. Find its angle of elevation from a point 75 m from its base.
15. A tree is 6·4 m tall. What is its angle of elevation 16 m from the foot of the trunk?
16. Find the angle of elevation of the top of a flagpole from a point 18 m from its base if the pole is 5·4 m tall.

7.3 THE SINE RATIO

The ratio $\dfrac{\text{opposite}}{\text{hypotenuse}}$ is called the sine

(or sin) of the angle.

$$\sin \hat{C} = \frac{AB}{BC}$$

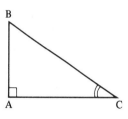

$$\sin \hat{X} = \frac{ZY}{ZX}$$

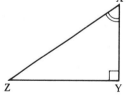

$$\sin \hat{Q} = \frac{RP}{RQ}$$

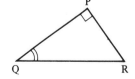

Exercise 7.3a

Write down the sine ratio of each angle marked with a small letter.

1.
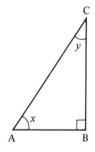

2.
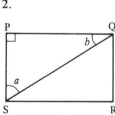

3.
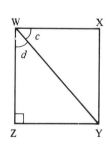

4.
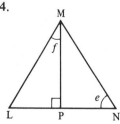

5.
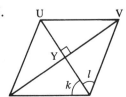

6.
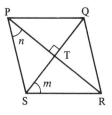

7.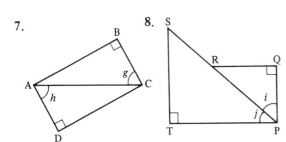

8.

9.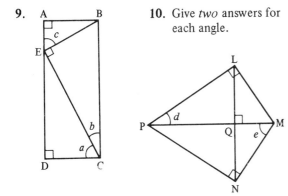

10. Give *two* answers for each angle.

Follow the steps below in using the sine ratio to find a missing side or a missing angle in a right-angled triangle.

1. Name the sides from the given angle (see Exercise 7.2a).
2. Write down the ratio (see Exercise 7.3a).
3. Solve the equation as shown in the following examples, using tables of natural sines.

(a) **Finding a missing side.**

Example 1

Find the length of AB in the diagram.

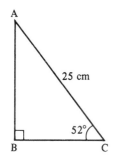

1. AB is the opposite side;
   AC is the hypotenuse.

2. $\dfrac{AB}{AC} = \sin \hat{C}$

3. $\dfrac{AB}{25} = \sin 52°$

   $\therefore AB = 25 \times \sin 52°$
   $= 25 \times 0{\cdot}7880 = 19{\cdot}7$ cm

Exercise 7.3b

In this exercise, triangle ABC is right-angled at B.

Find AB in questions **1** to **8**.

1. $AC = 8$ cm, $\hat{C} = 45°$
2. $AC = 4$ cm, $\hat{C} = 72°$
3. $AC = 6$ cm, $\hat{C} = 37°$
4. $AC = 7$ m, $\hat{C} = 18°$
5. $AC = 9$ m, $\hat{C} = 76°$
6. $AC = 12$ m, $\hat{C} = 50°$
7. $AC = 15$ mm, $\hat{C} = 27°$
8. $AC = 24$ mm, $\hat{C} = 11°$

Find BC in questions **9** to **15**.

9. $AC = 5$ cm, $\hat{A} = 60°$
10. $AC = 3$ cm, $\hat{A} = 23°$
11. $AC = 8$ cm, $\hat{A} = 67°$
12. $AC = 4$ m, $\hat{A} = 78°$
13. $AC = 11$ m, $\hat{A} = 31°$
14. $AC = 16$ mm, $\hat{A} = 20°$
15. $AC = 25$ mm, $\hat{A} = 43°$
16. How high up a vertical wall will a 5 m ladder reach if its foot makes an angle of 52° with the horizontal ground?
17. The supporting wire of a pole is 20 m long. If this wire makes an angle of 40° with the ground, how high is the pole?
18. The string of a kite is 50 m long and makes an angle of 60° with the ground. What is the vertical height of the kite?

(b) Finding a missing angle.

Example 2

Find the size of angle YZX in the diagram.

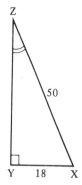

1. YX is the opposite side;
   XZ is the hypotenuse.

2. $\dfrac{YX}{XZ} = \sin Y\hat{Z}X$

3. $\dfrac{18}{50} = \sin Y\hat{Z}X$

$\therefore \sin Y\hat{Z}X = 0.36$

and $Y\hat{Z}X = 21°\ 6'$

Exercise 7.3c

In this exercise, triangle XYZ is right-angled at Y.

Find the size of $\hat{Z}$ in questions **1** to **8**.

1. XY = 56 mm, XZ = 112 mm
2. XY = 92 mm, XZ = 400 mm
3. XY = 132 mm, XZ = 200 mm
4. XY = 27 cm, XZ = 75 cm
5. XY = 24 cm, XZ = 120 cm
6. XY = 13·5 cm, XZ = 18 cm
7. XY = 17·5 cm, XZ = 50 cm
8. XY = 2·5 m, XZ = 4 m

Find the size of $\hat{X}$ in questions **9** to **15**.

9. ZY = 405 mm, XZ = 500 mm
10. ZY = 69 mm, XZ = 120 mm
11. ZY = 18 mm, XZ = 240 mm
12. ZY = 3·2 cm, XZ = 4 cm
13. ZY = 5·4 cm, XZ = 6 cm
14. ZY = 0·25 m, XZ = 2 m
15. ZY = 1·05 m, XZ = 1·2 m
16. A ladder 10 m long reaches 8 m up a vertical wall. What angle does the foot of the ladder make with the ground?
17. The two equal sides of an isosceles triangle are 20 cm long. If the vertical height of the triangle is 15 m, find the size of the equal angles.

7.4 THE COSINE RATIO

The ratio $\dfrac{\text{adjacent}}{\text{hypotenuse}}$ is called the cosine

(or cos) of the angle.

$\cos \hat{C} = \dfrac{CA}{CB}$

$\cos \hat{X} = \dfrac{XY}{XZ}$

$\cos \hat{Q} = \dfrac{QP}{QR}$

Exercise 7.4a

Write down the cosine ratio of each angle marked with a small letter.

1.   2.

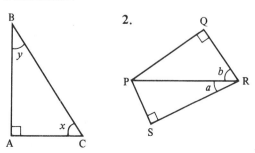

3.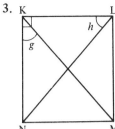

4.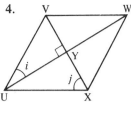

5.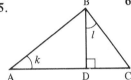

6.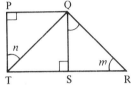

7.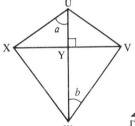

8. Give *two* answers for each angle.

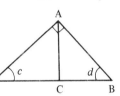

Follow the steps below in using the cosine ratio to find a missing side or a missing angle in a right-angled triangle.

1. Name the sides from the given angle (see Exercise 7.2a).
2. Write down the ratio (see Exercise 7.4a).
3. Solve the equation as shown in the following examples, using tables of natural cosines.

(a) Finding a missing side.

Example 1

Find the length of XY in the diagram.

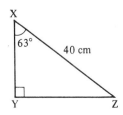

1. XY is the adjacent side; XZ is the hypotenuse.

2. $\dfrac{XY}{XZ} = \cos \hat{X}$

3. $\dfrac{XY}{40} = \cos 63°$

$\therefore XY = 40 \times \cos 63°$
$= 40 \times 0{\cdot}4540 = 18{\cdot}16$ cm

Exercise 7.4b

In this exercise, triangle XYZ is right angled at $\hat{Y}$. Find XY in questions **1** to **9**.

1. $XZ = 3$ cm, $\hat{X} = 23°$
2. $XZ = 6$ cm, $\hat{X} = 8°$
3. $XZ = 5$ cm, $\hat{X} = 14°$
4. $XZ = 9$ m, $\hat{X} = 67°$
5. $XZ = 3$ m, $\hat{X} = 83°$
6. $XZ = 6$ m, $\hat{X} = 42°$
7. $XZ = 20$ mm, $\hat{X} = 79°$
8. $XZ = 15$ mm, $\hat{X} = 84°$
9. $XZ = 24$ mm, $\hat{X} = 25°$

Find YZ in questions **10** to **15**.

10. $XZ = 5$ cm, $\hat{Z} = 70°$
11. $XZ = 4$ cm, $\hat{Z} = 63°$
12. $XZ = 8$ cm, $\hat{Z} = 47°$
13. $XZ = 30$ mm, $\hat{Z} = 16°$
14. $XZ = 11$ mm, $\hat{Z} = 36°$
15. $XZ = 18$ mm, $\hat{Z} = 27°$

16. A ladder 10 m long makes an angle of 30° with a vertical wall. How far up the wall does the ladder reach?
17. The diagonal of a rectangle makes an angle of 60° with the shorter side. If the diagonal is 13 cm long, how long is the shorter side?

(b) Finding the missing angle.

Example 2

Find the size of angle ABC in the diagram.

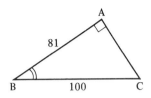

1. AB is the adjacent side;
   BC is the hypotenuse.

2. $\dfrac{AB}{BC} = \cos A\hat{B}C$

3. $\dfrac{81}{100} = \cos A\hat{B}C$

   $\therefore \cos A\hat{B}C = 0{\cdot}81$

   and $A\hat{B}C = 35° \, 54'$

Exercise 7.4c

In this exercise, triangle ABC is right-angled at A.
Find the size of $\hat{B}$ in questions **1** to **8**.

1. AB = 45 mm, BC = 125 mm
2. AB = 76 mm, BC = 80 mm
3. AB = 69 mm, BC = 300 mm
4. AB = 396 mm, BC = 400 mm
5. AB = 54 cm, BC = 240 cm
6. AB = 24·9 cm, BC = 30 cm
7. AB = 16 cm, BC = 80 cm
8. AB = 6·6 cm, BC = 12 cm

Find the size of $\hat{C}$ in questions **9** to **15**.

9. AC = 68 mm, BC = 136 mm
10. AC = 115 mm, BC = 200 mm
11. AC = 16·5 cm, BC = 25 cm
12. AC = 4·86 m, BC = 6 m
13. AC = 4·55 m, BC = 5 m
14. AC = 24 cm, BC = 60 cm
15. AC = 40 mm, BC = 64 mm

7.5 RIGHT-ANGLED TRIANGLES

In some questions, the side length of a triangle will have to be found by dividing the given length by the ratio. This is generally done using logarithms for the calculation.

Example 1

Find the length of AC from the diagram.

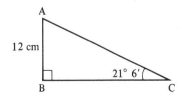

1. AB is the opposite side;
   AC is the hypotenuse.

2. $\dfrac{AB}{BC} = \sin A\hat{C}B$

3. $\dfrac{12}{AC} = \sin 21° \, 6' = 0{\cdot}36$

   $\therefore AC = \dfrac{12}{0{\cdot}36}$

   $= 33{\cdot}34$ cm

Questions using the cosine ratio can be done in a similar way.

Exercise 7.5a

In questions **1** to **8**, find AC in the triangle ABC where $\hat{B} = 90°$.

1. AB = 6 cm, $\hat{C} = 30°$
2. AB = 198 mm, $\hat{C} = 41° \, 18'$
3. AB = 220 mm, $\hat{C} = 33° \, 22'$
4. AB = 1·5 m, $\hat{C} = 48° \, 35'$
5. BC = 9 cm, $\hat{A} = 13°$

6. BC = 364 mm, $\hat{A}$ = 65° 30′

7. BC = 1·6 m, $\hat{A}$ = 11° 32′

8. BC = 2·4 m, $\hat{A}$ = 28° 41′

In questions **9** to **15**, find XZ in the triangle XYZ where $\hat{Y}$ = 90°.

9. YZ = 38 mm, $\hat{Z}$ = 18° 12′

10. YZ = 115 mm, $\hat{Z}$ = 76° 42′

11. YZ = 18 cm, $\hat{Z}$ = 66° 25′

12. YZ = 3·4 m, $\hat{Z}$ = 31° 47′

13. XY = 486 mm, $\hat{X}$ = 35° 54′

14. XY = 297 mm, $\hat{X}$ = 8° 6′

15. XY = 9·6 cm, $\hat{X}$ = 36° 52′

Example 2

A wire stay makes an angle of 15° with a vertical wall. How far up the wall does it reach if the other end of the stay is 8 m from the wall?

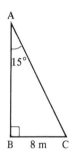

Method 1

1. BC is the opposite side:
   AB is the adjacent side.

2. $\dfrac{BC}{AB}$ = tan $B\hat{A}C$

3. $\dfrac{8}{AB}$ = tan 15° = 0·2679

   $\therefore$ AB = $\dfrac{8}{0·2679}$

   = 29·86 m

So the stay is fastened 29·9 m from the foot of the wall (to 3 s.f.).

Method 2

Angle ACB = 90° − 15° = 75°.
Using this angle:

$\dfrac{AB}{8}$ = tan 75°

AB = 8 × tan 75°
= 8 × 3·732 = 29·856 m

So the stay is fastened 29·9 m from the foot of the wall (to 3 s.f.).

Exercise 7.5b

1. In the diagram a ladder is shown resting against a vertical wall.

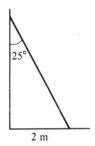

Find from the diagram:

(a) the height that the ladder reaches up the wall,
(b) the length of the ladder.

2. The diagram shows the jib of a crane and its hook.

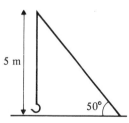

Find from the diagram:

(a) the distance of the hook from the base of the jib,
(b) the length of the jib.

3. The side of a lean-to shed is shown against a vertical wall in the diagram.

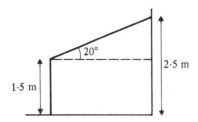

Find from the diagram the width of the side of the shed.

4. A canopy roller blind extends from the front of a shop as shown.

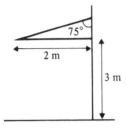

Find from the diagram:

(a) the height of the roller above the ground,
(b) the unrolled length of blind.

5. The plan of a bay window is shown in the diagram.

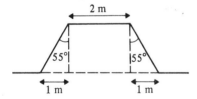

Find from the diagram:

(a) the distance the window projects from the wall of the room,
(b) the total length of window frame.

6. The side view of a road bridge is shown.

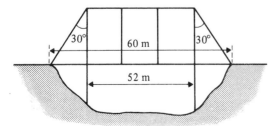

Find the height of the top of the bridge from the road.

7. The side of a terrace of houses is shown.

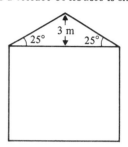

Find the width of the side of the houses.

8. Find the height of the church steeple from the diagram.

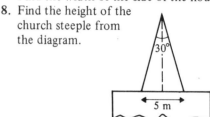

9. The diagram shows a children's playground slide.

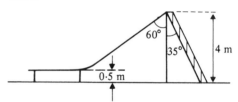

Find from the diagram the distance from the main support to:

(a) the foot of the steps,
(b) the end of the slope.

10. The frame of a boy's bicycle is shown in the diagram.

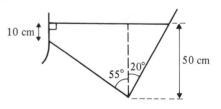

Find the length of the cross bar.

When a diagram is not given, always draw a suitable right-angled triangle and letter this triangle before attempting to answer the question.

Example 3

A ship is due South of a lighthouse and sails on a heading of 071° for a distance of 57 km until it is due East of the lighthouse. How far is it now from the lighthouse?

Drawing

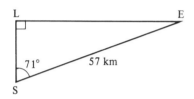

In triangle LSE

$$\sin 71° = \frac{LE}{57}$$

$$\therefore 57 \times \sin 71° = LE$$
$$57 \times 0{\cdot}9455 = LE$$
$$53{\cdot}89 = LE$$

So the ship is 53·9 km from the lighthouse (to 3 s.f.).

Exercise 7.5c

1. A ship is due South of a lighthouse. It sails on a heading of 050° for a distance of 60 km until it is due East of the lighthouse. How far is it now from the lighthouse?

2. A helicopter is due South of a beacon. It flies on a heading of 013° for a distance of 200 km until it is due East of the beacon.
   (a) How far is it now from the beacon?
   (b) How far was it originally from the beacon?

3. An aeroplane is due North of a beacon. It flies on a heading of 104° for a distance of 400 km until it is due East of the beacon. How far is it now from the beacon?

4. A ship is due North of a lighthouse. It sails on a heading of 162° for a distance of 90 km until it is due East of the lighthouse.
   (a) How far is it now from the lighthouse?
   (b) How far was it originally from the lighthouse?

5. A ship is due North of a lighthouse. It sails on a heading of 211° for a distance of 70 km until it is due West of the lighthouse, how far is it now from the lighthouse?

6. A desert traveller on a camel is due North of an oasis and he travels on a heading of 262° for a distance of 30 km until he is due West of the oasis.
   (a) How far is he now from the oasis?
   (b) How far was he originally from the oasis?

7. A helicopter is due South of a beacon and it flies on a heading of 340° for a distance of 120 km until it is due West of the beacon. How far is it now from the beacon?

8. Birmingham is due South of Bradford and Hull is due East of Bradford. Hull is 180 km from Birmingham and its bearing from Birmingham is 032°. Find:
   (a) the distance from Bradford to Hull,
   (b) the distance from Bradford to Birmingham.

9. Bristol is due South of Liverpool and Sheffield is due East of Liverpool. Sheffield is 250 km from Bristol and its bearing from Bristol is 024°. Find:
   (a) the distance from Liverpool to Sheffield,
   (b) the distance from Liverpool to Bristol.

10. Shrewsbury is due North of Ludlow and Kidderminster is due East of Ludlow. Kidderminster is 55 km from Shrewsbury and its bearing from Shrewsbury is 139°. Find:
    (a) the distance from Ludlow to Kidderminster,
    (b) the distance from Ludlow to Shrewsbury.

11. Bangor is due North of Swansea and Newport is due East of Swansea. Newport is 200 km from Bangor and its bearing from Bangor is 159°. Find:
    (a) the distance from Swansea to Newport,
    (b) the distance from Swansea to Bangor.

12. Dundee is due North of Carlisle and Stranraer is due West of Carlisle. Stranraer is 210 km from Dundee and its bearing from Dundee is 217°. Find:
    (a) the distance from Carlisle to Stranraer,
    (b) the distance from Carlisle to Dundee.

13. Filey is due North of Hull and Leeds is due West of Hull. Leeds is 90 km from Filey and its bearing from Filey is 238°. Find:
    (a) the distance from Hull to Leeds,
    (b) the distance from Hull to Filey.

In problems involving three dimensions, the following angles may have to be found.

1. Angle between a line and a plane.

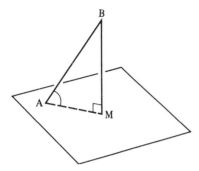

If a line AB meets a plane at A, and a perpendicular is drawn from B to the plane at M, the required angle is BÂM.

2. Perpendicular to a plane.

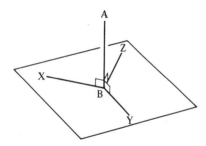

If AB is perpendicular to the plane, then any line drawn in this plane to meet the perpendicular at B is at right angles to it.

$$X\hat{B}A = Y\hat{B}A = Z\hat{B}A = 90°$$

3. Angle between two planes.

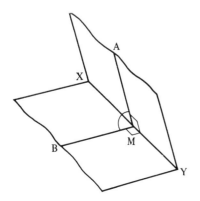

If XY is the line of intersection of the two planes, and AM and BM are perpendiculars to XY, then BMA is the angle between the two planes.

Example 1

In the solid illustrated below, name the angle between:

(a) AG and the base ABCD
(b) AG and the face GCDH
(c) face FGHE and plane FADG

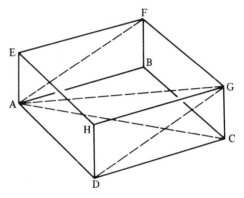

The required angles are:

(a) GÂC

(b) AĜD

(c) AF̂E or DĜH

Exercise 7.6a

1. In the cuboid illustrated, name the angle between:

(a) US and the base PQRS
(b) US and the face WVRS
(c) US and the face UTPQ
(d) the face TWSP and the plane PUVS

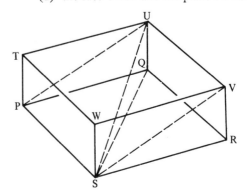

2. In the cuboid illustrated, name the angle between:

   (a) PM and the base KLMN
   (b) PM and the face SRMN
   (c) PM and the face PQLK
   (d) the base KLMN and the plane PSML

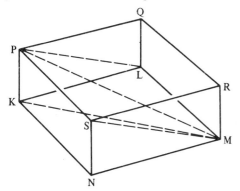

3. In the cube illustrated, name the angle between:

   (a) AG and the base ABCD
   (b) AG and the face FEAB
   (c) AG and the face GHDC
   (d) the face BFGC and the plane AFGD

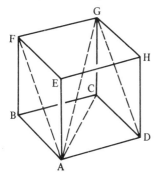

4. In the square-based pyramid shown, name the angle between:

   (a) FB and the base ABCD
   (b) EF and the face AFB
   (c) EG and the face AFB
   (d) the face AFB and the base ABCD

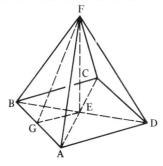

5. In the square-based pyramid shown, name the angle between:

   (a) QM and the base KLMN
   (b) PQ and the face LQM
   (c) PR and the face LQM
   (d) the face LQM and the base KLMN

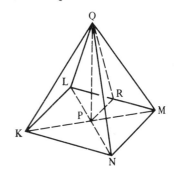

Example 2

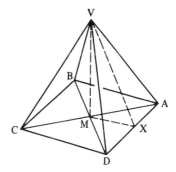

VACBD is a square-based pyramid of side 12 cm. If VA = VB = VC = VD = 15 cm, calculate:

(a) the length of AC
(b) the vertical height VM
(c) the angle between the edge VA and the base
(d) the angle between the face VAD and the base

(a) Length of AC. Use triangle ABC, right-angled at B.

$$AC^2 = AB^2 + BC^2$$
$$= 12^2 + 12^2$$
$$AC = \sqrt{288}$$
$$= 16 \cdot 98 \text{ cm}$$

(b) Vertical height VM. Use triangle VMA, right-angled at M.

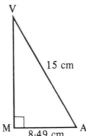

$MA = \frac{1}{2}AC = 8\cdot49$

$VM^2 = VA^2 - MA^2$
$\quad\quad = 15^2 - 8\cdot49^2$
$\quad\quad = 225 - 72 = 153$
$VM = \sqrt{153}$
$\quad\quad = 12\cdot37$ cm

(c) Angle between VA and base is $V\hat{A}M$

$\sin VAM = \dfrac{12\cdot37}{15} = 0\cdot8247$

$V\hat{A}M = 55°\ 33'$

(d) Angle between face VAD and base is $V\hat{X}M$

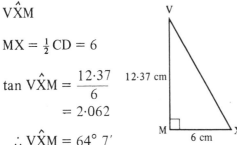

$MX = \frac{1}{2}CD = 6$

$\tan V\hat{X}M = \dfrac{12\cdot37}{6}$
$\quad\quad\quad = 2\cdot062$

$\therefore V\hat{X}M = 64°\ 7'$

Exercise 7.6b

1. In the cuboid shown, find:

   (a) the length of BD
   (b) the angle between FD and the base ABCD.

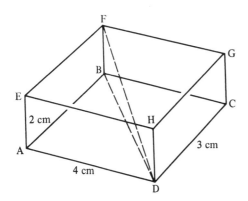

2. In the cube shown, find:

   (a) the length of QK
   (b) the length of RK
   (c) the angle between RK and the face QPKL.

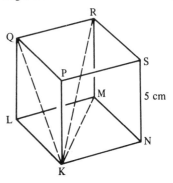

3. In the cuboid shown, find:

   (a) the length of TQ
   (b) the angle between TR and the face TUQP.

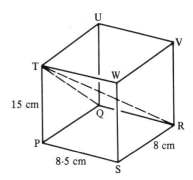

4. In the cube shown, find:

   (a) the length of BD
   (b) the angle between HB and the base ABCD
   (c) the angle between the base ABCD and the plane BEHC

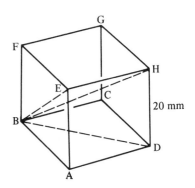

5. In the cuboid shown, find:

   (a) the angle between the base LMNP and the plane PQRN
   (b) the length of QP
   (c) the length of RP
   (d) the angle between RP and the face LQTP.

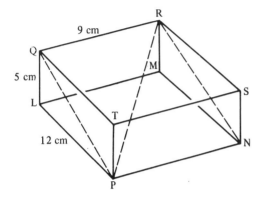

7. In the square-based pyramid shown, find:

   (a) the length of EG
   (b) the angle between EF and the face AFB
   (c) the angle between the base ABCD and the face AFB.

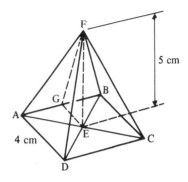

6. In the cuboid shown, find:

   (a) the angle between the base PQRS and the plane PQYZ
   (b) the length of PR
   (c) the angle between PY and the base PQRS.

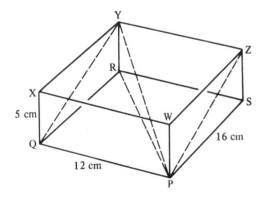

8. In the square-based pyramid shown, find:

   (a) the length of (i) PM; (ii) LM
   (b) the length of PQ
   (c) the length of PR
   (d) the angle between PQ and the face NQM
   (e) the angle between the base KLMN and the face NQM.

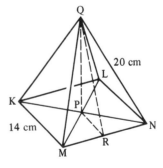

In any number base N the column headings are

$$\ldots N^4 \,; N^3 \,; N^2 \,; N\,; 1\,; \ldots$$

Example 1

Write the number 4322_{ten} in column form.

Power	10^3	10^2	10	1
Column value	1000's	100's	10's	units
$4322_{\text{ten}} =$	4 ·	3	2	2

Example 2

Write the number 110101_{two} in column form.

Power	2^5	2^4	2^3	2^2	2	1
Column value	32's	16's	8's	4's	2's	units
$110101_{\text{two}} =$	1	1	0	1	0	1

Exercise 8.1a

Write the following numbers in column form.

1. 3654_{ten} 2. 4037_{ten} 3. 905_{ten}

4. 2431_{five} 5. 4032_{five} 6. 340_{five}

7. 34_{five} 8. 101101_{two} 9. 10110_{two}

10. 1011_{two} 11. 111_{two} 12. 20121_{three}

13. 1202_{three} 14. 210_{three} 15. 3231_{four}

16. 2303_{four} 17. 320_{four} 18. 2547_{eight}

19. 650_{eight} 20. 4351_{six}

By writing a number in column form, it is easy to change a number written in any base to its equivalent in base ten.

Example 3

Change 4026_{eight} into base ten.

Power	8^3	8^2	8	1
Column value	512's	64's	8's	units
$4026_{\text{eight}} =$	4	0	2	6
Value	2048	0	16	6

Hence

$$\begin{aligned}
4026_{\text{eight}} &= (4 \times 512) + (0 \times 64) + \\
&\qquad (2 \times 8) + (6 \times 1) \\
&= 2048 + 0 + 16 + 6 \\
&= 2070_{\text{ten}}
\end{aligned}$$

Example 4

Find the 'odd value out' in base ten of
$32_{\text{eight}}\,; 101_{\text{five}}\,; 221_{\text{three}}$.

$$\begin{aligned}
32_{\text{eight}} &= (3 \times 8) + (2 \times 1) \\
&= 24 + 2 = 26
\end{aligned}$$

$$101_{\text{five}} = (1 \times 25) + (1 \times 1) = 25 + 1 = 26$$

$$\begin{aligned}
221_{\text{three}} &= (2 \times 9) + (2 \times 3) + (1 \times 1) \\
&= 18 + 6 + 1 = 25
\end{aligned}$$

So 221_{three} is the 'odd value out' because its base-ten value is 25.

Exercise 8.1b

In questions 1 to 10 change each number to base ten.

1. 1241_{five} 2. 2230_{five} 3. 312_{five}

4. 100101_{two} 5. 11010_{two} 6. 1110_{two}

7. 1231_{four} 8. 1042_{six} 9. 272_{eight}

10. 2021_{three}

In questions 11 to 20 change each number to base 10 and find the 'odd value out'.

11. (a) 10100_{two} (b) 102_{four} (c) 33_{five}

12. (a) 11101_{two} (b) 131_{four} (c) 102_{five}

13. (a) 110100_{two} (b) 302_{four} (c) 200_{five}

14. (a) 1102_{three} (b) 100111_{two} (c) 47_{eight}

15. (a) 1200_{three} (b) 134_{five} (c) 55_{eight}

16. (a) 2002_{three} (b) 321_{four} (c) 71_{eight}

17. (a) 52_{six} (b) 200_{four} (c) 11110_{two}

18. (a) 115_{six} (b) 101110_{two} (c) 142_{five}

19. (a) 140_{six} (b) 2022_{three} (c) 330_{four}

20. (a) 244_{five} (b) 1023_{four} (c) 2210_{three}

A number written in base ten is changed into another base by repeated division by the number of the required base.

Example 5

Change 222_{ten} into base eight.

$$8\,\overline{)\,222}$$
$$8\,\overline{)\,27} \quad + \quad 6 \text{ units (1)}$$
$$8\,\overline{)\,3} \quad + \quad 3 \text{ eights (8)}$$
$$0 \quad + \quad 3 \text{ sixty-fours } (8)^2$$

The remainders are read upwards to give the answer.

Hence $222_{ten} = 336_{eight}$

Exercise 8.1c

Change each base ten number into the required base.

1. 328 to base five	2. 556 to base five
3. 435 to base five	4. 270 to base five
5. 46 to base two	6. 39 to base two
7. 25 to base two	8. 19 to base two
9. 318 to base four	10. 431 to base four
11. 192 to base four	12. 725 to base six
13. 438 to base six	14. 157 to base six
15. 104 to base three	16. 51 to base three
17. 96 to base three	18. 624 to base eight
19. 336 to base eight	20. 211 to base eight

When changing a base ten number into base twelve, use T for 10 and E for 11.

21. 146 to base twelve	22. 432 to base twelve
23. 213 to base twelve	24. 132 to base twelve
25. 1728 to base twelve	26. 730 to base twelve.

To change a number in one base to another base, first change the number into base ten and then into the new base.

Example 6

Change 1003_{six} into base twelve.

1. 1003_{six} into base ten.

Power	6^3	6^2	6	1
Column value	216's	36's	6's	units
$1003_{six} =$	1	0	0	3

$$\therefore 1003_{six} = (1 \times 216) + (3 \times 1)$$
$$= 216 + 3 = 219_{ten}$$

2. Base ten to base twelve.

$$12\,\overline{)\,219}$$
$$12\,\overline{)\,18} \quad + \quad 3 \text{ units (1)}$$
$$12\,\overline{)\,1} \quad + \quad 6 \text{ twelves (12)}$$
$$0 \quad + \quad 1 \text{ one-hundred-and-forty-four } (12)^2$$

Reading upwards:

$$219_{ten} = 163_{twelve}$$

Hence $1003_{six} = 163_{twelve}$

Exercise 8.1d

1. Change 144_{five} to base two.

2. Change 132_{five} to base two.

3. Change 202_{five} to base two.

4. Change 111101_{two} to base five.

5. Change 100111_{two} to base five.

6. Change 101001_{two} to base five.

7. Change 2033_{four} to base six.

8. Change 2201_{four} to base six.

9. Change 1332_{four} to base six.

10. Change 2130_{four} to base eight.

11. Change 3002_{four} to base eight.

12. Change 332_{four} to base eight.

Shown below is a way of listing the members (or elements) of the set of odd numbers which are smaller than 12.

A = {1, 3, 5, 7, 9, 11}

Exercise 8.2a

List the members of the sets in questions **1** to **6**.

1. { odd numbers smaller than 10 }
2. { even numbers smaller than 12 }
3. { multiples of 5 smaller than 31 }
4. { square numbers less than 20 }
5. { days of the week beginning with T }
6. { vowels }

Describe the sets in questions **7** to **12**.

7. A = { 1, 3, 5, 7 }
8. B = { 3, 6, 9, 12, 15 }
9. C = { Saturday, Sunday }
10. D = { April, August }
11. E = { 2, 3, 5, 7, 11 }
12. F = { red, orange, yellow, green, blue, indigo, violet }

We can name all the possible days of the week as the *universal set* ($\mathcal{E}$)

$\mathcal{E}$ = { Sun, Mon, Tue, Wed, Thurs, Fri, Sat }

If A = { Sat, Sun }, then set A is called a *subset* of this universal set. This is written A ⊂ $\mathcal{E}$

Exercise 8.2b

List the elements in questions **1** to **6** if $\mathcal{E}$ = { months of the year }.

1. A = { months with 30 days }
2. B = { months whose first letter is J }
3. C = { months whose last letter is Y }
4. D = { months with five-letter names }
5. E = { the shortest month }
6. F = { months with six-letter names }

In questions **7** to **12**, give a suitable universal set to include the subset.

7. { Monday, Tuesday }
8. { April, May }
9. { a, e, i, o, u }
10. { apples, oranges, bananas }
11. { North, South, East }
12. { 1952. 1956, 1960, 1964 }

A set that contains no elements is called an empty set. Thus the subset F = { months with 29 days } is an empty set for three out of every four years.

This is written F = { }

or F = ϕ

Exercise 8.2c

State which of the following are empty sets.

1. { odd numbers divisible by 2 }
2. { prime numbers which are even }
3. { cubes with 8 faces }
4. { days with 1440 minutes }
5. { shapes whose angle sum is 400° }
6. { houses with flat roofs }

$$\text{If } \mathcal{E} = \{ 1, 2, 3, 4, 5, 6, 7, 8, 9, 10 \}$$
$$A = \{ 1, 3, 6, 10 \}$$
$$B = \{ 1, 3, 5, 7, 9 \}$$
$$\text{and } C = \{ 1, 4, 9 \}, \text{ then}$$

1. The *intersection* of set A and B is written A ∩ B = {1, 3}.
   This is the set of elements that are in both set A and set B.
2. The *union* of set B and set C is written B ∪ C = {1, 3, 4, 5, 7, 9}.
   This is the set of elements that are in set B or in set C or in both.
3. The *complement* of set C is written C' = { 2, 3, 5, 6, 7, 8, 10}.
   This is the set of elements that are not in set C.
4. The *number* of elements in set B is 5. This is written n(B) = 5.

Example

Using the above sets, list the elements in:

(a) $B \cap C$ (b) $A \cap B \cap C$ (c) $A \cup C$
(d) $A \cup B \cup C$ (e) $(A \cap B)'$

(a) $B \cap C = \{1, 9\}$

(b) $A \cap B \cap C = \{1\}$

(c) $A \cup C = \{1, 3, 4, 6, 9, 10\}$

(d) $A \cup B \cup C = \{1, 3, 4, 5, 6, 7, 9, 10\}$

(e) $(A \cap B) = \{1, 3\}$

      $\therefore (A \cap B)' = \{2, 4, 5, 6, 7, 8, 9, 10\}$

Exercise 8.2d

For questions **1** to **5**, list the elements of
(a) $A \cap B$,   (b) $A \cup B$.

  1. $A = \{1, 2, 3\}$, $B = \{2, 3, 5, 7\}$
  2. $A = \{1, 3, 5, 7\}$, $B = 2, 3, 5, 8\}$
  3. $A = \{$ multiples of 4 less than 20 $\}$,
     $B = \{$ multiples of 6 less than 20 $\}$
  4. $A = \{$ rectangle, square $\}$,
     $B = \{$ rhombus, square $\}$
  5. $A = \{1, 3, 5\}$, $B = \{2, 4, 6\}$

For questions **6** to **10**, find the value of
(a) $n(A)$ (b) $n(B)$ (c) $n(A \cap B)$ (d) $n(A \cup B)$

Also show, for each case, that
$n(A) + n(B) = n(A \cap B) + n(A \cup B)$

  6. $A = \{3, 6, 9, 12, 15\}$, $B = \{5, 10, 15\}$
  7. $A = \{1, 3, 6, 10\}$, $B = \{3, 6, 9, 12\}$
  8. $A = \{1, 4, 9, 16\}$, $B = \{1, 2, 4, 8, 16\}$
  9. $A = \{$ odd numbers less than 10$\}$,
     $B = \{$ prime numbers less than 10$\}$
  10. $A = \{1, 3, 5, 7\}$, $B = \{2, 4, 6, 8\}$

For questions **11** to **14**, list the elements of
(a) A' (b) B' (c) $(A \cap B)'$
(d) $(A \cup B)'$ (e) $A' \cap B'$ (f) $A' \cup B'$

Also show, for each case, that
(i) $(A \cap B)' = A' \cup B'$,   (ii) $(A \cup B)' = A' \cap B'$.

  11. $\&= \{1, 2, 3, 4\}$, $A = \{1, 4\}$, $B = \{1, 3\}$
  12. $\&= \{1, 2, 3, 4, 5, 6\}$, $A = \{1, 2, 4\}$,
      $B = \{1, 3, 6\}$

  13. $\&= \{$ London, Manchester, Cardiff,
         Edinburgh, Glasgow $\}$,
      $A = \{$ capital cities $\}$, $B = \{$ English cities $\}$
  14. $\&= \{1, 2, 3, 4\}$,
      $A = \{1, 2, 3\}$, $B = \{1, 2, 4\}$

For questions **15** to **18**, find the value of
(a) $n(A')$ (b) $n(B')$ (c) $n(A \cap B)'$
(d) $n(A \cup B)'$ (e) $n(A' \cap B')$ (f) $n(A' \cup B')$

Also show, for each case, that
(i) $n(A \cap B)' = n(A' \cup B')$, (ii) $n(A \cup B)' = n(A' \cap B')$.

  15. $\&= \{1, 2, 3, 4, 5\}, A = \{1, 3, 5\}, B = \{1, 4\}$
  16. $\&= \{1, 2, 3, 4, 5, 6, 7, 8\}$, $A = \{1, 2, 4, 7\}$,
      $B = \{2, 4, 6, 8\}$
  17. $\&= \{1, 2, 3, 4, 5\}$,
      $A = \{1, 2, 3, 4\}$, $B = \{1, 3, 5\}$
  18. $\&= \{$ iron, copper, silver, gold $\}$,
      $A = \{$ metals with four-letter names $\}$,
      $B = \{$ metals used in coinage $\}$

For questions **19, 20** and **21**, list the elements of:
(a) $A \cap B$ (b) $B \cap C$ (c) $A \cap C$
(d) $A \cap B \cap C$ (e) $A \cup B$ (f) $B \cup C$
(g) $A \cup C$ (h) $A \cup B \cup C$ (i) A'
(j) B' (k) C' (l) $(A \cap B \cap C)'$
(m) $(A \cup B \cup C)'$

  19. $\&= \{1, 2, 3, 4, 5, 6, 7, 8\}$, $A = \{1, 2, 3, 6\}$,
      $B = \{1, 2, 4, 7\}$ and $C = \{1, 3, 5, 7\}$
  20. $\&= \{1, 2, 3, 4, 5, 6\}$, $A = \{3, 6\}$,
      $B = \{1, 3, 5\}$ and $C = \{2, 5\}$
  21. $\&= \{1, 2, 3, 4\}$, $A = \{1, 2\}$, $B = \{1, 3\}$
      and $C = \{1, 4\}$

For questions **22, 23** and **24**, list the elements of:
(a) $(A \cap B)'$ (b) $(B \cap C)'$ (c) $(A \cap C)'$
(d) $(A \cup B)'$ (e) $(B \cup C)'$ (f) $(A \cup C)'$

  22. $\&= \{1, 2, 3, 4\}$, $A = \{1, 2, 3\}$
      $B = \{1, 2, 4\}$ and $C = \{2, 3, 4\}$
  23. $\&= \{$ bus, train, ship, aeroplane $\}$,
      $A = \{$ those which do not fly $\}$,
      $B = \{$ those which use wheels $\}$ and
      $C = \{$ those which cross oceans $\}$
  24. $\&= \{$ football, rugby, cricket, rounders $\}$,
      $A = \{$ games in which goals are scored $\}$,
      $B = \{$ games played with a round ball $\}$ and
      $C = \{$ games played with a bat $\}$

Sets can be shown pictorially on a Venn diagram. The universal set is usually represented by a rectangle, and subsets are shown as circles within the rectangle. When drawing Venn diagrams, the number of intersections must fit the given data.

Example 1

If & = { 1, 2, 3, 4, 5, 6 } ,

   A = { 1, 3, 5 } , and B = { 2, 3, 5 } ,

Illustrate the information on a Venn diagram.

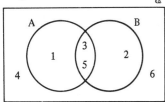

Exercise 8.3a

Show each group of sets on a Venn diagram.

1. & = { 1, 2, 3, 4, 5, 6 } ,
   A = { 1, 2, 6 } , B = { 2, 4, 6 } .
2. & = { 1, 2, 3, 4, 5, 6 } ,
   A = { 2, 4, 5 } , B = { 2, 3, 4 } .
3. & = { 1, 2, 3, 4, 5, 6 } ,
   A = { 3, 4, 6 } , B = { 2, 3, 6 } .
4. & = { 1, 2, 3, 4, 5, 6 } ,
   A = { 2, 5, 6 } , B = { 1, 2, 5 } .
5. & = { 1, 2, 3, 4, 5, 6, 7 } ,
   A = { 1, 3, 4 } , B = { 2, 3, 6 } .
6. & = { 1, 2, 3, 4, 5, 6, 7 }
   A = { 2, 3, 7 } , B = { 1, 2, 4 }
7. & = { 1, 2, 3, 4, 5, 6, 7 }
   A = { 2, 4, 5 } , B = { 3, 5, 7 }
8. & = { 1, 2, 3, 4, 5, 6, 7, 8 } ,
   A = { 1, 3, 5, 8 } , B = { 2, 3, 5, 6 } .
9. & = { 1, 2, 3, 4, 5, 6, 7, 8, } ,
   A = { 2, 3, 4, 7 } , B = { 3, 5, 7, 8 } .
10. & = { 1, 2, 3, 4, 5, 6, 7, 8 } ,
    A = { 3, 4, 6, 7 } , B = { 1, 4, 5, 7 } .

Example 2

If & = { 1, 2, 3, 4, 5, 6 }

   A = { 2, 4, 6 } , and B = { 2, 4 } ,

illustrate the information on a Venn diagram.

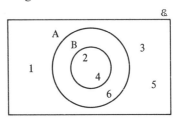

Exercise 8.3b

Illustrate each of the following sets on a Venn diagram.

1. & = { 1, 2, 3, 4, 5, 6 } , A = { 2, 3, 5 } ,
   B = { 2, 5 } .
2. & = { 1, 2, 3, 4, 5, 6 } , A = { 1, 3, 4, 6 } ,
   B = { 1, 4 } .
3. & = { 1, 2, 3, 4, 5, 6 } , A = { 1, 2, 3, 6 } ,
   B = { 2, 3, 6 } .
4. & = { 1, 2, 3, 4, 5, 6 } , A = { 1, 2, 4, 5, 6 } ,
   B = { 1, 2, 4 } .
5. & = { 1, 2, 3, 4, 5, 6, 7 } , A = { 2, 3, 5, 7 } ,
   B = { 2, 5 } .
6. & = { 1, 2, 3, 4, 5, 6, 7 } , A = { 1, 3, 4, 6 } ,
   B = { 1, 4, 6 } .
7. & = { 1, 2, 3, 4, 5, 6, 7 } , A = { 1, 2, 4, 5, 6 } ,
   B = { 1, 2, 5 } .
8. & = { 1, 2, 3, 4, 5, 6, 7, 8 } , A = { 1, 3, 4, 5, 8 }
   B = { 1, 3, 5 } .
9. & = { 1, 2, 3, 4, 5, 6, 7, 8 } ,
   A = { 2, 3, 4, 5, 7 } , B = { 2, 5, 7 } .
10. & = { 1, 2, 3, 4, 5, 6, 7, 8 } ,
    A = { 1, 3, 4, 5, 6, 7 } , B = { 1, 4, 5, 7 } .

Example 3

If $\varepsilon = \{\ 1, 2, 3, 4, 5, 6\ \}$,

$\quad$ A $= \{\ 1, 3, 5\ \}$, and B $= \{\ 2, 4\ \}$,

illustrate the information on a Venn diagram.

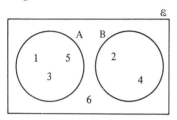

Exercise 8.3c

Illustrate each of the following sets on a Venn diagram.

1. $\varepsilon = \{\ 1, 2, 3, 4, 5, 6\ \}$, A $= \{\ 1, 2, 4\ \}$,
   B $= \{\ 3, 6\ \}$.
2. $\varepsilon = \{\ 1, 2, 3, 4, 5, 6\ \}$, A $= \{\ 2, 4, 6\ \}$,
   B $= \{\ 1, 5\ \}$.
3. $\varepsilon = \{\ 1, 2, 3, 4, 5, 6\ \}$, A $= \{\ 2, 6\ \}$,
   B $= \{\ 3, 4\ \}$.
4. $\varepsilon = \{\ 1, 2, 3, 4, 5, 6\ \}$, A $= \{\ 1, 5\ \}$,
   B $= \{\ 2, 3, 6\ \}$.
5. $\varepsilon = \{\ 1, 2, 3, 4, 5, 6, 7\ \}$, A $= \{\ 1, 3, 6\ \}$,
   B $= \{\ 2, 5, 7\ \}$.
6. $\varepsilon = \{\ 1, 2, 3, 4, 5, 6, 7\ \}$, A $= \{\ 1, 4, 5\ \}$,
   B $= \{\ 2, 6\ \}$.
7. $\varepsilon = \{\ 1, 2, 3, 4, 5, 6, 7\ \}$, A $= \{\ 2, 5\ \}$,
   B $= \{\ 1, 3, 4, 7\ \}$.
8. $\varepsilon = \{\ 1, 2, 3, 4, 5, 6, 7, 8\ \}$, A $= \{\ 1, 3, 6, 7\ \}$,
   B $= \{\ 2, 5, 8\ \}$.
9. $\varepsilon = \{\ 1, 2, 3, 4, 5, 6, 7, 8\ \}$, A $= \{\ 2, 4, 6, 8\ \}$,
   B $= \{\ 1, 7\ \}$.
10. $\varepsilon = \{\ 1, 2, 3, 4, 5, 6, 7, 8\ \}$, A $= \{\ 2, 7\ \}$,
    B $= \{\ 3, 4, 8\ \}$.

Example 4

Describe the shaded area in terms of the subsets for each Venn diagram.

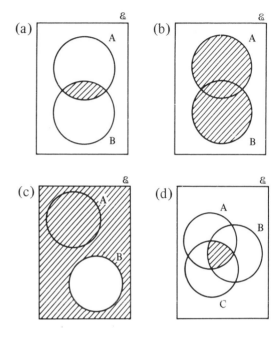

(a) A ∩ B   (b) A ∪ B
(c) B′   (d) A ∩ B ∩ C

Exercise 8.3d

In each of the following Venn diagrams describe the shaded area.

1. 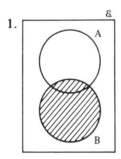   2.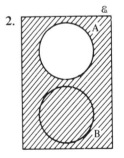

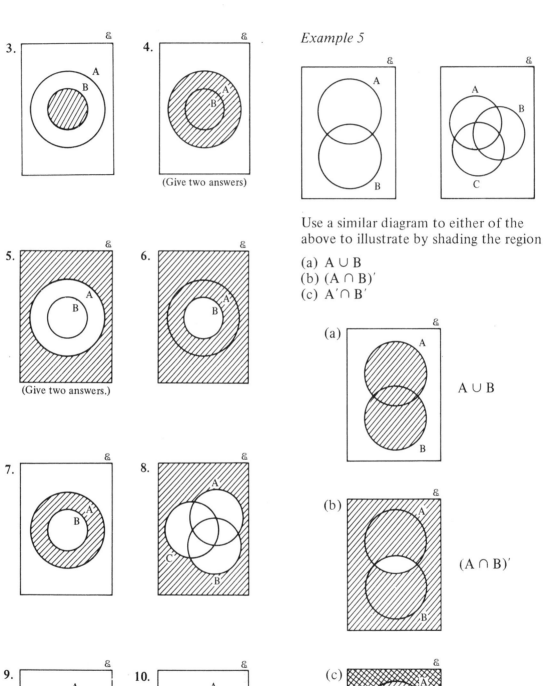

3.

4.

(Give two answers)

5.

(Give two answers.)

6.

7.

8.

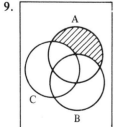

9.

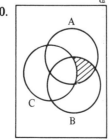

10.

Example 5

Use a similar diagram to either of the above to illustrate by shading the region

(a) A ∪ B
(b) (A ∩ B)′
(c) A′∩ B′

(a)

A ∪ B

(b)

(A ∩ B)′

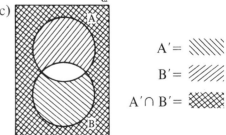

(c)

A′ = \\\\\\\
B′ = //////
A′∩ B′ = ▨▨▨

Exercise 8.3e

Use a similar diagram to either of the two in
example 5 to illustrate by shading the following
regions.

1. A' 2. B' 3. $(A \cup B)'$
4. $A \cap B'$ 5. $A' \cap B$ 6. $A \cap B' \cap C$
7. $A' \cap B \cap C$ 8. $A' \cap B \cap C'$ 9. $A' \cap B' \cap C$
10. $(A \cap B \cap C)'$

Example 6

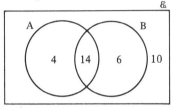

In the above Venn diagram

$\mathcal{E} = \{$ pupils in class III$\}$

i.e. $n(\mathcal{E}) = 4 + 14 + 6 + 10 = 34$

$A = \{$ pupils who play tennis $\}$

$B = \{$ pupils who play squash$\}$

How many pupils:

(a) play tennis?
(b) play both squash and tennis?
(c) play neither squash nor tennis?

(a) $n(A) = 4 + 14 = 18$
(b) $n(A \cap B) = 14$
(c) $n(A \cup B)' = 10$

Exercise 8.3f

1. The Venn diagram shows

$\mathcal{E} = \{$ girls in class 2A$\}$
$A = \{$ girls with blonde hair $\}$
$B = \{$ girls with blue eyes $\}$

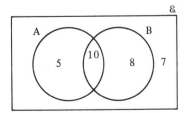

(a) How many girls have blonde hair?
(b) How many girls have blue eyes?
(c) How many girls have blue eyes and blonde hair?
(d) How many girls are in class 2A?
(e) How many girls have neither blue eyes nor blonde hair?

2. In the Venn diagram:

$\mathcal{E} = \{$ people at a meeting $\}$
$A = \{$ those who asked for tea $\}$
$B = \{$ those who voted $\}$

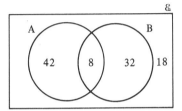

Write down the number who:

(a) asked for tea,
(b) asked for tea and voted,
(c) neither asked for tea nor voted,
(d) attended the meeting.

3. The Venn diagram shows:

$\mathcal{E} = \{$ pupils in the upper school$\}$
$A = \{$ those who saw the Cup Final$\}$
$B = \{$ those who saw the World Cup$\}$

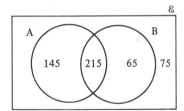

Write down the number who:

(a) watched the Cup Final,
(b) watched the World Cup,
(c) watched both,
(d) watched neither,
(e) are in the upper school.

4. In the Venn diagram,

$\mathscr{E}$ = { pupils in Westgate school }
A = { those who have a brother }
B = { those who have a sister }

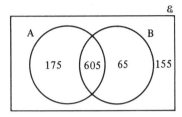

Write down the number who:

(a) have a brother,
(b) have a sister,
(c) have neither brother nor sister,
(d) are in the school altogether.

5. The Venn diagram shows quantities within the following sets.

$\mathscr{E}$ = { numbers less than 20 }
A = { prime numbers less than 20 }
B = { odd numbers less than 20 }

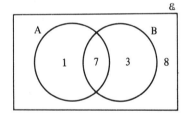

Redraw the diagram to show the actual elements. Are the numbers correct?

Example 7

If $n(A) = 30$, $n(B) = 20$ and $n(A \cap B) = 5$, draw a Venn diagram to find $n(A \cup B)$.

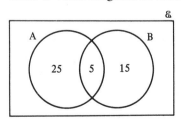

Because $n(A \cap B) = 5$
then $n(A) = 25 + 5 = 30$
$n(B) = 15 + 5 = 20$
$\therefore n(A \cup B) = 25 + 5 + 15 = 45$

Exercise 8.3g

1. If $n(\mathscr{E}) = 36$, $n(A) = 24$, $n(B) = 15$, and $n(A \cap B) = 10$, draw a Venn diagram to find:

(a) $n(A \cap B')$ (b) $n(A' \cap B)$
(c) $n(A \cup B)$ (d) $n(A \cup B)'$

2. If $n(\mathscr{E}) = 1000$, $n(A) = 840$, $n(B) = 720$, and $n(A \cap B) = 570$, draw a Venn diagram to find:

(a) $n(A \cap B')$ (b) $n(A' \cap B)$
(c) $n(A \cup B)$ (d) $n(A \cup B)'$

3. If $n(\mathscr{E}) = 1000$, $n(A) = 550$, $n(B) = 530$, and $n(A \cap B) = 385$, draw a Venn diagram to find:

(a) $n(A \cup B)$ (b) $n(A \cup B)'$

4. If $n(\mathscr{E}) = 500$, $n(A) = 360$, $n(B) = 290$ and $n(A \cap B) = 225$, draw a Venn diagram to find:

(a) $n(A \cup B)$ (b) $n(A \cup B)'$

5. If $\mathscr{E}$ = { all numbers less than 10 }
A = { all prime numbers less than 10 }
B = { all even numbers less than 10 }

draw a Venn diagram to find:

(a) $n(A \cup B)$ (b) $n(A \cap B)$
(c) $n(A \cup B)'$ (d) $n(A \cap B')$
(e) $n(A' \cap B)$

6. In a class of 30 pupils, 18 play squash and 19 play tennis. How many play both games, provided everyone plays at least one game?

7. In a sample of 200 people, 150 liked tea, 75 liked coffee and 11 liked neither. How many people liked both tea and coffee?

The following symbols may be used.

$<$ means *is less than*
$\leqslant$ means *is less than or equal to*
$>$ means *is greater than*
$\geqslant$ means *is greater than or equal to*

Example 1

Use the symbols $>$, $<$, or $=$ to connect the following.

(a) $3\frac{1}{2}$ and $3\cdot45$
(b) $9 + 7$ and $8 + 8$
(c) 10×0 and $10 \times 0\cdot1$
(d) $64 \div 8$ and $72 \div 9$
(e) $-3 + 4$ and $-6 - 7$

(a) $3\frac{1}{2} = 3\cdot5, \therefore 3\frac{1}{2} > 3\cdot45$
(b) $9 + 7 = 16, 8 + 8 = 16$
   $\therefore 9 + 7 = 8 + 8$
(c) $10 \times 0 = 0, 10 \times 0\cdot1 = 1$
   $\therefore 10 \times 0 < 10 \times 0\cdot1$
(d) $64 \div 8 = 8, 72 \div 9 = 8$
   $\therefore 64 \div 8 = 72 \div 9$
(e) $-3 + 4 = -1, -6 - 7 = -13$
   $\therefore -3 + 4 > -6 - 7$

Exercise 8.4a

Use $>$, $<$, or $=$ to connect the two parts of each question.

1. $2\frac{1}{4}$ and $2\cdot2$
2. $4\frac{3}{4}$ and $4\cdot85$
3. $3\frac{4}{5}$ and $3\cdot8$
4. $5 + 21$ and $3\cdot5 + 23\cdot5$
5. $8 + 7$ and $4\frac{1}{4} + 10\frac{3}{4}$
6. $2\cdot9 + 1\cdot6$ and $2\frac{2}{5} + 1\frac{4}{5}$
7. $-9 + 22$ and $-15 + 27$
8. $-11 + 30$ and $-38 + 57$
9. $-7 + 10\cdot6$ and $-4\cdot5 + 8$
10. $-16 + 8$ and $-23 + 16$
11. $-21 + 7$ and $-9 - 5$
12. $-12 - 3$ and $-7 - 9$
13. $-18 - 7$ and $-11 - 13$
14. 8×11 and 29×3
15. 11×12 and 19×7
16. 6×7 and $8\cdot4 \times 5$
17. $1\cdot75 \times 8$ and $2\cdot5 \times 6$
18. $144 \div 12$ and $104 \div 8$
19. $112 \div 7$ and $8 \div 0\cdot5$
20. $10\cdot8 \div 0\cdot9$ and $165 \div 15$

An equation such as $y = 5$ can only have one solution. In an inequality such as $a < 4$, a can have any value less than 4. Inequalities are solved in similar ways to equations.

Example 2

(a) Find the solution of $x + 4 > 6$
$$x > 6 - 4$$
$$\therefore x > 2$$
(b) Find the solution of $2x + 4 \leqslant 12$
$$2x \leqslant 12 - 4$$
$$2x \leqslant 8$$
$$x \leqslant 4$$

Exercise 8.4b

In questions **1** to **10**, find which part of the question has a different answer.

1. (a) $x + 15 > 18$
   (b) $20x > 60$
   (c) $2x + 15 > 23$
2. (a) $3x + 7 > 13$
   (b) $16x > 48$
   (c) $5x - 4 > 6$
3. (a) $6x - 5 < 25$
   (b) $x/3 < 2$
   (c) $7x + 15 < 50$
4. (a) $2y - 5 < 7$
   (b) $15y < 90$
   (c) $3y - 4 < 17$
5. (a) $3z + 2 \geqslant 17$
   (b) $5z - 6 \geqslant 14$
   (c) $14z \geqslant 70$
6. (a) $2p + 9 > 25$
   (b) $4p - 5 > 31$
   (c) $p/3 + 9 > 12$
7. (a) $6q - 7 > 35$
   (b) $q/4 + 11 > 13$
   (c) $q/2 - 3 > 1$
8. (a) $u/5 + 10 < 12$
   (b) $u/4 - 2 < 1$
   (c) $3u - 16 < 14$
9. (a) $v/2 - 3 < 1$
   (b) $5v + 12 < 42$
   (c) $v/3 + 19 < 21$
10. (a) $4t - 25 \leqslant 35$
    (b) $t/6 + 7 \leqslant 9$
    (c) $t/3 - 3 \leqslant 2$

If the unknown in an inequation is a variable on a given set, then the inequality has a *solution set*.

Example 4

If x is a positive integer, find the solution set for:

(a) $x < 4$ (b) $5x - 6 \leqslant 19$

(a) The solution set is $x = \{1, 2, 3\}$

(b) $5x - 6 \leqslant 19$
$$5x \leqslant 25$$
$$x \leqslant 5$$

So the solution set is $x = \{5, 4, 3, 2, 1\}$

Exercise 8.4c

Find the solution set for each question.

1. $x + 5 < 20$ where x is a positive prime number.
2. $x - 10 < 40$ where x is a square number.
3. $15x \leqslant 105$ where x is a positive odd number.
4. $18y \leqslant 144$ where y is a positive even number.
5. $\frac{z}{3} < 12$ where z is a positive integer divisible by 5.
6. $\frac{p}{4} < 10$ where p is a positive integer divisible by 6.
7. $4q + 11 < 35$ where q is any positive integer.
8. $2r + 15 \leqslant 51$ where r is a positive integer divisible by both 2 and 3.
9. $3u + 26 \leqslant 98$ where u is a positive integer divisible by both 3 and 4.
10. $2v - 17 < 79$ where v is a positive integer divisible by 8.
11. $3x - 16 \leqslant 29$ where $x > 12$.
12. $5y - 12 < 28$ where $y > 5$.
13. $\frac{z}{2} + 6 < 11$ where z is not a prime number.
14. $\frac{m}{5} - 3 < 7$ where m is a square number that is odd.
15. $\frac{n}{9} - 7 \leqslant 9$ where n is a square number that is even.

The solution of an inequality can be shown pictorially by means of a number line.

Example 5

Solve $x + 4 < 8$, and show the solution on a number line.

$$x + 4 < 8$$
$$\therefore \quad x < 4$$

The solution set is $x = \{3, 2, 1, 0, \ldots\}$

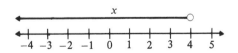

The symbol ○ means that 4 is not included in the solution set.

Example 6

Solve $3x + 2 \leqslant 8$, and show the solution on a number line.

$$3x + 2 \leqslant 8$$
$$3x \leqslant 6$$
$$x \leqslant 2$$

The solution set is $x = \{2, 1, 0, \ldots\}$

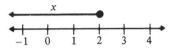

The symbol ● means that 2 is included in the solution set.

Exercise 8.4d

Solve the inequality and show the solution on a number line.

1. $x + 7 < 12$
2. $x + 16 \leqslant 22$
3. $x + 29 \leqslant 33$ for $x > 0$
4. $3y \leqslant 18$
5. $5z < 20$ for $z > 0$
6. $16t \leqslant 80$ for $t > 0$
7. $m/3 < 3$
8. $n - 3 > 2$
9. $4p \geqslant 12$
10. $2q + 17 < 21$
11. $5r - 2 \leqslant 13$
12. $12x - 17 \leqslant 7$ for $x > 0$

Inequalities can also be shown on a graph.

Example 7

(a) Sketch the graph of $x < 2$.

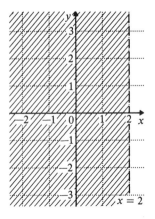

The shaded region to the left of the broken line is the required solution.
The broken line indicates that no point of the solution lies on the line.

(b) Sketch the graph of $y \geqslant -2$

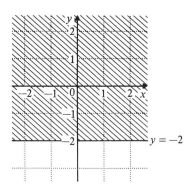

The shaded region above the line $y = -2$ is the required solution.
This line is unbroken, indicating that points of the solution also lie on the line.

Example 8

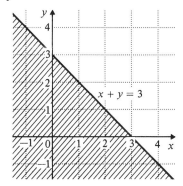

The set of points that lie on the line $x + y = 3$ is written as

$$\{(x, y): x + y = 3\}$$

The set of points in the shaded region is:

$$\{(x, y): x + y \leqslant 3\}$$

Exercise 8.4e

On a sketch graph, shade the region for the set of points that satisfy the conditions in questions **1** to **15**.

1. $x < 2$ and $y \leqslant 4$
2. $x \leqslant 3$ and $y < 1$
3. $x < -2$ and $y < 2$
4. $x < -1$ and $y \geqslant -3$
5. $x > 1$ and $y \geqslant 3$
6. $x \geqslant 4$ and $y > -2$

7. $x > -2$ and $y \geqslant -4$
8. $x + y < 4$ for $x > 0, y > 0$
9. $x + y \leqslant 6$ for $x > 0, y > 0$
10. $x + y \geqslant 5$ for $x > 0, y > 0$
11. $\{(x, y): x + y > 2\}$
12. $\{(x, y): x - y \leqslant 3\}$
13. $\{(x, y): x - y > 1\}$
14. $\{(x, y): y - x < 4\}$
15. $\{(x, y): y - x \leqslant 5\}$

Describe the shaded region in each of the following sketch graphs.

16.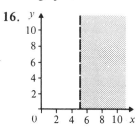

17.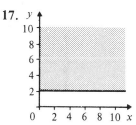

18.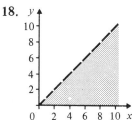

19.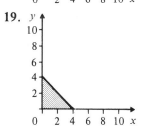

20.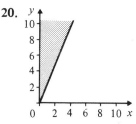

A *relation* is a pairing of elements of one set A with the elements of another set B. A *mapping* is a relation between two sets such that every element of the first set is paired off with one and only one element of the second set.

Example 1

State in each case whether the relation is a mapping from left to right, giving reasons.

(a)
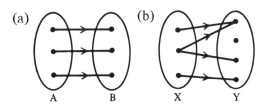

(b)
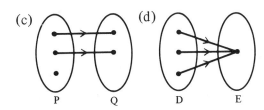

(c)
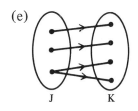

(d)

(e)

(a) This is a mapping because each element of A is paired with a unique element of B.

(b) This is not a mapping because one element of X is paired with two elements of Y.

(c) This is not a mapping because one element of P is not paired with an element of Q.

(d) This is a mapping because each element of D is paired with one element of E.

(e) This is not a mapping because one element of J is paired with two elements of K.

Exercise 8.5a

For each question state whether the relation A → B is a mapping, giving reasons.

1.
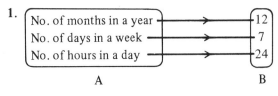

2.
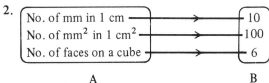

3.
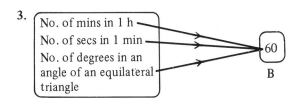

4.
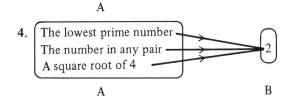

5.
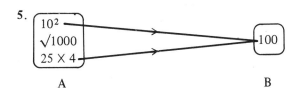

6.
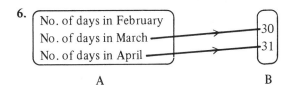

7.
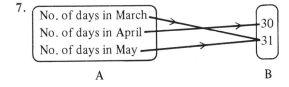

8.

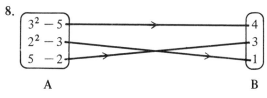

A   B

9.

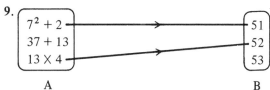

A   B

10.

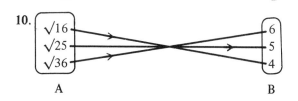

A   B

Example 2

Draw a diagram to illustrate the mapping $x \to 2x$ where x is an element of the set

$$A = \{ 0, 1, 2, 3, 4 \} .$$

Method 1

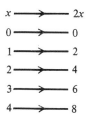

Method 2

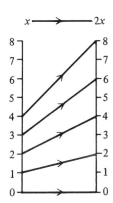

Exercise 8.5b

Draw a diagram to illustrate each mapping where x is an element of the set A using method 1 or method 2 above.

1. $x \to 3x$, $A = \{0, 1, 2, 3, 4\}$
2. $x \to \frac{1}{2}x$, $A = \{-6, -4, -2, 0, 2, 4, 6\}$
3. $x \to \frac{5}{2}x$, $A = \{0, 2, 4\}$
4. $x \to \frac{4}{3}x$, $A = \{-3, 0, 3, 6\}$
5. $x \to x + 3$, $A = \{-4, -3, -2, -1, 0, 1, 2, 3\}$
6. $x \to 2x + 1$, $A = \{-2, -1, 0, 1, 2\}$
7. $x \to 3x - 2$, $A = \{-1, 0, 1, 2, 3\}$
8. $x \to \frac{3}{2}x + 2$, $A = \{-2, 0, 2, 4\}$
9. $x \to \frac{2}{3}x - 3$, $A = \{-3, 0, 3, 6\}$
10. $x \to \frac{2}{x}$, $A = \{\frac{1}{4}, \frac{1}{2}, 1, 2, 4\}$
11. $x \to 7 - x$, $A = \{1, 2, 3, 4, 5\}$
12. $x \to 5 - \frac{1}{2}x$, $A = \{2, 4, 6, 8\}$
13. $x \to x^2 - 3$, $A = \{0, 1, 2, 3\}$
14. $x \to \frac{x^2}{4}$, $A = \{-2, 0, 2, 4\}$
15. $x \to x^2 - 6$, $A = \{-2, -1, 0, 1, 2, 3\}$

Example 3

The diagram below represents a mapping. Write this mapping in the form $x \to$

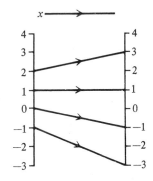

The mapping is $x \to 2x - 1$.

Exercise 8.5c

Write each mapping in the form $x \rightarrow$

1.
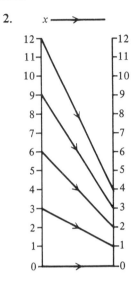

2.
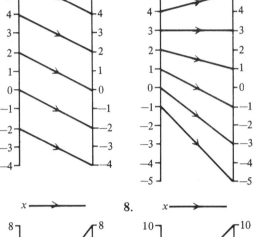

3.
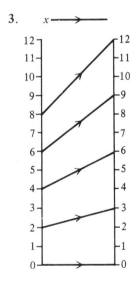

4.
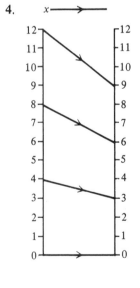

5.
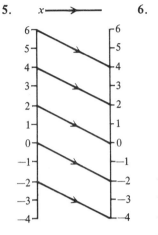

6.
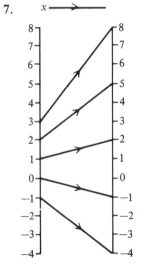

7.
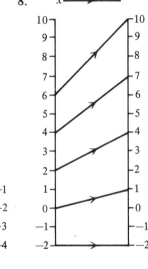

8.
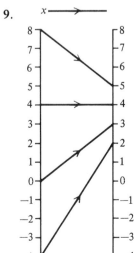

9.
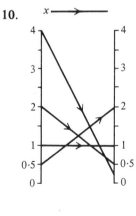

10.

11. 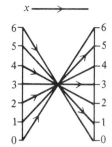 12.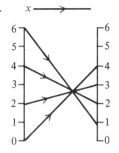

The mapping $x \to 2x - 1$ can be written as a set of ordered pairs:

$(-1, -3), (0, -1), (1, 1), (2, 3).$

This can be plotted to show the graph of the relation $y = 2x - 1$.

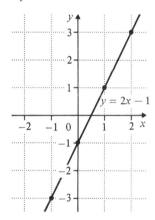

Example 4

Draw the graph of the mapping $x \to x^2 - 4$ where x is an element of the set

$A = \{ -3, -2, -1, 0, 1, 2, 3 \}$

x	-3	-2	-1	0	1	2	3
$x^2 - 4$	5	0	-3	-4	-3	0	5

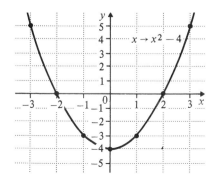

Exercise 8.5d

Draw a graph of each mapping of x where x is an element of the given set A.

1. $x \to \frac{1}{4}x$, A $= \{ -8, -4, 0, 4, 8 \}$
2. $x \to \frac{2}{3}x$, A $= \{ -9, -6, -3, 0, 3, 6, 9 \}$
3. $x \to 2x + 3$, A $= \{ -3, -2, -1, 0, 1, 2, 3, 4 \}$
4. $x \to 3x - 3$, A $= \{ -2, -1, 0, 1, 2, 3, 4 \}$
5. $x \to 4x - 5$, A $= \{ -1, 0, 1, 2, 3, 4 \}$
6. $x \to \frac{1}{2}x + 2$, A $= \{ -8, -6, -4, -2, 0, 2, 4, 6, 8 \}$
7. $x \to \frac{3}{2}x - 1$, A $= \{ -6, -4, -2, 0, 2, 4, 6 \}$
8. $x \to x^2 - 1$, A $= \{ -3, -2, -1, 0, 1, 2, 3 \}$
9. $x \to 2x^2$, A $= \{ -3, -2, -1, 0, 1, 2, 3 \}$
10. $x \to \frac{x^2}{2} - 2$, A $= \{ -4, -2, 0, 2, 4 \}$

Describe the mapping for x in each of the following graphs.

11.   12.

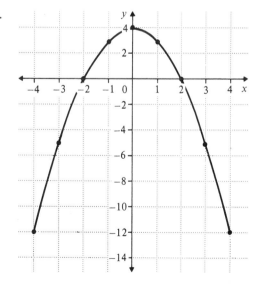

13.

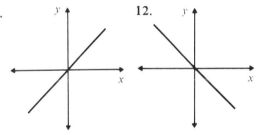

Reflection

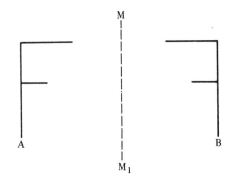

The object B is the mirror image of the object A. The line MM_1 is called the *axis of reflection* or the *mirror line*.

In a reflection each point on the image is the same perpendicular distance from the axis of reflection as the corresponding point on the object.

Rotation

Rotation is the movement of an object about a point through a given number of degrees in a clockwise (negative) or an anticlockwise (positive) direction.

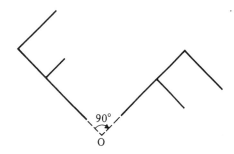

The object has been rotated through $-90°$ (clockwise) about the point O.

Translation

Translation is the movement along a straight line in a certain direction for a given distance of a point or an object, provided reflection or rotation is not involved.

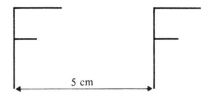

The object has been translated 5 cm to the right.

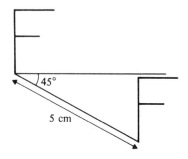

The object has been translated 5 cm at an angle of 45° to the horizontal.

Translations are sometimes given as *column vectors*, following the usual Cartesian notation.

e.g. $\begin{pmatrix} 2 \\ -1 \end{pmatrix}$ means 'move 2 units to the right and 1 unit down'.

Enlargement

An enlargement will enlarge or reduce the size of an object about a centre of enlargement to a given *scale factor*. The scale factor may be positive, negative, or fractional.

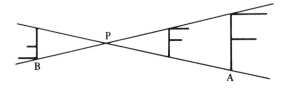

The centre of enlargement is P.
The image A has a scale factor of 2.
The image B has a scale factor of -1.

Exercise 8.6

1. An anticlockwise rotation is positive and a clockwise rotation is negative.
   Copy each drawing, then
   (i) draw the image reflected in the line AB,
   (ii) rotate the figure about the centre O.

(a) Rotate 90°.   (b) Rotate 120°

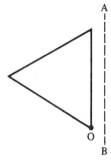

(c) Rotate − 135°   (d) Rotate 90°

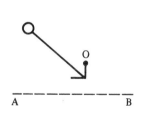

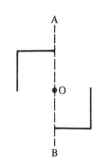

(e) Rotate 180°   (f) Rotate 180°

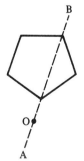

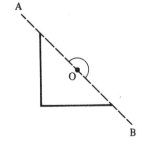

2. Copy the figure, then rotate it about the centre O in the given direction. Draw the image each time.

   (a) Rotate three times: (b) Rotate five times:
        90°, 180° and 270°.   60°, 120°, 180°, 240° and 300°.

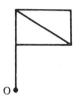

   (c) Rotate three times:
        90°, 180°, and −90°.

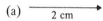

3. Copy the diagram to the size shown. Then sketch the position of the image after the given translation.

   (a) ⟶ 2 cm

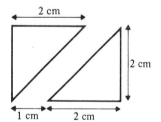

   (b) Distance AB   (c) ⟶ 3 cm

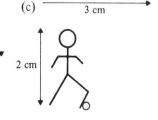

4. Copy each figure and enlarge with the centres of enlargement and scale factors given.

   (i) centre O, scale factor $+2$;
   (ii) centre O, scale factor -1;
   (iii) centre X, scale factor $+\frac{1}{2}$.

(a)

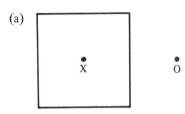

(b)

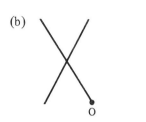

(c)

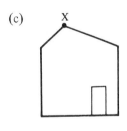

5. Copy each figure and rotate $180°$ about the axis XY. What other transformation would give the same result?

(a)   X   (b)   X

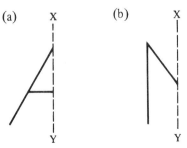

(c)   (d)

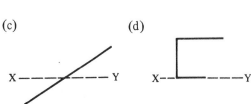

6. (a) Plot a trapezium ABCD with vertices A $(2, 1)$; B $(3, 1)$; C $(3, 2)$; D $(2, 3)$.
  (b) Plot on the diagram
     (i) the image $A_1 B_1 C_1 D_1$ of the trapezium ABCD under a rotation of $90°$ anti-clockwise about the origin O,
     (ii) the image $A_2 B_2 C_2 D_2$ of $A_1 B_1 C_1 D_1$ under a reflection in the x-axis,
     (iii) the image $A_3 B_3 C_3 D_3$ of $A_2 B_2 C_2 D_2$ under a rotation of $90°$ anticlockwise about the origin O.
  (c) Describe the single transformation which would map ABCD onto $A_3 B_3 C_3 D_3$.

7. Triangle ABC has coordinates A $(1, 1)$; B $(3, 1)$; C $(2\frac{1}{2}, 3)$.
  (a) Draw the image of ABC under the following transformations:
     (i) A reflection in the y-axis, label $A_1 B_1 C_1$,
     (ii) An enlargement, centre $(2, \frac{1}{2})$ and scale factor 2, label $A_2 B_2 C_2$,
     (iii) A rotation of $270°$ anticlockwise about $(0, 1)$, label $A_3 B_3 C_3$.
  (b) Describe a single transformation which will map $A_3 B_3 C_3$ onto ABC.

8. Ox and Oy are perpendicular coordinate axes.

  R denotes the translation $\begin{pmatrix} 0 \\ 4 \end{pmatrix}$

  S denotes a rotation of $180°$ about $(0, 0)$,

  T denotes a reflection in the x-axis.

  Find the image of the point $(2, 3)$ under each of the transformations below where RS (for example) means transformation S followed by transformation R.

  (a) **RS**   (b) **SR**   (c) **ST**
  (d) **TS**   (e) **RT**   (r) **TR**

  Which of the above transformations give the same result?

A set of numbers arranged in rows and columns as shown is called a *matrix*.

$$\begin{pmatrix} 1 & 2 & 3 \\ 6 & 5 & 4 \end{pmatrix}$$

This matrix has two rows and three columns. The *order* of this matrix is 2×3.

Exercise 8.7a

Give the order of the following matrices.

1. $\begin{pmatrix} 3 & 2 \\ 1 & 4 \\ 2 & 6 \end{pmatrix}$
2. $\begin{pmatrix} 3 & 2 & 5 & 1 \\ 1 & 3 & 7 & 4 \\ 5 & 2 & 2 & 1 \end{pmatrix}$

3. $\begin{pmatrix} 5 & 1 & 4 \\ 2 & 3 & 0 \\ 6 & 5 & 3 \end{pmatrix}$
4. $\begin{pmatrix} 3 & 1 & 5 & 2 \\ 4 & 0 & 6 & 7 \end{pmatrix}$

5. $\begin{pmatrix} 1 & 5 \\ 2 & 7 \\ 3 & 6 \\ 9 & 2 \end{pmatrix}$
6. $\begin{pmatrix} 5 & 0 & 4 \\ 3 & 2 & 6 \\ 1 & 7 & 0 \\ 6 & 3 & 1 \end{pmatrix}$

7. $\begin{pmatrix} 1 & 5 & 7 & 3 & 2 \\ 4 & 0 & 2 & 8 & 1 \end{pmatrix}$
8. $\begin{pmatrix} 1 & 5 & 3 & 6 \\ 4 & 0 & 4 & 5 \\ 6 & 3 & 2 & 2 \\ 1 & 7 & 5 & 3 \end{pmatrix}$

9. $\begin{pmatrix} 3 & 2 \\ 1 & 7 \\ 2 & 5 \\ 4 & 4 \\ 2 & 0 \end{pmatrix}$
10. $\begin{pmatrix} 3 & 6 & 1 \\ 2 & 5 & 3 \\ 4 & 1 & 8 \\ 7 & 1 & 3 \\ 6 & 5 & 4 \end{pmatrix}$

Matrices of the *same order* can be added or subtracted.

Example 1

Where possible, write each of the following as a single matrix.

(a) $\begin{pmatrix} 2 & 3 \\ 4 & 5 \end{pmatrix} + \begin{pmatrix} 0 & -1 \\ 2 & -6 \end{pmatrix}$

$= \begin{pmatrix} 2+0 & 3+-1 \\ 4+2 & 5+-6 \end{pmatrix}$

$= \begin{pmatrix} 2 & 2 \\ 6 & -1 \end{pmatrix}$

(b) $\begin{pmatrix} 1 & 4 \\ 2 & 5 \\ 3 & 6 \end{pmatrix} - \begin{pmatrix} 0 & -1 \\ 3 & 0 \\ 2 & 6 \end{pmatrix}$

$= \begin{pmatrix} 1-0 & 4--1 \\ 2-3 & 5-0 \\ 3-2 & 6-6 \end{pmatrix}$

$= \begin{pmatrix} 1 & 5 \\ -1 & 5 \\ 1 & 0 \end{pmatrix}$

(c) $\begin{pmatrix} 2 & 3 & 4 \\ 5 & 7 & 8 \end{pmatrix} + \begin{pmatrix} 2 & 5 \\ 3 & 7 \\ 4 & 8 \end{pmatrix}$

These two cannot be written as a single matrix.

Exercise 8.7b

Find the answer which is different for each question.

1. (a) $\begin{pmatrix} 1 & 2 \\ 3 & 5 \end{pmatrix} + \begin{pmatrix} 3 & 1 \\ 2 & 2 \end{pmatrix}$

   (b) $\begin{pmatrix} 2 & 0 \\ 1 & 2 \end{pmatrix} + \begin{pmatrix} 2 & 3 \\ 4 & 4 \end{pmatrix}$

   (c) $\begin{pmatrix} 6 & 7 \\ 8 & 9 \end{pmatrix} - \begin{pmatrix} 2 & 4 \\ 3 & 2 \end{pmatrix}$

2. (a) $\begin{pmatrix} 2 & 1 \\ 2 & 1 \\ 3 & 4 \end{pmatrix} + \begin{pmatrix} 3 & 3 \\ 4 & 2 \\ 4 & 1 \end{pmatrix}$

   (b) $\begin{pmatrix} 8 & 5 \\ 7 & 3 \\ 9 & 6 \end{pmatrix} - \begin{pmatrix} 3 & 2 \\ 1 & 0 \\ 2 & 1 \end{pmatrix}$

   (c) $\begin{pmatrix} 1 & 0 \\ 3 & 2 \\ 5 & 2 \end{pmatrix} + \begin{pmatrix} 4 & 3 \\ 3 & 1 \\ 2 & 3 \end{pmatrix}$

3. (a) $\begin{pmatrix} 1 & 2 & 3 \\ 2 & 3 & 2 \end{pmatrix} + \begin{pmatrix} 1 & 3 & 1 \\ 1 & 5 & 4 \end{pmatrix}$

   (b) $\begin{pmatrix} 5 & 8 & 9 \\ 9 & 9 & 8 \end{pmatrix} - \begin{pmatrix} 3 & 2 & 5 \\ 6 & 1 & 2 \end{pmatrix}$

   (c) $\begin{pmatrix} 7 & 9 & 5 \\ 5 & 8 & 9 \end{pmatrix} - \begin{pmatrix} 5 & 3 & 1 \\ 2 & 0 & 3 \end{pmatrix}$

4. (a) $\begin{pmatrix} 2 & 2 & 3 \\ 5 & 5 & 1 \\ 9 & 1 & 7 \end{pmatrix} + \begin{pmatrix} 3 & 2 & 0 \\ 2 & -3 & 4 \\ -6 & 5 & -3 \end{pmatrix}$

(b) $\begin{pmatrix} 1 & 7 & 2 \\ 3 & 8 & 9 \\ 1 & 7 & -1 \end{pmatrix} + \begin{pmatrix} 4 & -3 & 1 \\ 4 & -5 & -4 \\ 2 & -1 & 5 \end{pmatrix}$

(c) $\begin{pmatrix} 8 & 1 & 8 \\ 9 & 1 & 3 \\ -1 & 4 & -4 \end{pmatrix} + \begin{pmatrix} -3 & 3 & -5 \\ -2 & 1 & 2 \\ 4 & 2 & 8 \end{pmatrix}$

5. (a) $\begin{pmatrix} 1 & 3 \\ 7 & 2 \end{pmatrix} + \begin{pmatrix} 2 & -5 \\ -2 & -6 \end{pmatrix}$

(b) $\begin{pmatrix} 0 & 1 \\ 2 & 7 \end{pmatrix} + \begin{pmatrix} 3 & -3 \\ 3 & -3 \end{pmatrix}$

(c) $\begin{pmatrix} 3 & -6 \\ 4 & -9 \end{pmatrix} + \begin{pmatrix} 0 & 4 \\ 1 & 5 \end{pmatrix}$

6. (a) $\begin{pmatrix} 3 & 1 & 2 \\ 5 & 4 & 2 \\ 2 & 5 & 2 \end{pmatrix} + \begin{pmatrix} 2 & -3 & 2 \\ -2 & -6 & -8 \\ -6 & -4 & -5 \end{pmatrix}$

(b) $\begin{pmatrix} 4 & 4 & 1 \\ 2 & 1 & -9 \\ 3 & 0 & -1 \end{pmatrix} + \begin{pmatrix} 1 & -6 & 3 \\ 1 & -3 & 3 \\ -7 & 1 & -2 \end{pmatrix}$

(c) $\begin{pmatrix} 0 & 3 & 2 \\ 6 & 5 & -1 \\ -4 & 3 & 0 \end{pmatrix} + \begin{pmatrix} 5 & -5 & 2 \\ -3 & -7 & -5 \\ -1 & -2 & -3 \end{pmatrix}$

It is only possible to multiply two matrices together if the number of columns in the first matrix equals the number of rows in the second matrix.
The resulting matrix will have the same number of rows as the first matrix and the same number of columns as the second matrix. So a (4 × 2) matrix multiplied by a (2 × 3) matrix becomes a (4 × 3) matrix. The method of multiplication is shown below.

$$\begin{pmatrix} a & b \\ c & d \end{pmatrix} \begin{pmatrix} w & x \\ y & z \end{pmatrix} = \begin{pmatrix} (aw + by) & (ax + bz) \\ (cw + dy) & (cx + dz) \end{pmatrix}$$

Example 2

If $A = \begin{pmatrix} 1 & 4 \\ 2 & 5 \\ 3 & 6 \end{pmatrix}$; $B = \begin{pmatrix} 1 & 2 & 3 \\ 4 & 5 & 6 \end{pmatrix}$ and

$C = \begin{pmatrix} 2 & 1 \\ 1 & 3 \\ 2 & 1 \end{pmatrix}$, evaluate if possible

(a) AB   **(b) BC**   **(c) AC**

(a) $AB = \begin{pmatrix} 1 & 4 \\ 2 & 5 \\ 3 & 6 \end{pmatrix} \begin{pmatrix} 1 & 2 & 3 \\ 4 & 5 & 6 \end{pmatrix}$

$$= \begin{pmatrix} (1 \times 1) + (4 \times 4) & (1 \times 2) + (4 \times 5) & (1 \times 3) + (4 \times 6) \\ (2 \times 1) + (5 \times 4) & (2 \times 2) + (5 \times 5) & (2 \times 3) + (5 \times 6) \\ (3 \times 1) + (6 \times 4) & (3 \times 2) + (6 \times 5) & (3 \times 3) + (6 \times 6) \end{pmatrix}$$

$$= \begin{pmatrix} 17 & 22 & 27 \\ 22 & 29 & 36 \\ 27 & 36 & 45 \end{pmatrix}$$

(b) $BC = \begin{pmatrix} 1 & 2 & 3 \\ 4 & 5 & 6 \end{pmatrix} \begin{pmatrix} 2 & 1 \\ 1 & 3 \\ 2 & 1 \end{pmatrix}$

$$= \begin{pmatrix} 10 & 10 \\ 25 & 25 \end{pmatrix}$$

(c) $AC = \begin{pmatrix} 1 & 4 \\ 2 & 5 \\ 3 & 6 \end{pmatrix} \begin{pmatrix} 2 & 1 \\ 1 & 3 \\ 2 & 1 \end{pmatrix}$ This cannot be calculated.

Exercise 8.7c

Evaluate each product if possible.

1. $\begin{pmatrix} 1 & 3 \\ 2 & 4 \end{pmatrix} \begin{pmatrix} 2 & 1 \\ 3 & 2 \end{pmatrix}$

2. $\begin{pmatrix} 6 & 1 \\ 2 & 0 \end{pmatrix} \begin{pmatrix} 7 & 2 \\ 3 & 4 \end{pmatrix}$

3. $\begin{pmatrix} 2 & 1 \\ 3 & 2 \end{pmatrix} \begin{pmatrix} 1 & 3 \\ 2 & 4 \end{pmatrix}$

4. $\begin{pmatrix} 1 & 2 \\ 2 & 3 \end{pmatrix} \begin{pmatrix} 2 & 3 & 1 \\ 1 & 4 & 3 \end{pmatrix}$

5. $\begin{pmatrix} 4 & 1 \\ 2 & 0 \end{pmatrix} \begin{pmatrix} 1 & 5 & 6 \\ 3 & 2 & 2 \end{pmatrix}$

6. $\begin{pmatrix} 2 & 3 & 1 \\ 1 & 4 & 3 \end{pmatrix}\begin{pmatrix} 1 & 2 \\ 2 & 3 \end{pmatrix}$

7. $\begin{pmatrix} 2 & 3 \\ 1 & 5 \\ 2 & 2 \end{pmatrix}\begin{pmatrix} 4 & 2 \\ 3 & 3 \end{pmatrix}$

8. $\begin{pmatrix} 4 & 0 \\ 3 & 2 \\ 1 & 6 \end{pmatrix}\begin{pmatrix} 1 & 5 \\ 3 & 0 \end{pmatrix}$

9. $\begin{pmatrix} 4 & 2 \\ 3 & 3 \end{pmatrix}\begin{pmatrix} 2 & 3 \\ 1 & 5 \\ 2 & 2 \end{pmatrix}$

10. $\begin{pmatrix} 1 & 3 \\ 2 & 1 \\ 3 & 4 \end{pmatrix}\begin{pmatrix} 2 & 1 & 3 \\ 1 & 3 & 4 \end{pmatrix}$

11. $\begin{pmatrix} 2 & 1 \\ 3 & 0 \\ 1 & 4 \end{pmatrix}\begin{pmatrix} 1 & 0 & 2 \\ 0 & 1 & 3 \end{pmatrix}$

12. $\begin{pmatrix} 2 & 1 & 3 \\ 1 & 3 & 4 \end{pmatrix}\begin{pmatrix} 1 & 3 \\ 2 & 1 \\ 3 & 4 \end{pmatrix}$

13. $\begin{pmatrix} 1 & 0 & 2 \\ 0 & 1 & 3 \end{pmatrix}\begin{pmatrix} 2 & 1 \\ 3 & 0 \\ 1 & 4 \end{pmatrix}$

14. $\begin{pmatrix} 2 & 0 & 1 \\ 1 & 3 & 0 \\ 4 & 2 & 1 \end{pmatrix}\begin{pmatrix} 1 & 3 \\ 2 & 0 \\ 1 & 1 \end{pmatrix}$

15. $\begin{pmatrix} 1 & 3 \\ 2 & 0 \\ 1 & 1 \end{pmatrix}\begin{pmatrix} 2 & 0 & 1 \\ 1 & 3 & 0 \\ 4 & 2 & 1 \end{pmatrix}$

16. $\begin{pmatrix} 2 & 3 \\ 1 & 2 \end{pmatrix}\begin{pmatrix} 2 & -3 \\ -1 & 2 \end{pmatrix}$

17. $\begin{pmatrix} 1 & 2 \\ 2 & 3 \end{pmatrix}\begin{pmatrix} 3 & -2 \\ -2 & 1 \end{pmatrix}$

18. $\begin{pmatrix} 3 & 2 \\ 7 & 5 \end{pmatrix}\begin{pmatrix} 5 & -2 \\ -7 & 3 \end{pmatrix}$

19. $\begin{pmatrix} 3 & 4 \\ 5 & 6 \end{pmatrix}\begin{pmatrix} 6 & -4 \\ -5 & 3 \end{pmatrix}$

20. $\begin{pmatrix} 6 & 9 \\ 3 & 5 \end{pmatrix}\begin{pmatrix} 5 & -9 \\ -3 & 6 \end{pmatrix}$

21. If $\mathbf{A} = \begin{pmatrix} 3 & 4 \\ 5 & 6 \end{pmatrix}$, $\mathbf{B} = \begin{pmatrix} 1 & 2 \\ 3 & 4 \end{pmatrix}$, find

(a) $\mathbf{AB}$; (b) $\mathbf{BA}$. Does $\mathbf{AB} = \mathbf{BA}$?

22. If $\mathbf{A} = \begin{pmatrix} 2 & 3 \\ 1 & 2 \end{pmatrix}$, $\mathbf{B} = \begin{pmatrix} 2 & -3 \\ -1 & 2 \end{pmatrix}$, find

(a) $\mathbf{AB}$; (b) $\mathbf{BA}$. Does $\mathbf{AB} = \mathbf{BA}$?

23. If $\mathbf{I} = \begin{pmatrix} 1 & 0 \\ 0 & 1 \end{pmatrix}$, find (a) $\mathbf{I}^2$; (b) $\mathbf{I}^3$; (c) $\mathbf{I}^5$.

Note: $\mathbf{I}^2 = \mathbf{I} \times \mathbf{I}$

What is $\mathbf{I}^{10}$?

24. If $\mathbf{A} = \begin{pmatrix} 1 & 2 \\ 3 & 4 \end{pmatrix}$, $\mathbf{B} = \begin{pmatrix} 1 & 4 \\ 2 & 3 \end{pmatrix}$, $\mathbf{C} = \begin{pmatrix} 4 & 1 \\ 3 & 2 \end{pmatrix}$,

find (a) $\mathbf{AB}$; (b) $\mathbf{(AB)C}$; (c) $\mathbf{BC}$; (d) $\mathbf{A(BC)}$.

Does $\mathbf{(AB)C} = \mathbf{A(BC)}$?

25. Using the matrices in question 24,

find (a) $\mathbf{AB} + \mathbf{AC}$; (b) $\mathbf{A(B + C)}$.

Does $\mathbf{AB} + \mathbf{AC} = \mathbf{A(B + C)}$?

26. If $\begin{pmatrix} 2 & 1 \\ 4 & 3 \end{pmatrix}\begin{pmatrix} a \\ b \end{pmatrix} = \begin{pmatrix} 5 \\ 11 \end{pmatrix}$, find a and b.

27. If $\begin{pmatrix} 2 & -1 \\ -4 & 3 \end{pmatrix}\begin{pmatrix} a \\ b \end{pmatrix} = \begin{pmatrix} 0 \\ -10 \end{pmatrix}$, find a and b.

28. If $\mathbf{A} = \begin{pmatrix} 1 & 2 \\ 3 & 4 \end{pmatrix}$, $\mathbf{B} = \begin{pmatrix} 1 & 2 \\ 4 & 3 \end{pmatrix}$, find

(a) $\mathbf{A}^2$, i.e. $\mathbf{A} \times \mathbf{A}$; (b) $\mathbf{B}^2$;

(c) $\mathbf{(A + B)}^2$; (d) $\mathbf{AB}$; (e) $\mathbf{BA}$.

Does $\mathbf{(A + B)}^2 = \mathbf{A}^2 + \mathbf{AB} + \mathbf{BA} + \mathbf{B}^2$?

A vector has magnitude and direction: it can be represented by a straight line. For example, the movement from one point to another is a vector.

The movement (displacement) from A to B is shown above. This is written $\vec{AB}$ or **a**.

The reverse displacement would be either $\vec{BA}$ or $-\vec{AB}$ or $-$**a**.

Exercise 8.8a

Write each displacement in terms of **a** or **b**.

1. (a) $\vec{AB}$
   (b) $\vec{BA}$
   (c) $\vec{CB}$
   (d) $\vec{BC}$

2. (a) $\vec{XY}$
   (b) $\vec{YZ}$
   (c) $\vec{YX}$
   (d) $\vec{ZY}$

3. (a) $\vec{QP}$
   (b) $\vec{QR}$
   (c) $\vec{PQ}$
   (d) $\vec{RQ}$

4. (a) $\vec{LM}$
   (b) $\vec{NM}$
   (c) $\vec{ML}$
   (d) $\vec{MN}$

5. (a) $\vec{VU}$
   (b) $\vec{UV}$
   (c) $\vec{VW}$
   (d) $\vec{WV}$

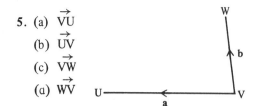

Vector addition

In the diagram, AB is produced to C so that $\vec{BC} = AB$

Then $\vec{AC} = 2\vec{AB} = 2$**a**.

If BC is drawn from B so that $\vec{BC} = $ **b**, as shown

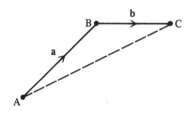

then $\vec{AC} = \vec{AB} + \vec{BC} = $ **a** + **b**
and $\vec{CA} = \vec{CB} + \vec{BA} = -$**b** $-$ **a**

Example 1

In the diagram $\vec{AB} = $ **a**, $\vec{DC} = 2$**a**, and $\vec{BC} = $ **b**.

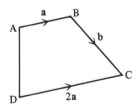

Find in terms of **a** and **b**

(a) $\vec{AC}$ (b) $\vec{BD}$ (c) $\vec{DA}$

(a) $\vec{AC} = \vec{AB} + \vec{BC} = $ **a** + **b**

(b) $\overrightarrow{BD} = \overrightarrow{BC} + \overrightarrow{CD} = \mathbf{b} - 2\mathbf{a}$

(c) $\overrightarrow{DA} = \overrightarrow{DC} + \overrightarrow{CB} + \overrightarrow{BA}$

$= 2\mathbf{a} - \mathbf{b} - \mathbf{a}$

$= \mathbf{a} - \mathbf{b}$

Exercise 8.8b

Write each displacement in terms of **a** and **b**.

1. (a) $\overrightarrow{AC}$
   (b) $\overrightarrow{BD}$
   (c) $\overrightarrow{AD}$

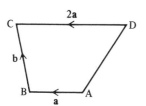

2. (a) $\overrightarrow{LN}$
   (b) $\overrightarrow{MP}$
   (c) $\overrightarrow{LP}$

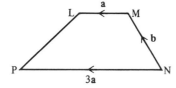

3. (a) $\overrightarrow{PR}$
   (b) $\overrightarrow{SQ}$
   (c) $\overrightarrow{SP}$

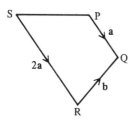

4. (a) $\overrightarrow{UW}$
   (b) $\overrightarrow{XV}$
   (c) $\overrightarrow{XU}$
   (d) $\overrightarrow{UX}$

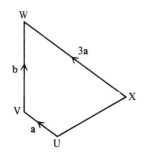

5. (a) $\overrightarrow{XZ}$
   (b) $\overrightarrow{YT}$
   (c) $\overrightarrow{XT}$

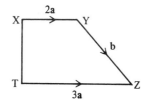

6. (a) $\overrightarrow{AC}$
   (b) $\overrightarrow{DB}$
   (c) $\overrightarrow{AB}$
   (d) $\overrightarrow{BA}$

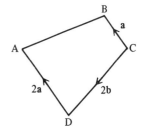

7. (a) $\overrightarrow{LN}$
   (b) $\overrightarrow{PM}$
   (c) $\overrightarrow{PL}$
   (d) $\overrightarrow{LP}$

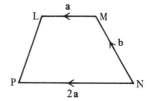

8. (a) $\overrightarrow{PU}$
   (b) $\overrightarrow{RS}$
   (c) $\overrightarrow{PS}$
   (d) $\overrightarrow{QT}$

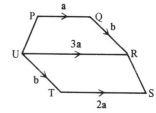

Example 2

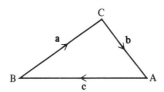

From the diagram rewrite the following as single vectors:

(a) $\mathbf{a} + \mathbf{b}$ (b) $\mathbf{b} + \mathbf{c}$ (c) $\mathbf{a} + \mathbf{b} + \mathbf{c}$

(a) $\mathbf{a} + \mathbf{b} = \overrightarrow{BC} + \overrightarrow{CA} = \overrightarrow{BA} = -\mathbf{c}$

(b) $\mathbf{b} + \mathbf{c} = \overrightarrow{CA} + \overrightarrow{AB} = \overrightarrow{CB} = -\mathbf{a}$

(c) $\mathbf{a} + \mathbf{b} + \mathbf{c} = \overrightarrow{BC} + \overrightarrow{CA} + \overrightarrow{AB} = 0$

In this case, the resultant displacement is zero because the movement is back to the starting point.

Exercise 8.8c

Write each part of the question as a single vector.

1. (a) $x + y$
   (b) $z + x$
   (c) $x + y + z$

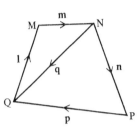

2. (a) $l + m + n + p$
   (b) $n + p$
   (c) $l + m$

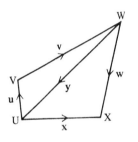

3. (a) $v + y$
   (b) $y + x$
   (c) $u + v + w$

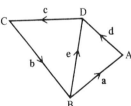

4. (a) $q + r$
   (b) $s + t$
   (c) $p + q + r$

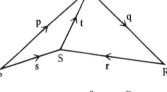

5. (a) $e + c$
   (b) $a + d$
   (c) $a + d + c + b$

6. (a) $v + w$
   (b) $y + z + x$
   (c) $u + v + y + z + x$

7. (a) $l + n$
   (b) $n + q + r$
   (c) $m + q + r + p$

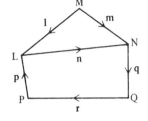

8. (a) $g + b$
   (b) $d + e$
   (c) $b + c + f$

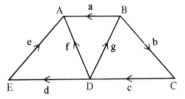

9. (a) $t + v$
   (b) $t + w + x$
   (c) $u + y + z$

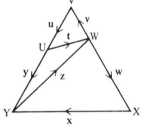

10. (a) $q + s$
    (b) $s + n$
    (c) $l + q + s$

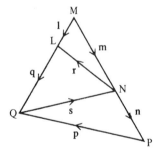

If two lines are equal and parallel, then their displacements are the same.

Example 3

OABC is a parallelogram in which $\overrightarrow{OA} = \mathbf{a}$ and $\overrightarrow{OC} = \mathbf{b}$.

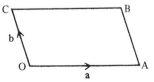

Express the following in terms of **a** and **b**.

(a) $\overrightarrow{CB}$ (b) $\overrightarrow{CO}$ (c) $\overrightarrow{BO}$ (d) $\overrightarrow{CA}$

(a) $\overrightarrow{CB} = \mathbf{a}$

(b) $\overrightarrow{CO} = -\mathbf{b}$

(c) $\overrightarrow{BO} = \overrightarrow{BA} + \overrightarrow{AO} = -\mathbf{b} - \mathbf{a}$

(d) $\overrightarrow{CA} = \overrightarrow{CB} + \overrightarrow{BA} = \mathbf{a} - \mathbf{b}$

Exercise 8.8d

In each question PQRS is a parallelogram.

Find the required vectors in terms of **a** and **b**, or **a**, **b** and **c**.

1. (a) $\overrightarrow{PS}$ (b) $\overrightarrow{SR}$
   (c) $\overrightarrow{QP}$ (d) $\overrightarrow{RQ}$
   (e) $\overrightarrow{PR}$ (f) $\overrightarrow{SQ}$

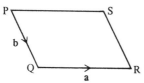

2. (a) $\overrightarrow{QR}$ (b) $\overrightarrow{PQ}$
   (c) $\overrightarrow{RS}$ (d) $\overrightarrow{SQ}$
   (e) $\overrightarrow{PR}$

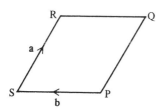

3. (a) $\overrightarrow{PS}$ (b) $\overrightarrow{PQ}$
   (c) $\overrightarrow{SP}$ (d) $\overrightarrow{QS}$
   (e) $\overrightarrow{RP}$

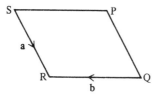

4. (a) $\overrightarrow{QR}$ (b) $\overrightarrow{QP}$
   (c) $\overrightarrow{PS}$ (d) $\overrightarrow{RP}$
   (e) $\overrightarrow{TQ}$

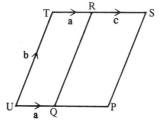

5. (a) $\overrightarrow{PQ}$ (b) $\overrightarrow{RQ}$
   (c) $\overrightarrow{QP}$ (d) $\overrightarrow{UT}$
   (e) $\overrightarrow{QS}$ (f) $\overrightarrow{QT}$

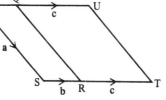

6. (a) $\overrightarrow{HE}$ (b) $\overrightarrow{EF}$
   (c) $\overrightarrow{QF}$ (d) $\overrightarrow{SP}$
   (e) $\overrightarrow{RQ}$ (f) $\overrightarrow{PQ}$
   (g) $\overrightarrow{SR}$

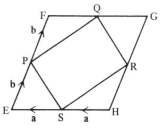

Column vectors

In the graph, the displacement $\overrightarrow{AB}$ can be expressed as the column vector $\begin{pmatrix} 4 \\ 3 \end{pmatrix}$

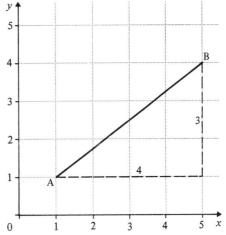

The distance AB can be found using the Theorem of Pythagoras.

$$AB^2 = 4^2 + 3^2 = 25$$
$$AB = \sqrt{25} = 5$$

Exercise 8.8e

Show on a graph one displacement represented by each column vector.
Using Pythagoras' Theorem, calculate the distance moved as a result of the displacement.

1. $\begin{pmatrix} 6 \\ 8 \end{pmatrix}$ 2. $\begin{pmatrix} 12 \\ 5 \end{pmatrix}$ 3. $\begin{pmatrix} 7 \\ 24 \end{pmatrix}$

4. $\begin{pmatrix} -15 \\ 8 \end{pmatrix}$ 5. $\begin{pmatrix} -20 \\ 21 \end{pmatrix}$ 6. $\begin{pmatrix} -40 \\ 9 \end{pmatrix}$

7. $\begin{pmatrix} 16 \\ -30 \end{pmatrix}$ 8. $\begin{pmatrix} 40 \\ -42 \end{pmatrix}$ 9. $\begin{pmatrix} -10 \\ -24 \end{pmatrix}$

10. $\begin{pmatrix} -48 \\ -14 \end{pmatrix}$

A column vector can also represent the Cartesian coordinates of a point. In the graph on page 221,

A is $\begin{pmatrix} 1 \\ 1 \end{pmatrix}$,   B is $\begin{pmatrix} 5 \\ 4 \end{pmatrix}$.

The displacement $\overrightarrow{AB}$ is obtained by subtracting the column vector of **A** from the column vector of **B**.

$$AB = \begin{pmatrix} 5 \\ 4 \end{pmatrix} - \begin{pmatrix} 1 \\ 1 \end{pmatrix} = \begin{pmatrix} 4 \\ 3 \end{pmatrix}$$

Example 4

Find $\overrightarrow{AB}$ if:

(a) B is (3, 4), A is (2, 3)
(b) B is (3, 4), A is (−2, 5)
(c) A is (3, 4), B is (2, 3)

(a) $\overrightarrow{AB} = \begin{pmatrix} 3 \\ 4 \end{pmatrix} - \begin{pmatrix} 2 \\ 3 \end{pmatrix} = \begin{pmatrix} 1 \\ 1 \end{pmatrix}$

(b) $\overrightarrow{AB} = \begin{pmatrix} 3 \\ 4 \end{pmatrix} - \begin{pmatrix} -2 \\ 5 \end{pmatrix} = \begin{pmatrix} 5 \\ -1 \end{pmatrix}$

(c) $\overrightarrow{AB} = \begin{pmatrix} 2 \\ 3 \end{pmatrix} - \begin{pmatrix} 3 \\ 4 \end{pmatrix} = \begin{pmatrix} -1 \\ -1 \end{pmatrix}$

Exercise 8.8f

For questions **1** to **5**, write down the column vectors of the numbered points on the graph.

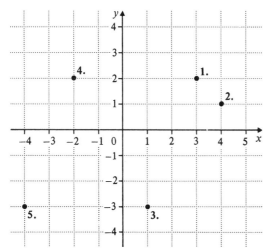

In questions **6** to **20**, write down the displacement $\overrightarrow{AB}$ as a column vector.

6. B is (5, 2), A is (3, 1)
7. B is (4, 3), A is (−1, 2)

8. B is (3, 1), A is (5, −3)
9. B is (2, −1), A is (4, 3)
10. B is (−3, 4), A is (2, 3)
11. A is (−2, 6), B is (5, −1)
12. A is (4, −3), B is (6, 0)
13. A is (−2, 1), B is (−1, 5)
14. A is (7, −3), B is (−3, 4)
15. A is (3, 1), B is (−2, −4)
16. B is (3, 4), A is (−2, −3)
17. B is (5, −2), A is (−1, −4)
18. B is (−2, 1), A is (−5, −2)
19. B is (−2, −1), A is (−3, 5)
20. B is (−1, −8), A is (4, −6)

Example 5

If $\overrightarrow{AB} = \begin{pmatrix} 4 \\ 3 \end{pmatrix}$ and $\overrightarrow{BC} = \begin{pmatrix} 2 \\ -1 \end{pmatrix}$, find $\overrightarrow{AC}$

where A is (1, 1).

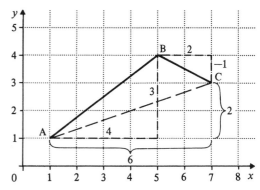

$\overrightarrow{AC} = \overrightarrow{AB} + \overrightarrow{BC}$

$$= \begin{pmatrix} 4 \\ 3 \end{pmatrix} + \begin{pmatrix} 2 \\ -1 \end{pmatrix} = \begin{pmatrix} 6 \\ 2 \end{pmatrix}$$

Exercise 8.8g

1. $\overrightarrow{AB} = \begin{pmatrix} 4 \\ 5 \end{pmatrix}$, $\overrightarrow{BC} = \begin{pmatrix} 2 \\ 3 \end{pmatrix}$; find $\overrightarrow{AC}$.

2. $\overrightarrow{LM} = \begin{pmatrix} 3 \\ 4 \end{pmatrix}$, $\overrightarrow{MN} = \begin{pmatrix} 2 \\ -2 \end{pmatrix}$; find $\overrightarrow{LN}$.

3. $\overrightarrow{PQ} = \begin{pmatrix} 5 \\ 3 \end{pmatrix}$, $\overrightarrow{QR} = \begin{pmatrix} -2 \\ 2 \end{pmatrix}$; find $\overrightarrow{PR}$.

4. $\overrightarrow{UV} = \begin{pmatrix} 3 \\ 2 \end{pmatrix}$, $\overrightarrow{VW} = \begin{pmatrix} -1 \\ 3 \end{pmatrix}$; find $\overrightarrow{UW}$.

5. $\overrightarrow{XY} = \begin{pmatrix} 2 \\ 1 \end{pmatrix}$, $\overrightarrow{YZ} = \begin{pmatrix} 5 \\ -3 \end{pmatrix}$; find $\overrightarrow{XZ}$.

6. $\overrightarrow{AB} = \begin{pmatrix} 3 \\ 3 \end{pmatrix}$, $\overrightarrow{BC} = \begin{pmatrix} -4 \\ 3 \end{pmatrix}$; find $\overrightarrow{AC}$.

A flow chart gives a program for finding a mathematical result.

Example 1

This chart gives a program for finding the average of the numbers 6, 4, and 11.

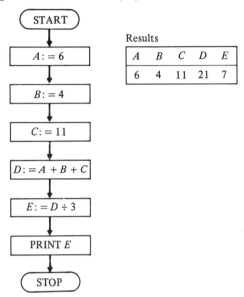

Results

A	B	C	D	E
6	4	11	21	7

Exercise 8.9a

Copy and complete the results table for each flow chart.

1.

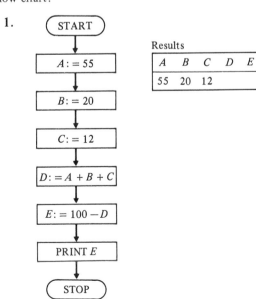

Results

A	B	C	D	E
55	20	12		

2.

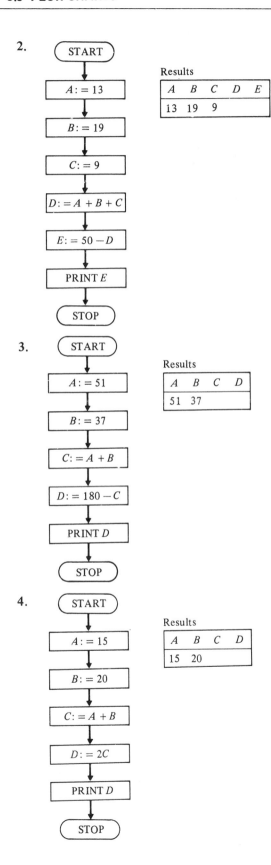

Results

A	B	C	D	E
13	19	9		

3.

Results

A	B	C	D
51	37		

4.

Results

A	B	C	D
15	20		

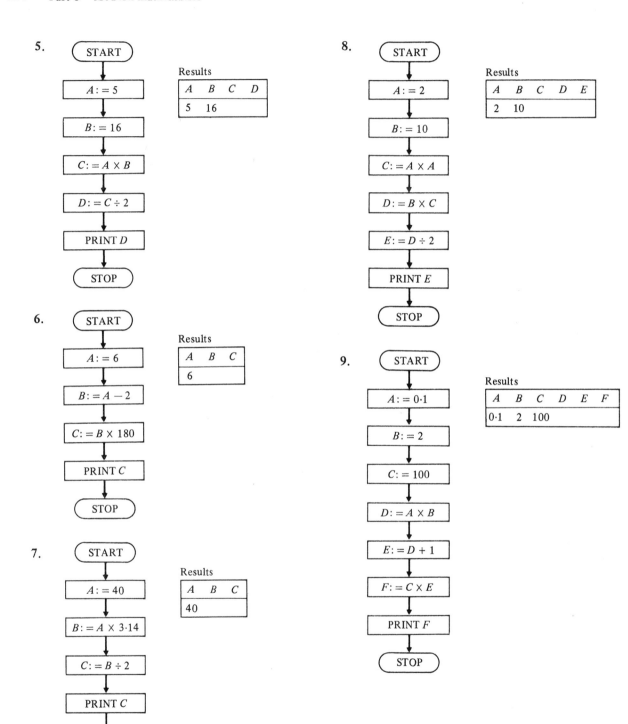

5.

START

$A := 5$

$B := 16$

$C := A \times B$

$D := C \div 2$

PRINT D

STOP

Results

A	B	C	D
5	16		

6.

START

$A := 6$

$B := A - 2$

$C := B \times 180$

PRINT C

STOP

Results

A	B	C
6		

7.

START

$A := 40$

$B := A \times 3 \cdot 14$

$C := B \div 2$

PRINT C

STOP

Results

A	B	C
40		

8.

START

$A := 2$

$B := 10$

$C := A \times A$

$D := B \times C$

$E := D \div 2$

PRINT E

STOP

Results

A	B	C	D	E
2	10			

9.

START

$A := 0 \cdot 1$

$B := 2$

$C := 100$

$D := A \times B$

$E := D + 1$

$F := C \times E$

PRINT F

STOP

Results

A	B	C	D	E	F
0·1	2	100			

10.

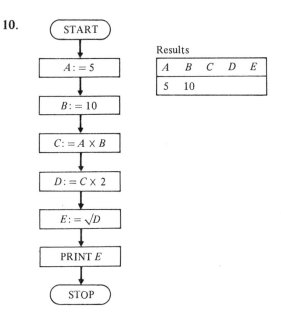

Results

A	B	C	D	E
5	10			

A more difficult flow chart is given below.

Example 2

This chart gives a program for finding the first eight square numbers.

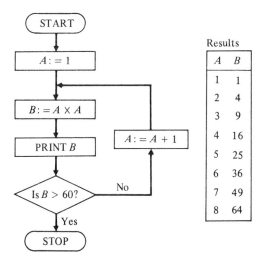

Results

A	B
1	1
2	4
3	9
4	16
5	25
6	36
7	49
8	64

Exercise 8.9b

Copy and complete the results table for each flow chart.

1.

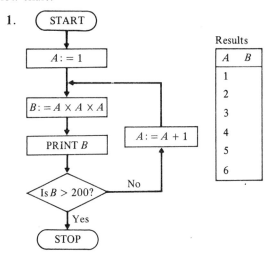

Results

A	B
1	
2	
3	
4	
5	
6	

2.

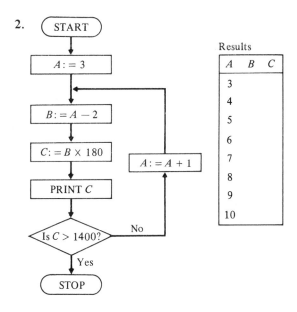

Results

A	B	C
3		
4		
5		
6		
7		
8		
9		
10		

3.

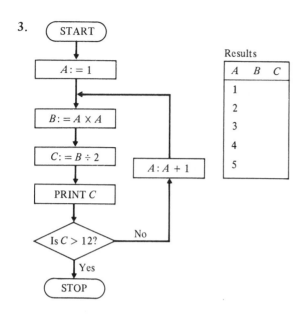

4.

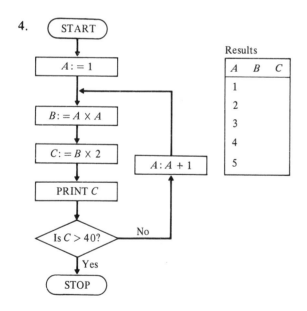

5.

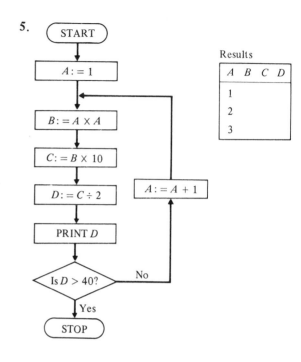

6.

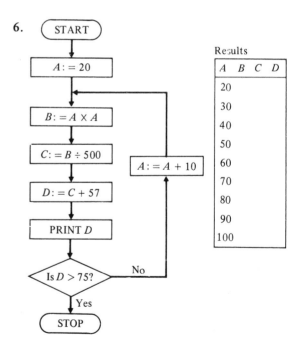

7.

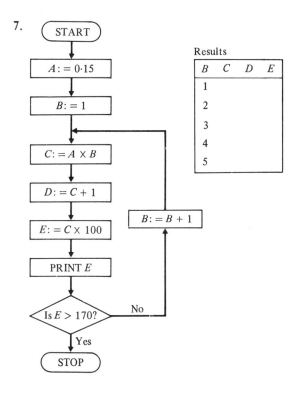

Results

B	C	D	E
1			
2			
3			
4			
5			

9.

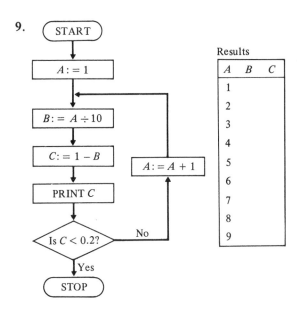

Results

A	B	C
1		
2		
3		
4		
5		
6		
7		
8		
9		

8.

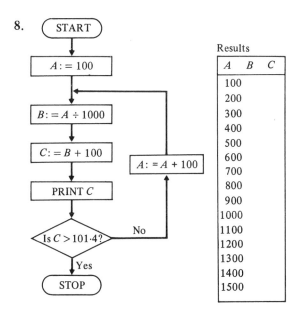

Results

A	B	C
100		
200		
300		
400		
500		
600		
700		
800		
900		
1000		
1100		
1200		
1300		
1400		
1500		

10.

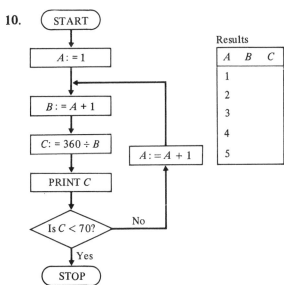

Results

A	B	C
1		
2		
3		
4		
5		

Example 3

The value of the expression $4 \cdot 22^2 - 2 \cdot 31^2$ is to be found using tables of logarithms. The instructions for carrying out this calculation are given below, but not in the correct order.

(a) Find the logarithm of $2 \cdot 31$.
(b) Find the logarithm of $4 \cdot 22$.
(c) Subtract the smaller antilogarithm from the larger.

(d) Multiply the logarithm of 4·22 by 2.
(e) Find the antilogarithm, having doubled log 4·22.
(f) Find the antilogarithm, having doubled log 2·31.
(g) Multiply the logarithm of 2·31 by 2.

Write the correct order for these instructions

The correct order is:

(b) → (d) → (e) → (a) → (g) → (f) → (c).

Exercise 8.9c

For each question, write out the correct order of instructions to achieve the required result.

1. To find the value of the expression $\sqrt{13} - \sqrt{5}$.

   (a) Find the logarithm of 5.
   (b) Find the logarithm of 13.
   (c) Divide the logarithm of 13 by 2.
   (d) Subtract the smaller antilogarithm from the larger.
   (e) Find the antilogarithm having divided log 5 by 2.
   (f) Divide the logarithm of 5 by 2.
   (g) Find the antilogarithm having divided log 13 by 2.

2. To find the value of the expression $5·1^3 - 1·3^3$.

   (a) Subtract the smaller antilogarithm from the larger.
   (b) Find the logarithm of 5·1.
   (c) Multiply the logarithm of 1·3 by 3.
   (d) Find the antilogarithm having trebled log 5·1.
   (e) Find the antilogarithm having trebled log 1·3.
   (f) Multiply the logarithm of 5·1 by 3.
   (g) Find the logarithm of 1·3.

3. To find the volume of the square-based pyramid shown.

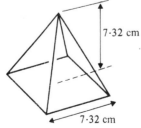

   (a) Find the logarithm of 7·32.
   (b) Divide by 3.
   (c) Look up the antilogarithm.
   (d) Multiply the logarithm of 7·32 by 3.

4. To find the area left when the smaller square shown is taken from the larger square.

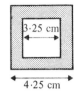

   (a) Find the logarithm of 4·25.
   (b) Multiply the logarithm of 3·25 by 2.
   (c) Subtract the smaller antilogarithm from the larger.
   (d) Multiply the logarithm of 4·25 by 2.
   (e) Find the antilogarithm having doubled log 3·25
   (f) Find the logarithm of 3·25.
   (g) Find the antilogarithm having doubled log 4·25

5. To find the perimeter of the rectangle shown, if its area is 1000 cm².

   (a) Having subtracted the logarithms, divide by 2.
   (b) Find the logarithm of 3.
   (c) Subtract log 3 from log 1000.
   (d) Find the logarithm of 1000.
   (e) Multiply by 8.
   (f) Find the antilogarithm.

6. To find the area of the shaded portion of the diagram.

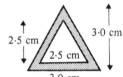

   (a) Square 2·5.
   (b) Divide by 2.
   (c) Subtract the smaller square from the larger square.
   (d) Square 3·0.

7. To find the perimeter of the right-angled triangle shown.

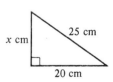

   (a) Find the difference between the two squares.
   (b) Square 20.
   (c) Add 20.
   (d) Find the square root.
   (e) Add 25.
   (f) Square 25.

8. To find the area of
    the shaded ring shown.

    21 cm   28 cm

    (a) Multiply by π.
    (b) Square 28.
    (c) Divide by 4.
    (d) Square 21.
    (e) Find the difference between the two squares.

9. To find the number of slabs
    measuring 0·5 m by 0·5 m to
    pave the border path shown.

    2 m

    3 m

    (a) Square 2.
    (b) Square 0·5.
    (c) Square 3.
    (d) Divide by $0·5^2$.
    (e) Subtract 3^2 from 2^2.

10. To find the average speed of a journey of 39 km
    from Birmingham to Leamington via Solihull if
    the 12 km to Solihull are covered at 48 km/h and
    the average speed of the journey from Solihull
    to Leamington is 54 km/h.

    (a) Add the times of the two halves of the
        journey.
    (b) Divide 12 by 48.
    (c) Divide the total distance by the total journey
        time.
    (d) Divide 27 by 54.

8.10 SOLIDS

A *net* of a solid is the outline of its faces
joined together from which a model of
the solid can be made.

Example 1

(a) The net of a cube.

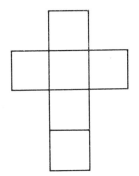

(b) The net of a square-based pyramid.

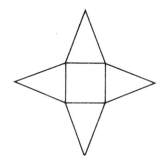

Exercise 8.10a

1. Draw the nets of the following solids.

   (a) cuboid   (b) triangular prism
   (c) tetrahedron   (d) cylinder
   (e) pentagonal prism
   (f) hexagonal-based pyramid.

2. Sketch the solids from the following nets.

   (a)   (b)

   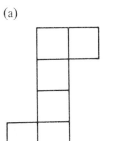   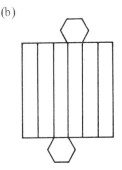

   (c)   (d)

   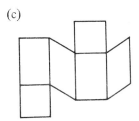   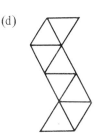

Euler's Theorem

For all polyhedra (solids with plane faces and straight edges):

$$V + F = E + 2$$

where V = the number of vertices
F = the number of faces
E = the number of edges.

Example 2

Cuboid

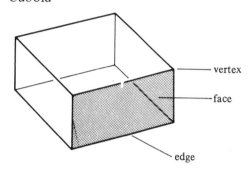

vertex

face

edge

$V = 8, F = 6, E = 12.$
$\therefore V + F = 8 + 6 = 14$
$\quad E + 2 = 12 + 2 = 14$
$\therefore V + F = E + 2.$

Example 3

Square-based pyramid

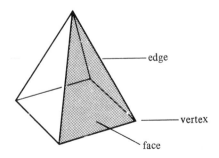

edge

vertex

face

$V = 5, F = 5, E = 8.$
$\therefore V + F = 5 + 5 = 10$
$\quad E + 2 = 8 + 2 = 10$
$\therefore V + F = E + 2$

Exercise 8.10b

In questions **1** to **6** show that Euler's Theorem is true for each solid.

1.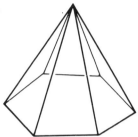

2.

3.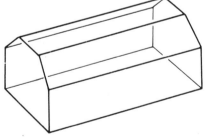

4.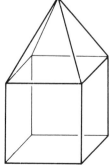

5.

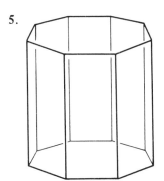

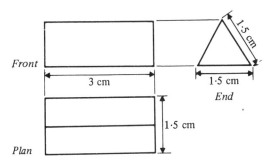

Front

3 cm   1·5 cm

End

Plan

1·5 cm

6.

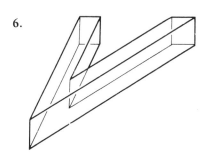

Exercise 8.10c

In questions **1** to **5** draw the front elevation, the end elevation and the plan for each solid.

In questions **7** to **12** sketch each solid and show that Euler's Theorem is true for each.

7. A triangular prism.
8. A hexagonal prism.
9. A pentagonal prism.
10. A tetrahedron (triangular-based pyramid)
11. A parallelopiped ('solid parallelogram').
12. An octahedron (two square-based pyramids with the same base).

1.

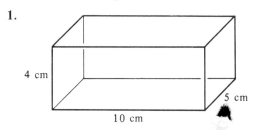

4 cm

10 cm

5 cm

2.

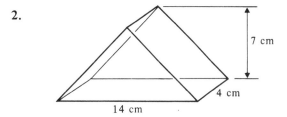

7 cm

14 cm

4 cm

Plans and elevations

A plan shows the shape and dimensions of an object as seen from above.
An elevation shows the shape and dimensions of an object as seen from the front, the end, or the side.
In all plans and elevations, a broken line is used to indicate a hidden edge.

Example 4

Draw the plan, the end elevation, and the side elevation of a triangular prism if it is 3 cm long and the length of each edge of the triangle is 1·5 cm.

3.

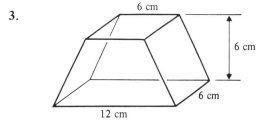

6 cm

6 cm

6 cm

12 cm

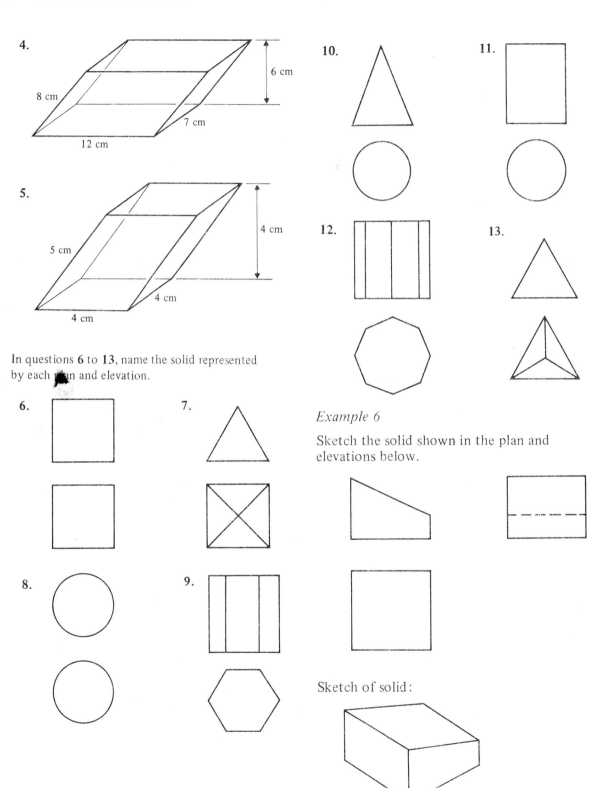

4.

8 cm

6 cm

7 cm

12 cm

5.

5 cm

4 cm

4 cm

4 cm

In questions **6** to **13**, name the solid represented by each plan and elevation.

6.

7.

8.

9.

10.

11.

12.

13.

Example 6

Sketch the solid shown in the plan and elevations below.

Sketch of solid:

Exercise 8.10d

Sketch the solid from each plan and elevation.

1.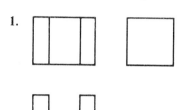

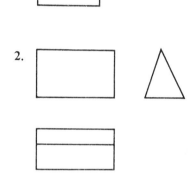

2.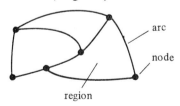

3.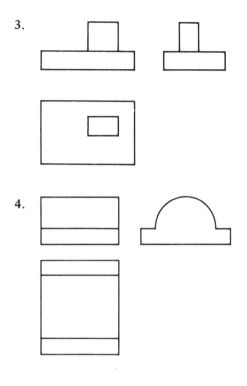

4.

8.11 NETWORKS

A network consists of lines (arcs) meeting at junctions (nodes or vertices).
The area outside the network and the areas bounded by the arcs are called regions.

Example 1

For the network below, write down the number of arcs, regions, and nodes.

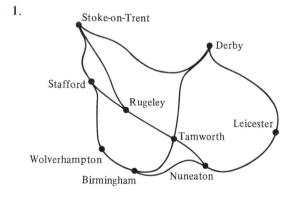

number of arcs = 8
number of regions = 4, i.e. 3 inside, 1 outside
number of nodes = 6

Exercise 8.11a

For each network, write down (a) the number of arcs, (b) the number of regions, (c) the number of nodes.

1.

2.

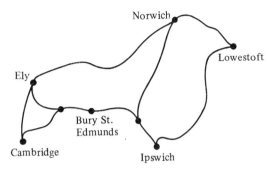

5.

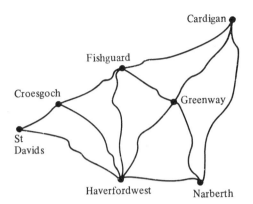

3.

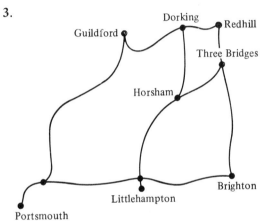

6.

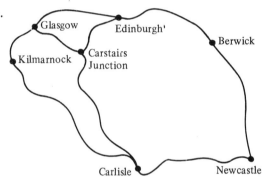

4.

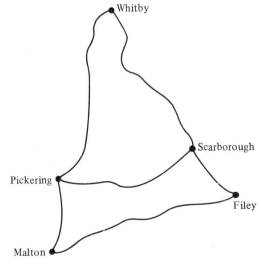

7.

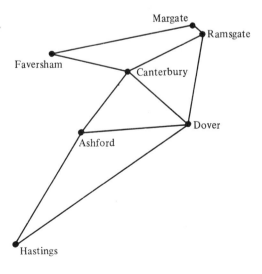

8.

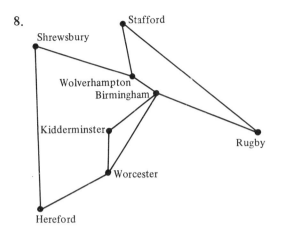

Stafford
Shrewsbury
Wolverhampton
Birmingham
Kidderminster
Rugby
Worcester
Hereford

Euler's Theorem

For all networks

$$R + N = A + 2$$

where R = the number of regions
N = the number of nodes
A = the number of arcs

Example 2

For the network below, show that Euler's Theorem is true.

$R = 4, N = 5, A = 7$
$\therefore R + N = 4 + 5 = 9$
$\quad A + 2 = 7 + 2 = 9$
$\therefore R + N = A + 2$

Example 3

A network has 27 arcs and 15 nodes. How many regions has it?

In any network $R + N = A + 2$
$$\therefore R + 15 = 27 + 2 = 29$$
$$R = 14$$
Hence the network has 14 regions.

Exercise 8.11b

For each network in questions **1** to **6**, show that Euler's Theorem is true.

1.    2.

3. 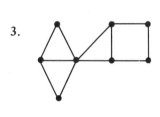   4.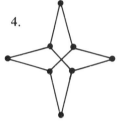

5.    6.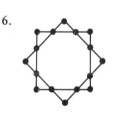

7. Find the number of regions in a network having 25 arcs and 12 nodes.
8. Find the number of regions in a network having 21 arcs and 18 nodes.
9. Find the number of regions in a network having 24 arcs and 13 nodes.
10. Find the number of nodes in a network having 26 arcs and 10 regions.
11. Find the number of nodes in a network having 27 arcs and 9 regions.
12. Find the number of nodes in a network having 23 arcs and 12 regions.
13. Find the number of arcs in a network having 11 regions and 15 nodes.
14. Find the number of arcs in a network having 8 regions and 14 nodes.
15. Find the number of arcs in a network having 14 regions and 16 nodes.
16. Find the number of arcs in a network having 10 regions and 16 nodes.

Traversability

If a network can be drawn without taking the pencil off the paper and without following the same line twice, then the network is said to be traversable.
A network is only traversable if
either (a) all its junctions are even.
or (b) only two of its junctions are odd.

Example 4

Find which of the networks are traversable.

(a)

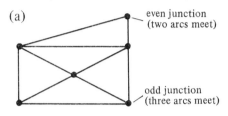

even junction
(two arcs meet)

odd junction
(three arcs meet)

(b)   (c)

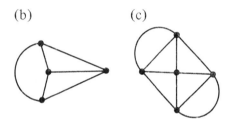

(a) no. of even junctions = 4;
    no. of odd junctions = 2.
    So network is traversable.
(b) no. of even junctions = 0;
    no. of odd junctions = 4.
    So network is *not* traversable.
(c) no. of even junctions = 5;
    no. of odd junctions = 0.
    So network is traversable.

Exercise 8.11c

State which networks are traversable, giving reasons.

1.   2.

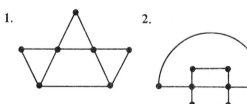

3.   4.

5.   6.

7.   8.

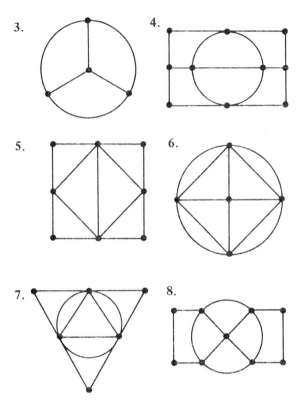

A network can be described by means of a matrix which gives the number of routes from node to node.

Example 5

1. *Single-stage* (direct routes)

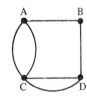

		to			
		A	B	C	D
from	A	0	1	2	0
	B	1	0	0	1
	C	2	0	0	2
	D	0	1	2	0

2. *Two-stage routes*

		to			
		A	B	C	D
from	A	5	0	0	5
	B	0	2	4	0
	C	0	4	8	0
	D	5	0	0	5

Exercise 8.11d

Write down (a) the single-stage matrix, (b) the two-stage matrix for each network.

1.

2.

3.

4.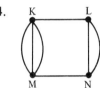

5.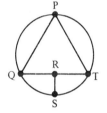

6.

7.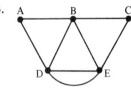

8.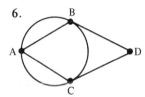

Example 6

The matrix below gives the single-stage routes on a network of roads.

$$
\begin{array}{c c c c c}
 & A & B & C & D \\
A & 0 & 0 & 2 & 1 \\
B & 0 & 0 & 1 & 0 \\
C & 2 & 1 & 0 & 2 \\
D & 1 & 0 & 2 & 0
\end{array}
$$

(a) Draw the network.
(b) Find four ways of reaching A from B by multi-stage routes.

(a)

(b) A can only be reached from B by the following routes. These are:

  (i) $B \rightarrow C \rightarrow A$ (two ways)
  (ii) $B \rightarrow C \rightarrow D \rightarrow A$ (two ways)

Exercise 8.11e

Each matrix below gives the single-stage routes on a network.

(a) Draw the network.
(b) List all routes from A to C.

1.
$$
\begin{array}{c c c c}
 & A & B & C \\
A & 0 & 2 & 1 \\
B & 2 & 0 & 1 \\
C & 1 & 1 & 0
\end{array}
$$

2.
$$
\begin{array}{c c c c}
 & A & B & C \\
A & 0 & 2 & 2 \\
B & 2 & 0 & 1 \\
C & 2 & 1 & 0
\end{array}
$$

3.
$$
\begin{array}{c c c c c}
 & A & B & C & D \\
A & 0 & 1 & 0 & 0 \\
B & 1 & 0 & 1 & 1 \\
C & 0 & 1 & 0 & 1 \\
D & 0 & 1 & 1 & 0
\end{array}
$$

4.
$$
\begin{array}{c c c c c}
 & A & B & C & D \\
A & 0 & 1 & 1 & 1 \\
B & 1 & 0 & 1 & 0 \\
C & 1 & 1 & 0 & 1 \\
D & 1 & 0 & 1 & 0
\end{array}
$$

5.
$$
\begin{array}{c c c c c}
 & A & B & C & D \\
A & 0 & 1 & 1 & 1 \\
B & 1 & 0 & 0 & 1 \\
C & 1 & 0 & 0 & 1 \\
D & 1 & 1 & 1 & 0
\end{array}
$$

6.
$$
\begin{array}{c c c c c}
 & A & B & C & D \\
A & 0 & 1 & 0 & 0 \\
B & 1 & 0 & 2 & 0 \\
C & 0 & 2 & 0 & 1 \\
D & 0 & 0 & 1 & 0
\end{array}
$$

7.
$$
\begin{array}{c c c c c}
 & A & B & C & D \\
A & 0 & 2 & 0 & 0 \\
B & 2 & 0 & 1 & 0 \\
C & 0 & 1 & 0 & 2 \\
D & 0 & 0 & 2 & 0
\end{array}
$$

8.
$$
\begin{array}{c c c c c}
 & A & B & C & D \\
A & 0 & 2 & 0 & 0 \\
B & 2 & 0 & 1 & 1 \\
C & 0 & 1 & 0 & 1 \\
D & 0 & 1 & 1 & 0
\end{array}
$$

9.
$$
\begin{array}{c c c c c}
 & A & B & C & D \\
A & 0 & 1 & 0 & 0 \\
B & 1 & 0 & 1 & 1 \\
C & 0 & 1 & 0 & 2 \\
D & 0 & 1 & 2 & 0
\end{array}
$$

10.
$$
\begin{array}{c c c c c}
 & A & B & C & D \\
A & 0 & 2 & 0 & 0 \\
B & 2 & 0 & 2 & 0 \\
C & 0 & 2 & 0 & 1 \\
D & 0 & 0 & 1 & 0
\end{array}
$$

A *tally chart* is a convenient way of recording the frequency of an event, as shown below.

choice	tally	frequency
mashed potatoes	⊦⊦⊦ ⊦⊦⊦ ⊦⊦⊦ ⊦⊦⊦ //	22
roast potatoes	⊦⊦⊦ ⊦⊦⊦ ⊦⊦⊦ ⊦⊦⊦ ⊦⊦⊦ ////	29
chipped potatoes	⊦⊦⊦ ⊦⊦⊦ ⊦⊦⊦ ⊦⊦⊦ ⊦⊦⊦ ⊦⊦⊦ ⊦⊦⊦ ⊦⊦⊦ ⊦⊦⊦ ////	49
	total	100

The tally chart shows the choice of potatoes by 100 pupils in a school canteen.

Example 1

A dice was rolled 50 times and the scores were recorded as follows.

2	5	1	3	1	3	4	1	6	4
4	4	4	2	5	5	5	3	5	4
3	5	2	3	4	2	3	2	6	6
1	6	6	1	6	6	2	3	2	5
2	6	5	4	4	1	4	5	6	1

Record this information on a suitable tally chart.

score	tally	frequency
6	⊦⊦⊦ ////	9
5	⊦⊦⊦ ////	9
4	⊦⊦⊦ ⊦⊦⊦	10
3	⊦⊦⊦ //	7
2	⊦⊦⊦ ///	8
1	⊦⊦⊦ //	7
	total	50

Exercise 9.1a

1. The details below show the first 30 names on a voting list for a certain district, together with how they voted.

name	vote	name	vote	name	vote
Adams P.	Conservative	Andrews M.	Conservative	Beasley O.	Conservative
Adams T.	Conservative	Andrews S.	Conservative	Benson D.	Liberal
Aitken R.	did not vote	Ashton B.	did not vote	Benson J.	Liberal
Alder C.	Labour	Ashton N.	did not vote	Benson P.	Labour
Alder W.	Labour	Atkinson H.	Labour	Brown P.	Conservative
Allerton V.	Labour	Atkinson M.	Labour	Brown V.	Conservative
Alsop L.	Conservative	Bailey W.	did not vote	Burns C.	Labour
Anderson G.	Liberal	Barker W.	Labour	Burns M.	did not vote
Anderson K.	Liberal	Bassett A.	Conservative	Butler J.	Conservative
Anderson R.	Labour	Bates L.	Labour	Butler L.	Conservative

Record the numbers of each kind of vote (including those who did not vote) on a tally chart.

2. A card was withdrawn from a pack forty times and replaced. The results were recorded as Picture, Ace, or Number. The details are given below.

Numbers	Number	Ace	Picture	Ace	Ace
Ace	Picture	Number	Number	Number	Number
Picture	Number	Ace	Number	Picture	Picture
Picture	Number	Picture	Number	Number	Number
Number	Picture	Ace	Number	Ace	Number
Number	Ace	Number	Picture	Picture	
Number	Number	Number	Number	Number	

Record this information on a suitable tally chart.

3. A football team played fifty matches in one season. The number of goals they scored in each match is shown below.

1	0	2	3	4	0	3	2	2	0
3	2	5	1	3	1	0	0	0	1
0	1	1	0	1	0	1	4	1	2
0	1	3	2	1	2	2	2	6	0
1	4	2	0	5	1	4	3	3	4

Show on a tally chart the frequency of goals scored.

4. The destination and time of departure of all inter-city trains leaving King's Cross station in London on a weekday is shown below.

4.05	Leeds	11.35	Newcastle	16.04	Bradford
5.50	Aberdeen	11.55	Aberdeen	16.08	York
7.40	Newcastle	12.04	Hull	16.42	Hull
7.45	Bradford	12.25	York	17.00	Edinburgh
8.00	Edinburgh	12.45	Bradford	17.04	Leeds
8.04	Hull	13.00	Edinburgh	17.35	Newcastle
8.25	Cleethorpes	13.04	Cleethorpes	18.00	Edinburgh
9.00	Edinburgh	13.25	Newcastle	18.04	Bradford
9.04	Leeds	14.00	Edinburgh	18.09	York
9.30	Newcastle	14.04	York	18.20	Cleethorpes
10.00	Edinburgh	14.45	Leeds	19.00	Newcastle
10.04	York	15.00	Aberdeen	19.04	Newcastle
10.45	Leeds	15.45	Leeds	19.40	Leeds
10.55	Edinburgh	16.00	Edinburgh		

Show on a tally chart the number of trains for each destination.

5. The year letter on the registration number plate of fifty cars that passed along a main road on a day in August was recorded as follows.

J	P	N	M	S	S	R	G	N
R	M	B	R	G	N	P	R	H
N	S	T	J	R	F	M	J	
L	G	R	N	P	R	R	S	
R	N	S	S	T	H	T	S	
S	S	N	J	P	N	S	M	

Draw up a tally chart to show how many cars of each year letter were recorded.

Statistical information can be represented by a bar chart, a pie chart, or a pictogram (see page 102). A frequency distribution (as in Example 1) is usually shown as a *histogram*.

In this book the height of each rectangle is a measure of the frequency of each event.

Example 2

Four coins are each tossed 64 times. The number of times a 'head' appeared on all four coins is shown on the tally chart. Show this information on a histogram.

'heads'	tally	frequency
4	///	3
3	ℍℍℓ ℍℍℓ ℍℍℓ /	16
2	ℍℍℓ ℍℍℓ ℍℍℓ ℍℍℓ ///	23
1	ℍℍℓ ℍℍℓ ℍℍℓ //	17
0	ℍℍℓ	5
	total	64

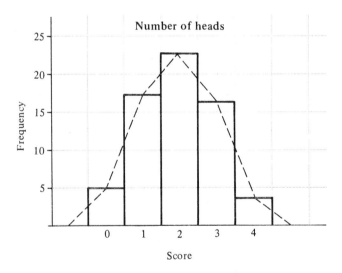

A *frequency polygon* is formed by joining the mid-points of each rectangle on the histogram by straight lines. This is shown by the broken lines on the graph above. The polygon is completed by continuing the lines to the horizontal axis as shown.

Exercise 9.1b

In questions **1** to **10**, illustrate the details on a histogram.

1. A farmer records on a tally chart for each of his chickens the number of eggs per week laid by the chicken. The chart for one of his chickens over a period of a year is shown below.

no. of eggs per week	tally	frequency (no. of weeks)
7	//	2
6	////	4
5	ＴＨＬ ＴＨＬ	10
4	ＴＨＬ ＴＨＬ ＴＨＬ /	16
3	ＴＨＬ ＴＨＬ /	11
2	////	4
1	///	3
0	//	2
	total	52

2. The tally chart shows the number of wet days in each week over a period of one year.

no. of wet days	tally	frequency (no. of weeks)
7	//	2
6	ＴＨＬ	5
5	ＴＨＬ //	7
4	ＴＨＬ ＴＨＬ //	12
3	ＴＨＬ ＴＨＬ	10
2	ＴＨＬ ////	9
1	ＴＨＬ /	6
0	/	1
	total	52

3. The number of minutes late that a certain bus arrives each day is recorded for a period of 90 days in the tally chart.

minutes late	tally	frequency (no. of days)
10	///	3
9	///	3
8	ＴＨＬ /	6
7	ＴＨＬ ////	9
6	ＴＨＬ ////	9
5	ＴＨＬ ＴＨＬ ＴＨＬ	15
4	ＴＨＬ ＴＨＬ ＴＨＬ ///	18
3	ＴＨＬ ＴＨＬ ＴＨＬ	15
2	ＴＨＬ /	6
1	///	3
0	///	3
	total	90

4. The finishing position of a sprinter in 50 different races is recorded in the tally chart. In each race there were eight runners.

position	tally	frequency
8th	////	4
7th	⧼⧽ /	6
6th	⧼⧽ ///	8
5th	⧼⧽ ⧼⧽ //	12
4th	⧼⧽ ⧼⧽	10
3rd	⧼⧽ //	7
2nd	//	2
1st	/	1
	total	50

5. Two dice are thrown together one hundred times. The scores are recorded in the tally chart.

score	tally	frequency
12	//	2
11	////	4
10	⧼⧽ /	6
9	⧼⧽ ///	8
8	⧼⧽ ⧼⧽ ⧼⧽ /	16
7	⧼⧽ ⧼⧽ ⧼⧽ ⧼⧽	20
6	⧼⧽ ⧼⧽ ////	14
5	⧼⧽ ⧼⧽ //	12
4	⧼⧽ ///	8
3	⧼⧽ /	6
2	////	4
	total	100

6. A football club plays 50 matches in a season. The number of goals they score for each match in one season is shown below.

No. of goals	0	1	2	3	4	5	6
No. of matches	6	10	11	12	6	3	2

7. The results in marks out of 10 of an English test for sixty children is shown below.

Marks scored	0	1	2	3	4	5	6	7	8	9	10
No. of pupils	1	2	3	6	10	11	9	8	5	3	2

8. The size of shoe worn by each of 30 children is shown in the table.

Size of shoe	2	3	4	5	6
No. of children	3	6	11	8	2

9. The money received in tips by a waitress over a period of 30 days serving in a restaurant is shown below.

Money collected	10p	20p	30p	40p	50p	60p	70p	80p	90p	£1	£1·10	£1·20	£1·30	£1·40	£1·50
No. of days	0	0	0	1	2	3	6	4	4	3	2	2	1	1	1

10. A table-tennis player wins 100 games. The number of points over those of his opponent is recorded in the table for each game.

Points difference	2	3	4	5	6	7	8	9	10	11	12	13	14	15	16	17	18	19	20	21
No. of games	2	2	3	4	5	6	8	15	14	13	10	7	5	2	2	1	1	0	0	0

In questions 11 to 15, illustrate the details on a frequency polygon.

11. A cricketer bowls in twenty matches in one season. The number of wickets he took in each match is shown in the table.

No. of wickets	0	1	2	3	4	5
No. of matches	3	5	6	3	2	1

12. A lorry driver can make a total of 6 deliveries between a depot and a shop in one working day. The actual number of deliveries he makes each day over a period of 30 days is shown in the table.

No. of deliveries	0	1	2	3	4	5	6
No. of days	1	3	4	7	8	5	2

13. The career of a batsman covered 25 seasons. The table records the number of centuries he scored over these seasons.

No. of centuries	0	1	2	3	4	5
No. of seasons	2	5	7	6	4	1

14. A football club entered a five-round knockout competition for 20 seasons. The club never won the trophy. The table shows the round in which the club was eliminated for each of the seasons.

Round when club was eliminated	1	2	3	4 (semi-final)	5 (final)
No. of seasons	4	5	6	3	2

15. A taxi has seating capacity for six passengers. The actual number of passengers carried for each of 40 journeys is shown.

No. of passengers	1	2	3	4	5	6
No. of journeys	7	8	10	9	4	2

The mode is 'the most popular' item,
i.e. the one result that occurs most
frequently in a set of results.

Example 1

Find the mode of the following:

4, 3, 2, 1, 4, 3, 3, 2, 1, 3, 4, 2, 3, 4, 3, 3,
2, 2, 2, 3, 4, 2, 3, 3, 1.

Number	Tally	Frequency
1	///	3
2	++++ //	7
3	++++ ++++	10
4	++++	5

∴ 3 is the mode.

Example 2

The number of goals scored by 15 teams
one Saturday was as follows. What is the
modal score?

Goals scored	Tally	Frequency
0	////	4
1	//	2
2	////	4
3	///	3
4	/	1
5	/	1

∴ there is no modal score as no result
occurs most often.

Exercise 9.2a

1. The temperature in °C on 20 winter days was:
   5, 3, 3, 2, 0, 0, 1, 2, 5, 4, 3, 1, 5, 4, 2,
   2, 4, 3, 4, 3.
   What was the modal temperature?

2. In a local football league, the goals scored one
   Saturday were 4, 0, 2, 6, 3, 0, 5, 2, 5, 7, 0, 1, 4,
   2, 1, 7, 6, 6, 5, 3, 2, 3, 7, 6, 2, 0, 1, 4.
   What is the modal score?

3. The times (in minutes) taken by a man going to
   work were 54, 57, 55, 57, 56, 58, 54, 53, 56,
   55, 56, 57, 53, 58, 54, 58, 54, 57, 55, 58.
   What is the modal time?

4. The shoe sizes of a class are 5, 7, 6, 7, 5, 5, 6, 6,
   7, 9, 9, 6, 6, 6, 8, 9, 8, 9, 5, 7, 7, 8, 5, 8.
   What is the modal size?

5. In various shops, 1 kg of tomatoes was priced in
   pence as follows.
   39, 39, 40, 43, 43, 44, 37, 44, 37, 39, 40, 42,
   43, 43, 44, 43, 44, 37, 39, 38, 39, 40, 44.
   What is the modal price?

6. The attendance of a class is as follows.
   30, 32, 28, 28, 29, 30, 31, 28, 27, 27, 31, 28,
   29, 32, 32, 28, 29, 30, 31, 30, 32, 31, 29, 29,
   30, 30, 28, 27, 28, 32.
   What is the modal attendance?

7. Two dice are thrown together 20 times and the
   results are shown below.

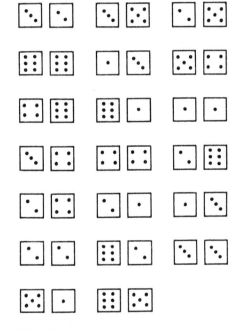

Find the modal score.

8. Three coins *A*, *B* and *C* are tossed simultaneously 20 times over. Each either lands Heads (H) or Tails (T) as shown in the table below.

A	B	C	A	B	C
H	H	T	T	H	T
H	T	T	T	T	H
T	T	H	H	T	H
H	T	T	H	H	H
T	H	T	T	H	T
T	T	T	T	T	T
H	H	T	H	T	H
H	T	H	T	H	T
H	T	T	T	H	H
H	H	H	T	T	T

Which is the modal set of appearances:
    3 Heads
    3 Tails
    2 Heads and 1 Tail
  or 1 Head and 2 Tails?

9. A hockey club takes part annually in a league tournament with five other clubs. The table below shows the record of this club over fifteen seasons. Find the modal number of league points.

	Games won (2 points)	Games drawn (1 point)	Games lost (0 point)
1964	3	4	3
1965	4	0	6
1966	8	0	2
1967	6	2	2
1968	4	2	4
1969	6	0	4
1970	2	4	4
1971	6	2	2
1972	6	1	3
1973	5	4	1
1974	5	1	4
1975	7	2	1
1976	5	0	5
1977	3	2	5
1978	7	0	3

10. The scoring record of a rugby club over fifteen matches is given in the table. Find the modal number of match points.

Tries (4 points)	Goals (3 points)	Conversions (2 points)
1	1	1
4	5	1
3	1	0
3	3	2
1	2	1
3	3	3
4	2	1
2	5	2
3	5	3
4	3	1
2	0	1
2	4	2
4	4	2
2	5	1
0	3	0

The *median* is the middle value when the results are placed in order of size.

Example 3

Find the median of:

(a) 2, 4, 6, 8, 10, 12, 14
(b) 3, 2, 5, 7, 4, 6, 9, 2

(a) In order of size, the numbers are:

    2, 4, 6, 8, 10, 12, 14

    Therefore the median is 8.

(b) In order of size, the numbers are:

    2, 2, 3, 4, 5, 6, 7, 9

    Therefore the median is $\dfrac{4+5}{2} = 4\frac{1}{2}$

Exercise 9.2b

1. The attendance of a class during one week was 32, 30, 29, 33, 34. What is the median attendance?

2. The number of potatoes needed to fill five 5-kg bags were 69, 84, 76, 71, 73. Find the median.

3. In one week the midday temperatures were 26°C, 25°C, 20°C, 19°C, 23°C, 22°C, 17°C. What is the median temperature?

4. In one week the daily tips received by a waitress were 75p, 120p, 90p, 95p, 60p, 70p, 80p. Find the median.

5. A 200-m sprinter ran in six races and his times were 23·7s, 23·0s, 23·5s, 24·1s, 22·9s, and 23·3s. What is his median time?

6. The times in minutes taken by a girl walking to work were 39, 37, 42, 40, 35, 36. What is the median time?

7. The price per $\frac{1}{2}$ kg of apples on various market stalls was 27p, 27p, 30p, 28p, 32p, 33p. What is the median price?

8. The weights in kg of eight boys were 40, 36, 33, 36, 45, 38, 41, 43. What is the median weight?

9. The waist measurements of eight girls measured in cm were 69, 68, 59, 70, 72, 57, 62, 64. What is the median of these measurements?

10. Two dice were thrown together five times and the results are shown below.

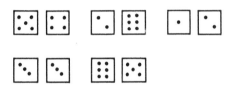

Find the median score.

The *mean* of a set of results is the sum of all the results divided by the number of results.

Example 4

The heights of six women in cm are 162, 156, 165, 153, 150, 156. What is the mean height?

$$\text{mean} = \frac{162 + 156 + 165 + 153 + 150 + 156}{6}$$

$$= \frac{942}{6} = 157 \text{ cm}$$

Exercise 9.2c

1. The number of wet days during four months were:
August, 11
September, 8
October, 13
November, 24
Find the mean number of wet days.

2. A salesman earns the following amounts over a four-week period: £53·32, £60·21, £55·27, £61·08. Find the mean wage.

3. The times in minutes of five railway journeys from Edinburgh to Glasgow were 43, 47, 49, 41, 50. Find the mean journey time.

4. The times of five coach journeys from London to Birmingham were 2 h 45 min, 2 h 33 min, 3 h 5 min, 2 h 32 min, 3 h 10 min. Find the mean journey time.

5. The number of spectators at a football club for five rounds of a cup competition was: 5200, 8130, 13 205, 18 055, 24 030. Find the mean attendance.

6. A small business spends the following on postal charges in one week: 98p, £1·10p, $86\frac{1}{2}$p, 79p, $95\frac{1}{2}$p, 35p. Find the mean.

7. A 400-m sprinter ran in six races and his times were 47·4s, 46·8s, 46·6s, 45·9s, 47·5s, 47·8s. Find the mean of these times.

8. Two dice were thrown together six times and the results are shown below.

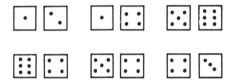

Find the mean of these scores.

9. The heights in cm of eight women are 163, 160, 166, 159, 167, 170, 162, 165. Find the mean of these heights.

10. The heights in cm of eight men are 185, 179, 178, 180, 186, 183, 180, 177. Find the mean of these heights.

Example 5

The frequency table shows the number of Heads when four coins are tossed 64 times. What is the mean?

Heads	frequency
4	3
3	16
2	23
1	17
0	5

$$\text{mean} = \frac{(4 \times 3) + (3 \times 16) + (2 \times 23) + (1 \times 17) + (0 \times 5)}{64}$$

$$= \frac{12 + 48 + 46 + 17 + 0}{64} = \frac{123}{64}$$

$$= 1·92 \text{ (to 3 s.f.)}$$

Exercise 9.2d

1. The table shows the different prices a woman paid for six eggs and the frequency of her purchases.

Price in p	27	26	25	24	23
No. of times paid	3	5	5	3	4

Find the mean price that she paid.

2. A girls' sports club has fifty members. The table shows the frequency of girls of each age.

Age in years	11	12	13	14	15	16
No. of girls	4	8	5	12	9	12

Find the mean age of the girls.

3. A class contains thirty-two pupils. The table shows the frequency of attendance over a period of 30 days.

No. present	32	31	30	29	28	27	26
No. of days	5	6	9	7	1	1	1

Find the mean attendance figure.

4. The table shows the number of passengers who left on the 15.30 bus from Stourbridge to Bridgnorth over a period of 60 days.

No. of passengers	8	7	6	5	4	3
No. of days	6	8	9	10	11	16

Find the mean number of passengers on this bus over the period.

5. A motorist buys petrol for his car daily. The table shows how many gallons he bought each day over a period of 42 days.

No. of gallons	6	5	4	3	2
No. of days	8	5	14	9	6

Find the mean number of gallons he bought each day.

6. The temperature at midday is recorded daily over a period of 30 days as shown in the table.

Temperature (°C)	24	23	22	21	20	19	18
No. of days	2	2	3	5	10	3	5

Find the mean temperature for this period.

7. The table shows the number of goals a football team scored for each of 50 matches.

No. of goals	6	5	4	3	2	1	0
No. of matches	1	2	4	8	10	12	13

Find the mean number of goals per match.

8. When four dice were thrown together a total of 200 times, the number of sixes scored per throw is shown in the table.

No. of sixes	4	3	2	1	0
No. of throws	1	2	7	40	150

Find the mean number of sixes scored per throw.

9. Four playing cards are removed from a pack at random. This is repeated 200 times. The number of picture cards drawn each time is shown in the table.

No. of picture cards	4	3	2	1	0
No. of draws	2	5	15	59	119

Find the mean number of picture cards removed per draw.

10. The table shows the atmospheric pressure in mm of mercury at midday over a period of 20 days.

Pressure	750	752	754	756	758	760
No. of days	1	0	3	2	1	3

Pressure	762	764	766	768	770
No. of days	3	2	0	3	2

Find the mean pressure over the period.

Example 6

Find a set of four numbers in ascending order starting with 5 each time, so that:

(a) the mode is 7
(b) the median is 7
(c) the mean is 7

(a) Here 7 is the result that occurs most frequently, so a possible set of numbers is 5, 6, 7, 7.

(b) The middle value has to be 7, so a possible set is 5, 6, 8, 9.

(c) The sum has to be 28, so a possible set is 5, 6, 7, 10.

Exercise 9.2e

Questions **1** to **10** give details of lists of numbers.
Match the details in each question with one of the
lettered sets of numbers below:

A = { 5, 7, 10, 7, 4, 2, 7}
B = { 10, 5, 11, 8, 12, 8}
C = { 3, 2, 6, 3, 1}
D = { 6, 2, 4, 3, 8, 3, 9}
E = { 5, 2, 9, 5, 4, 5}
F = { 6, 12, 4, 8, 2, 16, 4, 4}
G = { 12, 6, 1, 8, 15, 6}
H = { 7, 5, 10, 3, 5}
I = { 9, 8, 2, 10, 4, 7, 9}
J = { 6, 1, 8, 4, 6}

Each set may be used only once.

1. mode = 3; median = 4; mean = 5
2. mode = 9; median = 8; mean = 7
3. mode = median = 7; mean = 6
4. mode = median = 6; mean = 5
5. mode = median = 5; mean = 6
6. mode = median = mean = 3
7. mode = 6; median = 7; mean = 8
8. mode = 8; median = mean = 9
9. mode = median = mean = 5
10. mode = 4, median = 5, mean = 7

Exercise 9.2f

For each question, find which of the mode, the
median and the mean is different from the other
two.

1. A dice was thrown nine times and the scores
   were as follows:
   5, 3, 6, 5, 3, 1, 6, 4 and 3.
2. Two dice were thrown together nine times
   and the scores were as follows:
   7, 2, 5, 8, 12, 7, 5, 3 and 5.
3. On nine consecutive days the temperature in
   °C was:
   13, 11, 9, 7, 9, 13, 14, 10 and 13.
4. Over nine matches a football club's scoring
   record was:
   3, 1, 3, 5, 1, 2, 1, 0 and 2 goals.
5. Over seven matches a rugby club's scoring
   record was:
   16, 15, 4, 17, 15, 18 and 20 points.
6. During a seven-week period a man worked
   the following number of hours each week:
   37, 39, 41, 38, 51, 41 and 40.

7. During a seven-week period a man's weekly
   wages were as follows:
   £61, £68, £70, £64, £68, £65 and £66.
8. Seven light bulbs were used in the same
   socket. The life in days of each bulb was:
   75, 70, 75, 72, 75, 78 and 73.
9. A taxi-driver answers eleven calls. The number
   of passengers he carries are:
   4, 6, 3, 6, 1, 6, 3, 3, 5, 4 and 3.
10. The number of cars that used a ferry during
    a certain week was as follows:

Mon.	Tues.	Wed.	Thur.	Fri.	Sat.
14	16	13	12	16	25

11. On six consecutive days a train arrived at a
    station and was late by:
    6, 7, 3, 0, 5 and 3 minutes.
12. Over four cricket matches a batsman's
    scores were as follows:
    21, 11, 0, 11, 15, 14, 11 and 13 runs.
13. On a certain day the numbers of passengers
    using the buses from Worcester to Bromyard
    were as follows:

8.00 a.m.	9.00 a.m.	10.00 a.m.	11.00 a.m.
16	10	12	12

2.00 p.m.	4.00 p.m.	5.00 p.m.	6.00 p.m.
16	16	35	11

14. Over a two-week period a class attendance
    record was:

Mon.	Tues.	Wed.	Thur.	Fri.
28	26	25	28	30
32	28	25	24	24

15. At a factory the number of absent workers
    over a two-week period was:

Mon.	Tues.	Wed.	Thur.	Fri.
5	0	5	5	7
1	1	5	0	1

A scatter diagram is a graph which is plotted to see if there is any relationship between two sets of data.

Example

The table gives the height in cm and shoe size of 10 children.

Child	A	B	C	D	E	F	G	H	I	J
Height (cm)	130	135	135	155	150	160	120	170	140	125
Shoe size	2	4	3	6	5	7	1	9	4	2

Illustrate the information on a scatter diagram.

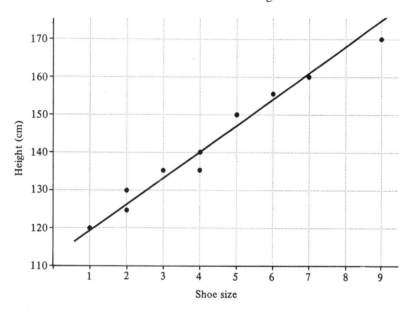

If a straight line can be drawn so that the points plotted line on or about it, then this line is called a *line of best fit*.

(a) Scatter diagram showing indirect correlation.

Correlation

The scatter diagram in the above example shows that people who are small wear small-sized shoes, and taller people wear larger shoes. In this case, there is a *direct* (positive) correlation between the heights and the shoe sizes.

In general, the nearer the points lie to the line of best fit, the higher is the degree of correlation.

Negative (indirect) correlation.

If the scatter diagram shows that high values in one set of data correspond to low values in the other set, then there is negative (indirect) correlation between the two sets of data, graph (a).
If there is no correlation between the two sets of data, then the scatter diagram will be similar to graph (b).

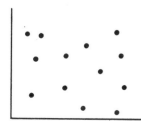

(b) Scatter diagram showing no correlation.

Exercise 9.3

For questions **1** to **10**, draw a scatter diagram to a suitable scale and describe the type of correlation shown.

1. The marks for ten pupils studying both Mathematics and Physics were as shown.

Maths	54	22	65	68	69	40	46	30	50	79
Physics	63	36	69	70	71	49	54	40	53	82

2. Ten samples of home-made jam were judged separately by Mr Brown and Mr Smith, and awarded points as follows.

Mr Brown	25	20	21	15	16	35	30	32	23	22
Mr Smith	30	22	25	17	17	39	33	38	27	33

3. The time of running of ten pupils was recorded.
(a) over 100 m in seconds,
(b) over a cross-country course in minutes.
The results are shown in the table.

100 m	14·6	15	14·9	13	13·5	13·8	14·4	15·1	13·9	14·9
Cross-country	22	25	26	24	26	28	24	23	26	25

4. The table shows the marks obtained by 12 pupils (a) in a French examination, (b) in a History examination.

French	11	59	14	30	41	45	21	49	81	24	35	70
History	73	40	73	60	54	50	68	46	21	65	56	30

5. The table shows (a) the batting averages, (b) the bowling averages (runs per wicket) of a cricketer over ten seasons.

Batting	35	40	25	50	60	27	15	15	12	10
Bowling	18	19	12	24	31	14	8	7	8	7

6. The table shows (a) the number of penalties awarded and (b) the number of goals scored from these penalties by a football team during ten weeks of the season.

Penalties	20	18	12	15	5	10	14	16	18	13
Goals	15	12	8	10	4	7	8	9	12	9

7. The table shows (a) the engine size in cm^3 of various cars and (b) the distance travelled in km on one litre of petrol.

Engine size	1000	1100	1300	1500	1600	1800	2000	2200	2500	4200
Distance	11·0	10·9	10·0	8·8	8·5	7·5	6·9	6·4	6·0	3·3

8. The table shows (a) the weight in kg of 10 men and (b) the size of shoe that they wear.

Weight	65	70	75	78	80	83	90	92	95	100
Shoe size	5	5	7	6	10	9	8	11	10	12

9. The table shows (a) the waist measurement in cm of 8 men and (b) their weight in kg.

Waist	60	69	75	83	90	99	110	125
Weight	64	78	80	88	88	94	112	120

10. The table shows (a) the number of cornets sold from an ice-cream van on each of seven days and (b) the temperature at midday in °C.

No. of cornets sold	40	70	150	50	100	170	200
Temperature	16	19	22	18	20	24	26

For questions 11 to 15, draw a scatter diagram to a suitable scale, put in the line of best fit, and use your diagram to answer the questions.

11. The marks for ten pupils studying science were:

Physics	35	24	37	41	39	30	32	27	31	41
Chemistry	30	17	35	39	38	24	27	24	29	40

If a pupil missed the Chemistry exam but obtained 36 marks in the Physics exam, what would be the expected mark in Chemistry?

12. Ten flower arrangements were assessed by two judges X and Y and awarded marks as follows:

Judge X	18	7	12	3	19	5	1	20	14	7
Judge Y	15	6	11	4	15	6	2	17	13	7

A late entry was awarded 15 marks by Judge X. Estimate the mark that might have been given by Judge Y.

13. The number of therms of gas used monthly in a house together with the average monthly temperature in °C is given in the table.

Month	Jan.	Feb.	Mar.	Apr.	May	June	July	Aug.	Sept.	Oct.	Nov.	Dec.
Therms used	46	42	35	32	23	8	6	13	16	17	26	41
Temperature (°C)	2·1	3·6	6·5	7·5	11·1	17·0	17·4	15·8	13·8	12·1	9·5	3·3

(a) Estimate the number of therms used when the average temperature is 10 °C.

(b) Estimate the average temperature in a month when 20 therms were used.

14. Over a period of 14 days the number of visitors to a museum and the number of hours of sunshine was as follows.

Hours sunshine	12·5	7	6	2	1·5	12	11	10·5	9	8·5	1	0·5	3	5
No. of visitors	95	340	350	560	605	100	140	206	220	300	620	684	550	410

Estimate the number of visitors when the number of hours of sunshine was:
(a) 4 hours   (b) 10 hours.

15. The marks of 10 pupils in the two parts of a mathematics examination were as follows:

Paper I	35	40	45	50	60	65	75	80	85	90
Paper II	15	21	28	34	47	54	66	73	80	85

If a pupil who scored 55 marks on Paper I was absent for Paper II, estimate his expected mark.

The probability p of an event happening is:

$$p = \frac{\text{number of 'successful' results}}{\text{total number of all possible results}}$$

This is usually written as a fraction, except that a certainty is 1 and an impossibility is 0.

Example 1

Seven counters numbered 1, 2, 3, 4, 5, 6, 7, are placed in a box. If one counter is drawn out at random, what is the probability that it is a counter with a number divisible by 3?
number of 'successes' = 2: these are counters 3 and 6.
total number of possibilities = 7

$$\therefore p = \frac{2}{7}$$

Exercise 9.4a

1. Six counters numbered 1, 3, 4, 5, 8, 9, are placed in a box. If one counter is drawn out at random, what is the probability that it is a counter

 (a) with an odd number,
 (b) with an even number?

2. If a dice is thrown, what is the probability that the score is

 (a) a prime number,
 (b) a square number?

3. Twelve counters labelled A, B, C, D, E, F, G, H, I, J, K, L, are placed in a box. If one counter is drawn out at random, what is the probability that it is a counter

 (a) with a consonant letter,
 (b) with a vowel letter?

4. If a card is withdrawn at random from a pack of 52 playing cards, what is the probability that it is:

 (a) an ace,
 (b) a picture card,
 (c) any number from 2 to 10?

5. A class contains 15 boys and 9 girls. If the class votes for one pupil to be the class captain and all of them stand an equal chance of election, what is the probability that the elected captain

will be (a) a boy,
       (b) a girl?

6. A farmer has 40 white sheep and 32 black sheep. If they are rounded up in any order for shearing, what is the probability that the first to be sheared is (a) a white sheep,
       (b) a black sheep?

7. A class of 30 boys contains 18 with dark hair, 8 with blonde hair and 4 with red hair. If the class proceeds to the assembly hall in random order, what is the probability that the first to enter the hall has

 (a) dark hair,
 (b) blonde hair,
 (c) red hair?

8. A television set has four channel-selectors, one for BBC 1, one for BBC 2, one for ITV and one which is not tuned to any channel. If I press any one of the four selectors at random, what is the probability

 (a) that I will receive a programme,
 (b) that I will receive a programme broadcast by the BBC?

9. A note-paper manufacturer makes pads in four colours: blue, white, pink, and green. In a pack of 36, there are always 12 blue pads, 10 white pads, 8 pink pads and 6 green pads. If I open a new pack, what is the probability that the pad is:

 (a) blue,
 (b) white,
 (c) pink,
 (d) green?

10. If two coins are tossed simultaneously, what is the probability of:

 (a) two heads,
 (b) two tails,
 (c) one head and one tail?

Combined probabilities

Addition rule If events A and B are *mutually exclusive* (i.e. they cannot happen at the same time) then

$$p(A \text{ } or \text{ } B) = p(A) + p(B)$$

Example 2

A box contains 6 blue balls, 4 red balls, and 5 white balls. What is the probability of picking out from the box either a red ball or a blue ball?

$$p \text{ (red)} = \frac{4}{15}; \ p \text{ (blue)} = \frac{6}{15}$$

$$\therefore p \text{ (red } or \text{ blue)} = \frac{4}{15} + \frac{6}{15} = \frac{10}{15} = \frac{2}{3}$$

Note:

The probability of picking out either a red ball or a blue ball or a white ball

$$= \frac{4}{15} + \frac{6}{15} + \frac{5}{15} = 1$$

Exercise 9.4b

1. A bag contains 8 red discs, 4 blue discs, and 1 white disc. What is the probability of picking out:
   (a) either a red disc or a blue disc,
   (b) either a red disc or a white disc,
   (c) either a blue disc or a white disc,
   (d) either a blue disc or a white disc or a red disc?
2. A small box of chocolates contains 4 hard centres, 6 soft centres and 2 foil-wrapped chocolates. What is the probability of picking out:
   (a) either a hard centre or a soft centre,
   (b) either a hard centre or a foil-wrapped chocolate,
   (c) either a soft centre or a foil-wrapped chocolate,
   (d) not a soft centre?
3. On a shelf in a supermarket, there are 8 packets of Sudso, 10 packets of Foamo and 12 packets of Washo soap powder. What is the probability that a housewife takes off the shelf:
   (a) either a packet of Foamo or a packet of Washo,
   (b) either a packet of Sudso or a packet of Foamo,
   (c) not a packet of Foamo?
4. A £1 cash bag contains four 10p coins, eight 5p coins and ten 2p coins. If one coin is picked out, what is the probability that the coin is:
   (a) a 10p coin,
   (b) not a 5p coin,
   (c) not a copper coin,
   (d) either a 5p coin or a 2p coin?
5. In a class of 30 boys, 15 choose Rugby, 12 choose Soccer and 3 choose Hockey to play during the winter term. What is the probability that the youngest:
   (a) does not play Rugby,
   (b) does not play Soccer,
   (c) does not play Hockey?

Multiplication rule If A and B are *independent* events (i.e. they do not affect each other) then

$$p \text{ (A } and \text{ B)} = p \text{ (A)} \times p \text{ (B)}$$

Example 3

A bag contains 6 blue balls and 4 red balls. A ball is taken out at random and replaced; then a second ball is taken out at random. What is the probability that:
(a) both balls are red,
(b) both balls are a different colour,
(c) both are red if the first ball is not replaced?

(a) p (first ball is red) $= \dfrac{4}{10}$

p (second ball is red) $= \dfrac{4}{10}$

$\therefore p$ (both are red) $= \dfrac{4}{10} \times \dfrac{4}{10}$

$\qquad\qquad\qquad = \dfrac{16}{100} = \dfrac{4}{25}$

(b) Method 1.

p (both are blue) $= \dfrac{6}{10} \times \dfrac{6}{10}$

$\qquad\qquad\quad = \dfrac{36}{100} = \dfrac{9}{25}$

p (both are red) $= \dfrac{4}{10} \times \dfrac{4}{10}$

$\qquad\qquad\quad = \dfrac{16}{100} = \dfrac{4}{25}$

$\therefore p$ (both are different) $= 1 - \left(\dfrac{9}{25} + \dfrac{4}{25} \right)$

$\qquad\qquad\qquad\qquad = \dfrac{12}{25}$

Method 2.

p (first ball is red) $= \dfrac{4}{10}$

p (second ball is blue) $= \dfrac{6}{10}$

$$\therefore p \text{ (first red, second blue)} = \frac{4}{10} \times \frac{6}{10}$$

$$= \frac{24}{100} = \frac{6}{25}$$

Similarly p (first blue,

$$\text{second red}) = \frac{6}{25}$$

$$\therefore p \text{ (both are different)} = \frac{6}{25} + \frac{6}{25}$$

$$= \frac{12}{25}$$

(c) p (first ball is red) $= \frac{4}{10}$

p (second ball is red) $= \frac{3}{9}$

$$\therefore p \text{ (both are red)} = \frac{4}{10} \times \frac{3}{9}$$

$$= \frac{12}{90} = \frac{2}{15}$$

Exercise 9.4c

1. A box contains 6 red socks and 4 white socks. A sock is taken out at random and put back in the box; a second sock is then taken out. What is the probability that:
   (a) both socks are red,
   (b) both socks are white,
   (c) both socks are a different colour,
   (d) both are red if the first is not replaced,
   (e) both are white if the first is not replaced,
   (f) both are different colours if the first is not replaced?
2. Eight counters numbered 1, 2, 3, 4, 5, 6, 7, 8, are placed in a box. One is taken out at random and replaced; a second one is then taken out. What is the probability that:
   (a) both are number 5,
   (b) both are even numbers,
   (c) both are prime numbers,
   (d) both are divisible by 3,
   (e) both are divisible by 3 if the first is not replaced,
   (f) the sum of both counters is 15,
   (g) the sum of both counters is 15 if the first is not replaced?

3. Two dice are thrown. What is the probability that:
   (a) one is a three and one is a four,
   (b) the total score is 12,
   (c) only one is a six,
   (d) both are even numbers,
   (e) the total score is 7,
   (f) neither is a 1?
4. In a geography quiz, cards are made with the names LONDON, LIVERPOOL, LEEDS, MANCHESTER, GLASGOW, EDINBURGH printed on them. If a card is picked at random and replaced, then a second is picked, what is the probability that:
   (a) both cards have names beginning with L,
   (b) both cards have names of cities in England,
   (c) both cards have names of cities in Scotland,
   (d) one has an English city and the other a Scottish city,
   (e) both cards have the names of English cities if the first is not replaced,
   (f) one has an English city and one has a Scottish city if the first is not replaced?
5. From a pack of 52 playing cards, a card is chosen and its suit is noted; it is then put back and a second card is chosen. What is the probability that:
   (a) both cards are Hearts,
   (b) both cards are Spades,
   (c) both cards are black suits,
   (d) neither card is a Diamond,
   (e) the first card is a red suit and the second is a black suit,
   (f) one card is a Heart and the second is a Spade?
6. A domino is chosen at random from the following group; it is then replaced and a second is chosen.

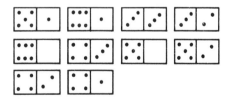

What is the probability that:
   (a) both dominoes have six dots,
   (b) both dominoes have seven dots,
   (c) both dominoes have five dots,
   (d) the first domino has six dots and the second five,
   (e) both dominoes have six dots if the first is not replaced?

Probability problems may be solved with the help of *tree diagrams.*

Example 4

A small box of chocolates contains 8 soft centres (S) and 5 hard centres (H). One chocolate is eaten and then another chocolate is taken out and eaten. Illustrate this by means of a tree diagram.

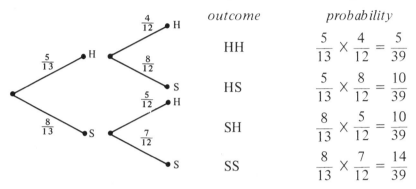

outcome	probability
HH	$\frac{5}{13} \times \frac{4}{12} = \frac{5}{39}$
HS	$\frac{5}{13} \times \frac{8}{12} = \frac{10}{39}$
SH	$\frac{8}{13} \times \frac{5}{12} = \frac{10}{39}$
SS	$\frac{8}{13} \times \frac{7}{12} = \frac{14}{39}$

To check, remember that the sum of all the probabilities must be 1.

In this case, $\dfrac{5}{39} + \dfrac{10}{39} + \dfrac{10}{39} + \dfrac{14}{39} = \dfrac{39}{39} = 1$

The tree diagram can then be used to find particular probabilities, e.g.

(a) the probability of eating two hard centres in succession $= \dfrac{5}{39}$

(b) the probability of eating a hard centre followed

by a soft centre $= \dfrac{10}{39}$

(c) the probability of eating one of each if the

order is unimportant $= \dfrac{10}{39} + \dfrac{10}{39} = \dfrac{20}{39}$

(d) the probability of *not* eating two soft

centres in succession $= 1 - \dfrac{14}{39} = \dfrac{25}{39}$

Exercise 9.4d

1. A coin is tossed twice. Draw a tree diagram to show the different results of Heads and Tails obtained.
2. A bag contains three balls, one red, one white and one blue. If they were drawn out one after the other, draw a tree diagram to show all the possibilities.

3. A small box of chocolates contains 5 soft centres and 4 hard centres. One chocolate is eaten and then another chocolate is eaten. Illustrate the possibilities by means of a tree diagram.
4. A box contains 3 red ties and 2 white ties. John picks a tie and puts it on and then David picks a tie to put it on. Draw a tree diagram to show the possible outcomes.

5. A box contains 4 red, 3 yellow and 2 black counters. A counter is taken from the box and not replaced; then a second counter is taken from the box. Draw a tree diagram to show the possible outcomes.

6. On a shelf in a supermarket, there are 10 boxes containing brown eggs and 5 containing white eggs. One box is taken off the shelf followed by another. Use a tree diagram to determine the probability of:
   (a) picking up 2 boxes containing brown eggs,
   (b) picking up 2 boxes containing white eggs,
   (c) picking up a box containing white eggs followed by one containing brown eggs.

7. In a class there are 16 boys and 9 girls. The head master picks two pupils at random from this class to answer a questionnaire. Use a tree diagram to determine the probability that the headmaster will pick:
   (a) two boys.
   (b) two girls,
   (c) a boy first and a girl second,
   (d) a girl first and a boy second,
   (e) a girl and a boy regardless of order.

8. I have to make two sandwiches using baps (bread buns). If I have 7 brown and 9 white baps and I pick up two at random, use a tree diagram to determine the probability that:
   (a) both will be brown baps,
   (b) both will be white baps,
   (c) one will be brown and the other will be white.

9. A box contains 21 red apples and 28 green ones. Two apples are removed from the box at random. Use a tree diagram to determine the probability that:
   (a) both apples are red,
   (b) both apples are green,
   (c) a red apple is removed followed by a green one,
   (d) one of each colour is removed.

10. A farmer has 33 white sheep and 12 black sheep; any two at a time are taken out for shearing. Use a tree diagram to determine the probability that:
    (a) both are white,
    (b) both are black,
    (c) one is white and the other is black.

11. On a supermarket shelf there are 15 bags of plain flour and 6 bags of self-raising flour. A housewife picks any two bags at random. Use a tree diagram to determine the probability of:
    (a) picking 2 bags of plain flour,
    (b) picking 2 bags of self-raising flour,
    (c) picking up a bag of plain flour followed by one of self-raising flour,
    (d) not picking up 2 bags of plain flour.

12. On a supermarket shelf there are 21 packets of plain crisps and 7 packets of cheese-and-onion crisps. A boy picks any two packets at random. Use a tree diagram to determine the probability of:
    (a) picking 2 packets of plain crisps,
    (b) picking 2 packets of cheese-and-onion crisps,
    (c) picking two packets which are different,
    (d) not picking 2 packets of plain crisps.

13. Nine counters are lettered A, B, C, D, E, F, G, H, and I. A counter is chosen at random and not replaced; then a second counter is also taken at random. Use a tree diagram to determine the probability that:
    (a) both letters are consonants,
    (b) both letters are vowels,
    (c) one letter is a consonant and the other is a vowel,
    (d) two vowels are not chosen.

14. A £1 cash bag contains 11 five-pence coins and 45 one-penny coins. Two coins are withdrawn from the bag together. Use a tree diagram to determine the probability that:
    (a) both coins are silver,
    (b) two silver coins are not withdrawn together,
    (c) two copper coins are not withdrawn together.

15. A box contains 36 tennis balls, 28 of which are white and 8 of which are yellow. One ball is taken from the box at random and not replaced. A second ball is then removed, the choice again being random. Use a tree diagram to determine the probability that:
    (a) both balls are white,
    (b) two white balls are not removed,
    (c) two yellow balls are not removed.

Exercise 1.1a page 1

1. (c)   **2.** (b)   **3.** (a)   **4.** (a)   **5.** (c)   **6.** (b)   **7.** (b)   **8.** (c)
9. (a)   **10.** (b)   **11.** (c)   **12.** (a)   **13.** (b)   **14.** (c)   **15.** (a)

Exercise 1.1b page 1

1. (a)   **2.** (a)   **3.** (b)   **4.** (c)   **5.** (a)   **6.** (a)   **7.** (b)   **8.** (b)
9. (b)   **10.** (c)   **11.** (a)   **12.** (c)   **13.** (c)   **14.** (b)   **15.** (a)

Exercise 1.1c page 2

1. (b)   **2.** (b)   **3.** (c)   **4.** (c)   **5.** (b)   **6.** (c)   **7.** (b)   **8.** (c)
9. (a)   **10.** (a)   **11.** (b)   **12.** (a)   **13.** (b)   **14.** (b)   **15.** (c)

Exercise 1.1d page 2

1. (a)   **2.** (b)   **3.** (b)   **4.** (a)   **5.** (c)   **6.** (a)   **7.** (a)   **8.** (b)
9. (a)   **10.** (c)   **11.** (a)   **12.** (a)   **13.** (a)   **14.** (b)   **15.** (a)

Exercise 1.1e page 3

1. 105, 45   **2.** 177 cm, 109 cm   **3.** 178 m, 315 m   **4.** 1900, 5450
5. 533 m, 453 m

Exercise 1.2a page 3

1. 70, 700   **2.** 140, 1400   **3.** 250, 2500   **4.** 370, 3700
5. 100, 1000   **6.** 700, 7000   **7.** 2200, 22 000   **8.** 4000, 40 000
9. 1000, 10 000   **10.** 7860, 78 600   **11.** 80   **12.** 16
13. 3200   **14.** 100   **15.** 49   **16.** 100
17. 120   **18.** 5000   **19.** 900   **20.** 300

Exercise 1.2b page 4

1. 660, 66   **2.** 980, 98   **3.** 1050, 105   **4.** 1200, 120
5. 2000, 200   **6.** 12 500, 1250   **7.** 30 100, 3010   **8.** 100, 10
9. 20 000, 2000   **10.** 1000, 100   **11.** 72   **12.** 360
13. 28   **14.** 100   **15.** 6500   **16.** 100
17. 18 000   **18.** 400   **19.** 3000   **20.** 90 000

Exercise 1.2c page 4

1. (c)   **2.** (b)   **3.** (a)   **4.** (b)   **5.** (b)   **6.** (a)   **7.** (b)   **8.** (b)
9. (b)   **10.** (c)   **11.** (c)   **12.** (b)

Exercise 1.2d page 4

1. (b)   **2.** (c)   **3.** (a)   **4.** (b)   **5.** (c)   **6.** (a)   **7.** (b)   **8.** (a)
9. (a)   **10.** (c)   **11.** (b)   **12.** (a)

Exercise 1.2e page 5

1. (b)   **2.** (c)   **3.** (b)   **4.** (c)   **5.** (b)   **6.** (a)   **7.** (c)   **8.** (c)
9. (c)   **10.** (a)   **11.** (a)   **12.** (b)

Exercise 1.2f page 5

1. 84 kg, 28 kg   **2.** 15 kg, 75 kg   **3.** 144 km, 18 km   **4.** 18, 90, 38

Exercise 1.3a page 6

1. 2, 4, 6, 8, 10   **2.** 5, 10, 15, 20, 25
3. 8, 16, 24, 32, 40   **4.** 7, 14, 21, 28, 35
5. 11, 22, 33, 44, 55   **6.** 20, 40, 60, 80, 100
7. 30, 60, 90, 120, 150   **8.** 60, 120, 180, 240, 300
9. 15, 30, 45, 60, 75   **10.** 25, 50, 75, 100, 125
11. 16, 32, 48, 64, 80   **12.** 18, 36, 54, 72, 90
13. 14, 28, 42, 56, 70   **14.** 13, 26, 39, 52, 65
15. 24, 48, 72, 96, 120   **16.** 21, 42, 63, 84, 105
17. 45, 90, 135, 180, 225   **18.** 36, 72, 108, 144, 180
19. 51, 102, 153, 204, 255   **20.** 72, 144, 216, 288, 360

Exercise 1.3b page 6

1. 1, 3   **2.** 1, 2, 4, 8   **3.** 1, 2, 5, 10
4. 1, 2, 3, 4, 6, 12   **5.** 1, 3, 5, 15   **6.** 1, 2, 3, 6, 9, 18
7. 1, 2, 3, 5, 6, 10, 15, 30   **8.** 1, 3, 9, 27   **9.** 1, 2, 3, 4, 6, 8, 12, 24
10. 1, 2, 4, 8, 16, 32   **11.** 1, 3, 5, 9, 15, 45 **12.** 1, 2, 4, 5, 8, 10, 20, 40
13. 1, 2, 3, 6, 9, 18, 27, 54   **14.** 1, 2, 3, 6, 7, 14, 21, 42
15. 1, 2, 3, 4, 5, 6, 10, 12, 15, 20, 30, 60 **16.** 1, 2, 3, 4, 6, 8, 12, 16, 24, 48
17. 1, 3, 7, 9, 21, 63
18. 1, 2, 3, 4, 6, 7, 12, 14, 21, 28, 42, 84 **19.** 1, 2, 3, 6, 11, 22, 33, 66
20. 1, 2, 3, 4, 6, 8, 9, 12, 18, 24, 36, 72

Exercise 1.3c page 6

1. 1, 2   **2.** 1, 5
3. 1, 2, 3, 6   **4.** 1, 7
5. 1, 11   **6.** 1, 2, 4, 5, 10, 20
7. 1, 5, 7, 35   **8.** 1, 2, 4; square number
9. 1, 2, 5, 10, 25, 50   **10.** 1, 5, 25; square number
11. 1, 2, 4, 8, 16; square number   **12.** 1, 2, 11, 22
13. 1, 2, 7, 14   **14.** 1, 13
15. 1, 2, 4, 7, 14, 28   **16.** 1, 3, 7, 21
17. 1, 7, 49; square number   **18.** 1, 2, 3, 4, 6, 9, 12, 18, 36; square number
19. 1, 3, 17, 51   **20.** 1, 2, 5, 7, 10, 14, 35, 70

Exercise 1.3d page 6

1. 1, 17; prime number   **2.** 1, 3, 11, 33
3. 1, 2, 13, 26   **4.** 1, 29; prime number
5. 1, 3, 13, 39   **6.** 1, 3, 19, 57
7. 1, 37; prime number   **8.** 1, 7, 13, 91
9. 1, 3, 29, 87   **10.** 1, 53; prime number
11. 1, 3, 37, 111   **12.** 1, 97; prime number
13. 1, 61; prime number   **14.** 1, 67; prime number
15. 1, 7, 17, 119   **16.** 1, 103, prime number
17. 1, 73; prime number   **18.** 1, 3, 41, 123
19. 1, 101; prime number   **20.** 1, 3, 9, 13, 39, 117

Exercise 1.3e page 6

1. 2×5 **2.** 3×5
3. $2 \times 2 \times 7$ **4.** $2 \times 2 \times 2 \times 3$
5. $2 \times 2 \times 3 \times 3$ **6.** $2 \times 2 \times 2 \times 5$
7. $3 \times 3 \times 5$ **8.** $2 \times 3 \times 3 \times 3$
9. $2 \times 5 \times 7$ **10.** $3 \times 3 \times 7$
11. $2 \times 3 \times 13$ **12.** $2 \times 2 \times 3 \times 5$
13. $2 \times 3 \times 11$ **14.** $2 \times 2 \times 2 \times 2 \times 7$
15. $2 \times 2 \times 2 \times 2 \times 2 \times 3$ **16.** $2 \times 3 \times 3 \times 3 \times 3$
17. $2 \times 7 \times 11$ **18.** $2 \times 2 \times 3 \times 3 \times 5$
19. $2 \times 2 \times 2 \times 3 \times 7$ **20.** $2 \times 2 \times 2 \times 3 \times 3 \times 3$

Exercise 1.3f page 7

1. 2   **2.** 6   **3.** 6   **4.** 9   **5.** 14   **6.** 24   **7.** 18   **8.** 21
9. 15   **10.** 16 **11.** 6   **12.** 4   **13.** 14 **14.** 16 **15.** 8   **16.** 6
17. 12 **18.** 13 **19.** 16 **20.** 4

Exercise 1.3g page 7

1. 24   **2.** 30   **3.** 24   **4.** 18   **5.** 36   **6.** 60   **7.** 30   **8.** 40
9. 45   **10.** 48 **11.** 24 **12.** 36 **13.** 40 **14.** 60 **15.** 36 **16.** 60
17. 144 **18.** 120 **19.** 90 **20.** 144

Exercise 1.3h page 7

1. (b)   2. (a)   3. (c)   4. (c)   5. (a)   6. (c)   7. (a)   8. (a)
9. (b)   10. (a)

Exercise 1.4a page 8

1. $\frac{7}{10}$

2. $10, \frac{4}{100}$

3. $500, \frac{9}{10}$

4. $\frac{3}{10}, \frac{5}{1000}$

5. $\frac{3}{100}, \frac{6}{1000}$

6. $30, \frac{5}{10}, \frac{7}{1000}$

7. $5, \frac{3}{10}, \frac{2}{100}$

8. $80, \frac{4}{10}$

9. $5, \frac{2}{10}, \frac{2}{1000}$

10. $200, \frac{3}{1000}$

Exercise 1.4b page 8

1. 37·6   2. 31·2   3. 35·85   4. 72·11   5. 87·834
6. 43·185   7. 38·16   8. 61·385   9. 16·935   10. 24·995
11. 1·4   12. 1·8   13. 3·85   14. 3·72   15. 3·75
16. 7·15   17. 4·38   18. 18·85   19. 106·65   20. 41·89
21. (b)   22. (c)   23. (a)   24. (a)   25. (b)
26. (b)   27. (c)   28. (a)   29. (b)   30. (a)

Exercise 1.4c page 9

1. 24, 240   2. 122, 1220   3. 37·5, 375   4. 153·6, 1536
5. 21·35, 213·5   6. 185·76, 1857·6   7. 8·5, 85   8. 7, 70
9. 2·368, 23·68   10. 0·139, 1·39   11. 36   12. 4·5
13. 290   14. 3·2   15. 100   16. 90
17. 0·8   18. 100   19. 0·04   20. 0·0015

Exercise 1.4d page 9

1. 2·53, 0·253   2. 3·816, 0·3816   3. 0·625, 0·0625   4. 0·735, 0·0735
5. 3·6, 0·36   6. 6, 0·6   7. 2·04, 0·204   8. 10·03, 1·003
9. 0·085, 0·0085 10. 0·0032, 0·00032 11. 0·42   12. 5·1
13. 0·36   14. 480   15. 100   16. 0·6
17. 30   18. 100   19. 0·009   20. 0·012

Exercise 1.4e page 9

1. 3·6   2. 1·6   3. 15·2   4. 34·4   5. 37·5
6. 15·12   7. 28·08   8. 61·8   9. 1·92   10. 21·76
11. 35·1   12. 14·72   13. 0·135   14. 0·126   15. 0·14
16. 0·0288   17. 0·027   18. 7·2   19. 0·9   20. 0·39

Exercise 1.4f page 10

1. 2·22   2. 2·76   3. 6·02   4. 2·496   5. 3·3
6. 0·7572   7. 46·35   8. 19·44   9. 10·38   10. 0·5525
11. 0·8192   12. 0·648   13. 3·248   14. 7·815   15. 0·0984
16. 0·0904   17. 0·062   18. 0·063   19. 0·0951   20. 0·07712

Exercise 1.4g page 10

1. 0·3   2. 0·2   3. 0·24   4. 0·14   5. 0·143
6. 0·182   7. 0·272   8. 0·245   9. 0·252   10. 0·221
11. 0·086   12. 0·083   13. 0·019   14. 0·0124   15. 0·042
16. 0·093   17. 0·045   18. 0·042   19. 0·013   20. 0·011

Exercise 1.4h page 10

1. 21·2   2. 9·3   3. 11·7   4. 1·5   5. 10·9
6. 18   7. 13   8. 14   9. 14   10. 65
11. 630   12. 730   13. 920   14. 6500   15. 1·9
16. 1·4   17. 12   18. 27   19. 140   20. 130

Exercise 1.4i page 10

1. 2·64   2. 1·34   3. 17·6   4. 42·8   5. 1·734
6. 1·563   7. 13·7   8. 25·38   9. 8·05   10. 24·0
11. 5·10   12. 4·51   13. 27·1   14. 9·51   15. 6·31
16. 3·90   17. 5·10   18. 14·80   19. 14·0   20. 2·00

Exercise 1.4j page 11

1. 4	2. 3	3. 3	4. 2	5. 3
6. 4	7. 2	8. 1	9. 2	10. 3
11. 3	12. 4	13. 3·2	14. 7·6	15. 13·4
16. 17·2	17. 31	18. 37	19. 15	20. 20
21. 3570	22. 4290	23. 5500	24. 20·0	25. 4·0
26. 10·0	27. 0·064	28. 0·0072	29. 0·009	30. 0·008

Exercise 1.5a page 11

1. 3·57 m	2. 53·29 m	3. 37·6 m	4. 0·49 m	5. 0·6 m
6. 0·09 m	7. 5·276 m	8. 0·752 m	9. 0·08 m	10. 0·007 m
11. 9·137 km	12. 0·83 km	13. 3372 m	14. 2490 m	15. 19600 m
16. 345 cm	17. 920 cm	18. 5936 mm	19. 8210 mm	20. 7900 mm

Exercise 1.5b page 11

1. 1·328 g	2. 0·536 g	3. 0·78 g	4. 0·09 g	5. 0·008 g
6. 1·5 kg	7. 0·59 kg	8. 0·03 kg	9. 0·002 kg	10. 1·32 t
11. 0·8 t	12. 4536 mg	13. 8980 mg	14. 3400 mg	15. 5260 g
16. 8500 g	17. 700 g	18. 3710 kg	19. 5600 kg	20. 300 kg

Exercise 1.5c page 12

1. 3·278 l	2. 8·25 l	3. 9·3 l	4. 6·035 l	5. 5·02 l
6. 1332 ml	7. 7600 ml	8. 755 ml	9. 320 ml	10. 100 ml

Exercise 1.5d page 12

1. 20p	2. 15p	3. 18p	4. 21p	5. £1.20
6. 75p	7. 36p	8. 3p	9. 15p	10. 26p
11. 14p	12. 3p	13. £20	14. £42	15. £3·75

Exercise 1.6a page 12

1. $\frac{2}{12}$ 2. $\frac{10}{16}$ 3. $\frac{2}{3}$ 4. $\frac{9}{13}$ 5. $\frac{1}{3}$ 6. $\frac{2}{3}$ 7. $\frac{1}{2}$ 8. $\frac{1}{3}$

9. $\frac{1}{4}$ 10. $\frac{2}{3}$ 11. $\frac{3}{4}$ 12. $\frac{3}{4}$ 13. $\frac{3}{5}$ 14. $\frac{6}{7}$ 15. $\frac{2}{3}$

Exercise 1.6b page 13

1. $1\frac{1}{2}$ 2. $1\frac{1}{3}$ 3. $1\frac{3}{4}$ 4. $2\frac{1}{2}$ 5. $2\frac{1}{3}$ 6. $2\frac{4}{7}$ 7. $1\frac{1}{2}$ 8. $1\frac{1}{3}$

9. $\frac{5}{4}$ 10. $\frac{6}{5}$ 11. $\frac{10}{7}$ 12. $\frac{7}{5}$ 13. $\frac{9}{4}$ 14. $\frac{13}{4}$ 15. $\frac{25}{4}$ 16. $\frac{7}{2}$

17. $\frac{11}{3}$ 18. $\frac{38}{5}$ 19. $\frac{26}{7}$ 20. $\frac{23}{3}$

Exercise 1.6c page 13

1. $\frac{2}{5}$ 2. $\frac{3}{5}$ 3. $\frac{7}{10}$ 4. $\frac{1}{2}$ 5. $1\frac{1}{3}$ 6. 1 7. $1\frac{4}{7}$ 8. $1\frac{1}{4}$

9. $\frac{1}{3}$ 10. $\frac{2}{5}$ 11. $\frac{3}{7}$ 12. $\frac{3}{4}$ 13. $\frac{3}{5}$ 14. $\frac{1}{3}$ 15. $\frac{3}{8}$

Exercise 1.6d page 13

1. $\frac{8}{15}$ 2. $\frac{7}{12}$ 3. $\frac{3}{10}$ 4. $\frac{11}{12}$ 5. $\frac{7}{8}$ 6. $\frac{1}{2}$ 7. $\frac{1}{3}$ 8. $\frac{3}{4}$

9. $\frac{1}{2}$ 10. $\frac{9}{16}$ 11. $\frac{1}{4}$ 12. $\frac{1}{10}$ 13. $\frac{1}{6}$ 14. $\frac{1}{20}$ 15. $\frac{5}{12}$ 16. $\frac{1}{8}$

17. $\frac{1}{3}$ 18. $\frac{1}{6}$ 19. $\frac{7}{12}$ 20. $\frac{1}{6}$

Exercise 1.6e page 13

1. $3\frac{1}{2}$ 2. $4\frac{8}{15}$ 3. $5\frac{7}{12}$ 4. $5\frac{3}{10}$ 5. $7\frac{7}{8}$ 6. $3\frac{3}{10}$ 7. $5\frac{5}{12}$ 8. $6\frac{1}{10}$

9. $3\frac{1}{2}$ 10. $2\frac{3}{4}$ 11. $3\frac{1}{3}$ 12. $4\frac{2}{5}$ 13. $1\frac{1}{4}$ 14. $1\frac{1}{6}$ 15. $1\frac{3}{8}$ 16. $5\frac{1}{10}$

17. $5\frac{1}{4}$ 18. $\frac{5}{6}$ 19. $1\frac{3}{5}$ 20. $2\frac{1}{2}$ 21. $\frac{2}{3}$ 22. $2\frac{5}{6}$ 23. $\frac{29}{30}$ 24. $1\frac{5}{12}$

25. $2\frac{8}{9}$

Exercise 1.6f page 14

1. $\frac{1}{6}$ 2. $\frac{1}{10}$ 3. $\frac{3}{8}$ 4. $\frac{3}{14}$ 5. $\frac{8}{15}$ 6. $\frac{1}{5}$ 7. $\frac{2}{7}$ 8. $\frac{2}{7}$

9. $\frac{3}{8}$ 10. $\frac{15}{22}$ 11. $\frac{4}{39}$ 12. $\frac{1}{14}$ 13. $\frac{3}{40}$ 14. $\frac{2}{3}$ 15. $\frac{3}{4}$

Exercise 1.6g page 14

1. $2\frac{11}{12}$ 2. $2\frac{1}{12}$ 3. $6\frac{1}{4}$ 4. $2\frac{11}{12}$ 5. $3\frac{9}{10}$ 6. $3\frac{1}{3}$ 7. $2\frac{2}{3}$ 8. $4\frac{1}{2}$

9. $12\frac{1}{2}$ 10. 30 11. $2\frac{6}{7}$ 12. $8\frac{5}{8}$ 13. 12 14. 12 15. $2\frac{1}{2}$

Exercise 1.6h page 14
    1. $\frac{3}{4}$ 2. $1\frac{2}{5}$ 3. $1\frac{1}{15}$ 4. $1\frac{1}{14}$ 5. $\frac{25}{36}$ 6. $\frac{3}{5}$ 7. $\frac{4}{9}$ 8. $\frac{6}{11}$
    9. $\frac{2}{3}$ 10. $\frac{1}{2}$ 11. 4   12. 4   13. 3   14. 2   15. $\frac{11}{12}$ 16. $\frac{11}{18}$
    17. $2\frac{2}{3}$ 18. $1\frac{1}{2}$ 19. $\frac{4}{9}$ 20. $1\frac{3}{5}$

Exercise 1.6i page 14
    1. (b)   2. (c)   3. (b)   4. (a)   5. (c)   6. (c)   7. (b)   8. (a)
    9. (b)   10. (c)   11. $1\frac{1}{5}$ 12. $1\frac{3}{5}$ 13. $\frac{1}{3}$ 14. $\frac{1}{6}$ 15. $\frac{1}{10}$ 16. $\frac{7}{15}$
    17. $\frac{9}{20}$ 18. $\frac{2}{3}$ 19. (a)   20. (a)   21. (b)   22. (b)   23. (a)   24. (a)
    25. (c)   26. (b)   27. (c)   28. (b)   29. $3\frac{1}{2}$ 30. $4\frac{1}{2}$ 31. $4\frac{1}{2}$ 32. 6
    33. 40   34. 18

Exercise 1.7a page 16
    1. $\frac{7}{10}$ 2. $\frac{1}{2}$ 3. $\frac{2}{5}$ 4. $\frac{9}{10}$ 5. $\frac{9}{20}$ 6. $\frac{13}{50}$
    7. $\frac{17}{20}$ 8. $\frac{8}{25}$ 9. $\frac{29}{50}$ 10. $\frac{18}{25}$ 11. $\frac{7}{8}$ 12. $\frac{13}{40}$
    13. $\frac{5}{8}$ 14. $\frac{19}{40}$ 15. $\frac{3}{100}$ 16. $\frac{2}{25}$ 17. $\frac{1}{20}$ 18. $\frac{7}{100}$
    19. $\frac{11}{200}$ 20. $\frac{3}{125}$ 21. $\frac{31}{500}$ 22. $\frac{11}{400}$ 23. $\frac{1}{16}$ 24. $\frac{3}{1000}$
    25. $\frac{3}{500}$ 26. $\frac{1}{500}$ 27. $\frac{7}{2000}$ 28. $\frac{3}{625}$ 29. $3\frac{1}{5}$ 30. $8\frac{1}{4}$
    31. $12\frac{4}{25}$ 32. $41\frac{1}{50}$ 33. $5\frac{9}{500}$ 34. $13\frac{3}{250}$ 35. $2\frac{1}{80}$ 36. $8\frac{7}{400}$
    37. $14\frac{1}{250}$ 38. $11\frac{1}{200}$ 39. $16\frac{1}{400}$ 40. $21\frac{3}{2000}$

Exercise 1.7b page 16
    1. 0·375   2. 0·125   3. 0·25   4. 0·15   5. 0·55   6. 0·35
    7. 0·65   8. 0·95   9. 0·6   10. 0·2   11. 0·8   12. 0·275
    13. 0·225   14. 0·425   15. 0·525   16. 0·1875   17. 0·6875   18. 0·14
    19. 0·18   20. 0·62

Exercise 1.7c page 17
    1. 0·$\dot{6}$ 2. 0·8$\dot{3}$ 3. 0·1$\dot{6}$ 4. 0·41$\dot{6}$ 5. 0·58$\dot{3}$ 6. 0·91$\dot{6}$
    7. 0·4$\dot{5}$ 8. 0·8$\dot{1}$ 9. 0·2$\dot{7}$ 10. 0·6$\dot{3}$ 11. 0·2$\dot{6}$ 12. 0·7$\dot{3}$
    13. 0·4$\dot{6}$ 14. 0·8$\dot{6}$ 15. 0·5$\dot{3}$ 16. 0·2$\dot{3}$ 17. 0·3$\dot{6}$ 18. 0·4$\dot{3}$
    19. 0·2$\dot{7}$ 20. 0·3$\dot{8}$

Exercise 1.7d page 17
    1. 0·714   2. 0·857   3. 0·571   4. 0·286   5. 0·143   6. 0·429
    7. 0·538   8. 0·769   9. 0·692   10. 0·846   11. 0·476   12. 0·095
    13. 0·762   14. 0·529   15. 0·235   16. 0·176   17. 0·353   18. 0·158
    19. 0·263   20. 0·368

Exercise 1.7e page 17
    1. $\frac{5}{8}$, 0·625   2. $\frac{7}{20}$, 0·35   3. $\frac{9}{10}$, 0·9   4. $\frac{39}{100}$, 0·39
    5. $\frac{1}{10}$, 0·1   6. $\frac{9}{10}$, 0·9   7. $\frac{49}{100}$, 0·49   8. $\frac{1}{10}$, 0·1
    9. $\frac{7}{10}$, 0·7   10. $\frac{3}{10}$, 0·3   11. (a) by $\frac{1}{100}$ or 0·01
    12. (b) by $\frac{1}{100}$ or 0·01   13. (b) by $\frac{1}{100}$ or 0·01
    14. (b) by $\frac{3}{50}$ or 0·06   15. (a) by $\frac{1}{25}$ or 0·04

Exercise 1.8a page 18
    1. 1:4   2. 1:6   3. 1:3   4. 2:5   5. 2:3   6. 3:4
    7. 5:6   8. 4:7   9. 5:2   10. 8:3   11. 7:4   12. 9:5
    13. 12:7   14. 3:20   15. 2:15   16. 3:50   17. 9:20   18. 1:8
    19. 3:40   20. 2:25   21. 3:8   22. 20:3   23. 15:4   24. 50:9
    25. 40:3   26. 25:2   27. 3:4   28. 7:8   29. 3:7   30. 1:6
    31. 4:11   32. 12:5   33. 9:7   34. 15:2   35. 9:4   36. 5:8
    37. 3:20   38. 5:7   39. 9:2   40. 12:5

Exercise 1.8b page 18   **1.** 2:3   **2.** 5:6   **3.** 3:10   **4.** 4:7   **5.** 9:16   **6.** 5:3
                            7. 8:7   **8.** 9:2   **9.** 7:5   **10.** 4:3

Exercise 1.8c page 19   **1.** £27 and £63   **2.** £24 and £60   **3.** £20 and £25
                            4. 25 and 35   **5.** 18 and 30 kg   **6.** £30 and £36
                            7. £30, £40 and £50   **8.** £16, £24 and £32   **9.** £24, £36 and £72
                            10. £7, £14 and £35   **11.** 10, 15 and 35   **12.** 5, 15 and 25
                            13. 8, 40 and 48 kg   **14.** £50, £40 and £20   **15.** 12, 36 and 72

Exercise 1.8d page 19   **1.** £45, £50 and £65   **2.** £32, £40 and £48   **3.** £45, £39 and £24
                            4. £64, £44 and £32   **5.** £55, £35 and £30   **6.** 21, 24 and 27 kg
                            7. 135, 165 and 240 kg   **8.** £2·50, £3 and £4·50   **9.** 56p, 64p and 80p
                            10. $3\frac{1}{2}$, $5\frac{1}{2}$ and 6 t

Exercise 1.8e page 20   **1.** £1, 20p   **2.** £1·10, 66p   **3.** £63, £27
                            4. 91p, £3·90, £1·17, £2·60   **5.** 1 h 50 min, 33 min
                            6. 30, 9   **7.** 28, 10   **8.** 44 m², 12 m²
                            9. 32, 22, 10, 9   **10.** Huddersfield, Glossop

Exercise 1.8f page 20   **1.** 90 min, 36 min   **2.** 4 h 10 min, 1 h 15 min   **3.** 50 h, 18 h
                            4. 24°, 16°   **5.** 24°, 15°   **6.** 20 min, 8 min
                            7. 15 min, 6 min   **8.** 12 cm, 8 cm   **9.** 36 min, 48 min
                            10. 30 min, 42 min

Exercise 1.9a page 21   **1.** $\frac{9}{100}$ **2.** $\frac{3}{20}$ **3.** $\frac{9}{20}$ **4.** $\frac{1}{5}$ **5.** $\frac{7}{10}$ **6.** $\frac{8}{25}$
                            7. $\frac{17}{25}$ **8.** $\frac{14}{25}$ **9.** $\frac{21}{50}$ **10.** $\frac{3}{50}$ **11.** $1\frac{2}{5}$ **12.** $1\frac{4}{5}$
                            13. $1\frac{3}{10}$ **14.** $2\frac{1}{10}$ **15.** $2\frac{3}{4}$ **16.** $3\frac{7}{20}$ **17.** $2\frac{9}{25}$ **18.** $3\frac{2}{25}$
                            19. $4\frac{1}{20}$ **20.** $4\frac{13}{20}$ **21.** 0·15   **22.** 0·35   **23.** 0·9   **24.** 0·8
                            25. 0·6   **26.** 0·05   **27.** 0·16   **28.** 0·72   **29.** 0·44   **30.** 0·58
                            31. 1·25   **32.** 1·56   **33.** 1·7   **34.** 2·3   **35.** 2·25   **36.** 3·4
                            37. 2·12   **38.** 3·04   **39.** 4·08   **40.** 4·01

Exercise 1.9b page 22   **1.** $\frac{7}{40}$ **2.** $\frac{3}{16}$ **3.** $\frac{19}{30}$ **4.** $\frac{8}{15}$ **5.** $\frac{5}{24}$ **6.** $\frac{5}{12}$
                            7. $\frac{17}{40}$ **8.** $\frac{1}{18}$ **9.** $\frac{11}{18}$ **10.** $\frac{7}{32}$ **11.** $1\frac{1}{40}$ **12.** $1\frac{19}{40}$
                            13. $1\frac{7}{30}$ **14.** $2\frac{1}{24}$ **15.** $4\frac{11}{40}$ **16.** $3\frac{1}{12}$ **17.** $3\frac{2}{15}$ **18.** $2\frac{1}{15}$
                            19. $1\frac{7}{24}$ **20.** $1\frac{5}{32}$ **21.** 0·625   **22.** 0·875   **23.** 0·325   **24.** 0·475
                            25. 0·5625   **26.** 0·8125   **27.** 0·09375   **28.** 0·6̇   **29.** 0·1̇   **30.** 0·83̇
                            31. 1·125   **32.** 1·375   **33.** 1·075   **34.** 2·225   **35.** 1·3125   **36.** 3·0625
                            37. 2·03125   **38.** 3·3̇   **39.** 4·03̇   **40.** 2·1̇6̇

Exercise 1.9c page 22   **1.** (a)   **2.** (b)   **3.** (b)   **4.** (c)   **5.** (a)   **6.** (b)   **7.** (a)   **8.** (b)
                            9. (c)   **10.** (a)   **11.** (b)   **12.** (b)   **13.** (a)   **14.** (b)   **15.** (a)

Exercise 1.9d page 23   **1.** (b)   **2.** (a)   **3.** (c)   **4.** (a)   **5.** (b)   **6.** (a)   **7.** (a)   **8.** (b)
                            9. (b)   **10.** (c)   **11.** (a)   **12.** (b)   **13.** (a)   **14.** (a)   **15.** (c)

Exercise 1.9e page 23   **1.** 120   **2.** 240   **3.** 74   **4.** 600, 360, 960, 480
                            5. 1080, 1020, 1104, 1176, 1140   **6.** 80%   **7.** 25%   **8.** 75%   **9.** 15%
                            10. 40%, 25%, 30%, 5%

Exercise 2.1a page 24
  1. £480, £1920   **2.** £90, £510   **3.** £36, £364   **4.** £20, £230
  5. £42, £308   **6.** £24, £126   **7.** £15, £110   **8.** £12, £68
  9. £45, £135   **10.** £17, £68   **11.** £4·80, £55·20   **12.** £2·70, £51·30
  13. £3·78, £59·22   **14.** £1·10, £20·90   **15.** 60p, £14·40

Exercise 2.1b page 25
  1. £6   **2.** £9   **3.** £50, £44, £37, £31, £41   **4.** £24, £16·80, £19·20
  5. £9, £14·40, £12·70   **6.** £180, £235, £305, £363

Exercise 2.1c page 25
  1. £15, £60, £210   **2.** £18, £72, £192   **3.** £30, £90, £340
  4. £56, £112, £462   **5.** £15, £60, £180   **6.** £45, £135, £375
  7. £34, £102, £262   **8.** £31·50, £126, £336
  9. £37·80, £75·60, £390·60   **10.** £11·25, £22·50, £147·50
  11. £21, £84, £434   **12.** £14, £56, £336   **13.** £34, £102, £527
  14. £27, £81, £441   **15.** £7, £35, £147   **16.** £8, £32, £288
  17. £15, £75, £235   **18.** £40·50, £162, £612
  19. £10·80, £21·60, £156·60   **20.** £14·40, £28·80, £388·80

Exercise 2.1d page 26
  1. £30, £45, £295   **2.** £48, £60, £380   **3.** £72, £162, £612
  4. £18, £24, £174   **5.** £63, £147, £497   **6.** £27, £45, £225
  7. £36, £42, £267   **8.** £24, £52, £212   **9.** £12, £30, £180
  10. £32, £56, £376   **11.** £6, £21, £141   **12.** £60, £200, £950
  13. £15, £40, £290   **14.** £36, £66, £786   **15.** £36, £114, £564

Exercise 2.1e page 27
  1. £622·08, £322·08   **2.** £608·35, £208·35   **3.** £752·64, £152·64
  4. £672·80, £172·80   **5.** £278·48, £78·48   **6.** £732·05, £232·05
  7. £463·05, £63·05   **8.** £1123·60, £123·60   **9.** £237·62, £37·62
  10. £540·80, £40·80

Exercise 2.1f page27
  1. £102, £12   **2.** £63, £3   **3.** £81, £6   **4.** £147, £12
  5. £252, £12   **6.** £504, £54   **7.** £402, £52   **8.** £2655, £155
  9. £945, £45   **10.** £1968, £168

Exercise 2.1g page 28
  1. £96, £2784   **2.** £153, £2106   **3.** £42, £1368   **4.** £39, £1176
  5. £18, £672   **6.** £13, £372   **7.** £9, £138   **8.** £3, £87
  9. £7, £109   **10.** £21, £342

Exercise 2.1h page 29
  1. +3p   **2.** +2p   **3.** +1p   **4.** −1p   **5.** +2p   **6.** +2p   **7.** +4p   **8.** +6p
  9. +10p **10.** −6p   **11.** −2p   **12.** −20p **13.** +12p **14.** −4p   **15.** +60p **16.** −30
  17. +4p   **18.** +6p   **19.** +3p   **20.** +6p

Exercise 2.1i page 30
  1. $+33\frac{1}{3}\%$ **2.** +25%   **3.** $+12\frac{1}{2}\%$ **4.** $-16\frac{2}{3}\%$ **5.** $+33\frac{1}{3}\%$ **6.** +4%
  7. $+11\frac{1}{9}\%$ **8.** $+11\frac{1}{9}\%$ **9.** +25%   **10.** $-16\frac{2}{3}\%$ **11.** $-6\frac{2}{3}\%$ **12.** $-13\frac{1}{3}\%$
  13. $+11\frac{1}{9}\%$ **14.** $-16\frac{2}{3}\%$ **15.** $+12\frac{1}{2}\%$ **16.** −20%   **17.** +50%   **18.** $+11\frac{1}{9}\%$
  19. +20%   **20.** +150%

Exercise 2.2a page 31
  1. 54 kg   **2.** 34 kg   **3.** 179 cm   **4.** 157 cm   **5.** 34   **6.** 330
  7. 38   **8.** 14   **9.** 15 min   **10.** 36 min   **11.** 1 h 4 min   **12.** £39·63
  13. 84 cm   **14.** 62 cm   **15.** 8   **16.** 17 h   **17.** 48p   **18.** £29·39
  19. £176, £22   **20.** £408, £34

Exercise 2.2b page 32
  1. 608   **2.** £336   **3.** £5850   **4.** £247·92   **5.** 8 h 30 min
  6. 376 800   **7.** 3 years   **8.** 30 m   **9.** 54   **10.** 6 h 30 min

Exercise 2.2c page 32   **1.** 173 cm   **2.** 35 kg   **3.** 25   **4.** 18 min   **5.** 136   **6.** 29°
7. 74 000   **8.** £37·36   **9.** 5   **10.** 1·2 h

Exercise 2.2d page 33   **1.** 632 km/h   **2.** 488 km/h   **3.** 720 km/h   **4.** 864 km/h   **5.** 632 km/h
6. 40 km/h   **7.** 24 km/h   **8.** 40 km/h   **9.** 42 km/h   **10.** 44 km/h
11. 128 km/h   **12.** 96 km/h   **13.** 100 km/h   **14.** 120 km/h   **15.** 104 km/h
16. 76 km/h   **17.** 68 km/h   **18.** 63 km/h   **19.** 68 km/h   **20.** 75 km/h

Exercise 2.2e page 33   **1.** 90 km/h   **2.** 36 km/h   **3.** 50 km/h   **4.** 48 km/h   **5.** 99 km/h
6. 117 km/h   **7.** 24 km/h   **8.** 30 km/h   **9.** 85 km/h   **10.** 36 km/h

Exercise 2.3a page 34   **1.** 4   **2.** 25   **3.** 81   **4.** 400   **5.** 3600
6. 4900   **7.** 160 000   **8.** 640 000   **9.** 10 000   **10.** 4 000 000
11. 25 000 000 **12.** 36 000 000 **13.** 0·25   **14.** 0·64   **15.** 0·01
16. 0·0009   **17.** 0·0004   **18.** 0·0036   **19.** $\frac{1}{16}$ **20.** $\frac{1}{49}$
21. $\frac{1}{144}$ **22.** $\frac{9}{16}$ **23.** $\frac{4}{25}$ **24.** $\frac{16}{121}$ **25.** $1\frac{11}{25}$
26. $1\frac{7}{9}$ **27.** $1\frac{9}{16}$ **28.** $2\frac{7}{9}$ **29.** $4\frac{21}{25}$ **30.** $11\frac{1}{9}$
31. $30\frac{1}{4}$ **32.** $12\frac{1}{4}$ **33.** $20\frac{1}{4}$ **34.** $1\frac{21}{100}$ **35.** $13\frac{4}{9}$
36. $3\frac{6}{25}$ **37.** $5\frac{19}{25}$ **38.** $2\frac{14}{25}$ **39.** $1\frac{40}{81}$ **40.** $1\frac{23}{121}$
41. 169   **42.** 256   **43.** 529   **44.** 841   **45.** 961
46. 2·89   **47.** 3·61   **48.** 6·76   **49.** 12·25   **50.** 16·81

Exercise 2.3b page 34   **1.** 4   **2.** 6   **3.** 7   **4.** 12   **5.** 30
6. 10   **7.** 50   **8.** 110   **9.** 600   **10.** 900
11. 700   **12.** 500   **13.** 0·6   **14.** 0·7   **15.** 0·2
16. 0·09   **17.** 0·08   **18.** 0·04   **19.** $\frac{1}{3}$ **20.** $\frac{1}{8}$
21. $\frac{1}{11}$ **22.** $\frac{3}{5}$ **23.** $\frac{7}{8}$ **24.** $\frac{5}{12}$ **25.** $2\frac{1}{2}$
26. $2\frac{2}{3}$ **27.** $1\frac{3}{4}$ **28.** $1\frac{2}{5}$ **29.** $2\frac{1}{3}$ **30.** $1\frac{5}{6}$
31. $1\frac{1}{6}$ **32.** $1\frac{1}{9}$ **33.** $1\frac{3}{7}$ **34.** $1\frac{1}{8}$ **35.** $1\frac{1}{7}$
36. $2\frac{3}{4}$ **37.** $6\frac{2}{3}$ **38.** 18   **39.** 14   **40.** 24
41. 21   **42.** 27   **43.** 33   **44.** 15   **45.** 25
46. 45   **47.** 1·2   **48.** 1·1   **49.** 1·6   **50.** 2·2

Exercise 2.3c page 34   **1.** +5   **2.** +3   **3.** −11   **4.** +9   **5.** −0·01
6. +0·26   **7.** −0·01   **8.** +0·29   **9.** +0·05   **10.** +0·14
11. − 0·04   **12.** + 0·11   **13.** − 0·04   **14.** − 0·01   **15.** − 0·16
16. − 0·0025   **17.** − 0·11   **18.** − 0·24   **19.** + 1·99   **20.** − 0·41
21. + 0·31   **22.** + 0·64   **23.** − 0·25   **24.** − 4   **25.** − 1
26. − 1   **27.** − 5   **28.** + 11   **29.** + 4   **30.** + 1

Exercise 2.3d page 35   **1.** 27   **2.** 125   **3.** 729   **4.** 1331
5. 1728   **6.** 3375   **7.** 27 000   **8.** 216 000
9. $\frac{1}{8}$ **10.** $\frac{1}{64}$ **11.** $\frac{1}{512}$ **12.** $\frac{1}{343}$
13. $\frac{8}{27}$ **14.** $\frac{64}{125}$ **15.** $3\frac{3}{8}$ **16.** $15\frac{5}{8}$
17. 0·001   **18.** 0·064   **19.** 0·729   **20.** 0·216

Exercise 2.3e page 35

1. 2	**2.** 4	**3.** 8	**4.** 6	**5.** 10
6. 20	**7.** 40	**8.** 50	**9.** $\frac{1}{6}$	**10.** $\frac{1}{7}$
11. $\frac{1}{9}$	**12.** $\frac{3}{4}$	**13.** $\frac{2}{9}$	**14.** $\frac{5}{6}$	**15.** $1\frac{1}{4}$
16. $1\frac{3}{5}$	**17.** 0·2	**18.** 0·3	**19.** 0·5	**20.** 0·7

Exercise 2.3f page 35

1. −4	**2.** +3	**3.** −25	**4.** −0·029
5. −0·016		**6.** +0·057	

Exercise 2.3g page 35

1. 8×10^5	**2.** $3\cdot6 \times 10^5$	**3.** $5\cdot48 \times 10^5$
4. 5×10^4	**5.** $3\cdot5 \times 10^4$	**6.** $2\cdot24 \times 10^4$
7. 9×10^3	**8.** $7\cdot5 \times 10^3$	**9.** $2\cdot84 \times 10^3$
10. $1\cdot563 \times 10^3$	**11.** 7×10^2	**12.** $2\cdot9 \times 10^2$
13. $3\cdot42 \times 10^2$	**14.** $1\cdot863 \times 10^2$	**15.** 8×10^1
16. $3\cdot6 \times 10^1$	**17.** $2\cdot98 \times 10^1$	**18.** 4×10^0
19. $8\cdot2 \times 10^0$	**20.** $7\cdot36 \times 10^0$	**21.** $1\cdot23 \times 10^0$
22. $4\cdot274 \times 10^2$	**23.** $1\cdot1692 \times 10^3$	**24.** $4\cdot0006 \times 10^3$
25. $1\cdot111\ 111 \times 10^6$	**26.** $2\cdot020\ 206 \times 10^6$	**27.** $1\cdot000\ 0001 \times 10^7$
28. $4\cdot006\ 0007 \times 10^7$	**29.** $1\cdot86 \times 10^5$	**30.** $1\cdot000\ 01 \times 10^1$

Exercise 2.3h page 35

1. 700 000	**2.** 825 000	**3.** 966 300	**4.** 500 100	**5.** 80 000
6. 63 000	**7.** 75 060	**8.** 6000	**9.** 3910	**10.** 5376
11. 450	**12.** 938	**13.** 883·7	**14.** 620·6	**15.** 50
16. 72·8	**17.** 15·32	**18.** 50·86	**19.** 9·51	**20.** 4·537
21. 1·6	**22.** 10·6	**23.** 482	**24.** 400 700	**25.** 2700 000
26. 991 000		**27.** 10 200 000		**28.** 94 360 000
29. 842 000 000		**30.** 96 300 000 000		

Exercise 2.4a page 36

1. 212, 210	**2.** 156, 160	**3.** 122, 120
4. 172, 170	**5.** 195, 200	**6.** 116, 120
7. 1650, 1600	**8.** 1930, 1900	**9.** 1348, 1300
10. 1077, 1000	**11.** 1461, 1500	**12.** 1448, 1500
13. 1506, 1500	**14.** 1011, 1000	**15.** 1593, 1600
16. 1389, 1390	**17.** 681, 690	**18.** 1018, 1040
19. 780, 800	**20.** 1075, 1110	**21.** 470, 500
22. 407, 400	**23.** 288, 300	**24.** 481, 500
25. 343, 300	**26.** 286, 300	**27.** 829, 810
28. 670, 700	**29.** 488, 500	**30.** 964, 1000

Exercise 2.4b page 37

1. 7·9, 8	**2.** 12·3, 12	**3.** 18·5, 19
4. 21·4, 22	**5.** 62·4, 60	**6.** 116·9, 120
7. 133·8, 130	**8.** 141·8, 140	**9.** 197·3, 200
10. 178·5, 180	**11.** 155·8, 160	**12.** 13·38, 13
13. 13·86, 14	**14.** 11·36, 11	**15.** 17·88, 18
16. 20·19, 20	**17.** 84·39, 80	**18.** 85·97, 80
19. 135·20, 140	**20.** 131·25, 130	**21.** 167·32, 170
22. 79·52, 83	**23.** 65·57, 64	**24.** 63·99, 65
25. 84·71, 78	**26.** 27·62, 30	**27.** 43·55, 40
28. 33·46, 30	**29.** 51·25, 50	**30.** 37·37, 40
31. 5·733, 5	**32.** 5·542, 6	**33.** 49·178, 51
34. 86·784, 85	**35.** 53·34, 53	**36.** 73·186, 71
37. 4·164, 4	**38.** 4·088, 4	**39.** 7·747, 8
40. 4·402, 4		

Exercise 2.4c *page 37*

1. 638 000	**2.** 484 000	**3.** 254 000	**4.** 108 000
5. 119 000	**6.** 194 000	**7.** 215 000	**8.** 198 000
9. 146 000	**10.** 63 900	**11.** 96 100	**12.** 97 400
13. 62 300	**14.** 80 100	**15.** 71 900	**16.** 53 200
17. 6270	**18.** 1820	**19.** 5160	**20.** 3290
21. 2380	**22.** 4350	**23.** 326	**24.** 111
25. 562	**26.** 782	**27.** 351	**28.** 215
29. 137	**30.** 255	**31.** 423	**32.** 199
33. 43·7	**34.** 29·5	**35.** 35·2	**36.** 17·4
37. 32·5	**38.** 56·9	**39.** 22·2	**40.** 35·5
41. 19·3	**42.** 42·3	**43.** 4·33	**44.** 8·55
45. 7·78	**46.** 5·39	**47.** 8·81	**48.** 6·01
49. 5·43	**50.** 6·38		

Exercise 2.4d *page 38*

1. 0·24	**2.** 0·32	**3.** 0·72	**4.** 0·08
5. 0·42	**6.** 0·66	**7.** 0·98	**8.** 0·06
9. 0·288	**10.** 0·512	**11.** 0·896	**12.** 0·008
13. 0·162	**14.** 0·722	**15.** 0·15625	**16.** 0·84375
17. 0·03125	**18.** 1·16	**19.** 1·64	**20.** 1·04
21. 1·18	**22.** 1·256	**23.** 2·12	**24.** 2·54

Exercise 2.4e *page 38*

1. 0·5$\dot{6}$	**2.** 0·9$\dot{6}$	**3.** 0·9$\dot{3}$	**4.** 0·1$\dot{3}$
5. 0·0$\dot{3}$	**6.** 0·0$\dot{6}$	**7.** 0·7$\dot{2}$	**8.** 0·6$\dot{1}$
9. 0·9$\dot{4}$	**10.** 0·42$\dot{6}$	**11.** 0·74$\dot{6}$	**12.** 0·17$\dot{3}$
13. 0·32$\dot{6}$	**14.** 0·80$\dot{6}$	**15.** 0·541$\dot{6}$	**16.** 0·291$\dot{6}$
17. 0·208$\dot{3}$	**18.** 0·347$\dot{2}$	**19.** 0·013$\dot{8}$	**20.** 1·3$\dot{6}$
21. 1·2$\dot{3}$	**22.** 1·52$\dot{7}$	**23.** 1·85$\dot{3}$	**24.** 2·5$\dot{3}$

Exercise 2.4f *page 38*

1. 3·873	**2.** 4·243	**3.** 2·646	**4.** 1·414	**5.** 3·162
6. 1·732	**7.** 2·236	**8.** 7·616	**9.** 9·487	**10.** 9·274
11. 2·214	**12.** 1·897	**13.** 1·581	**14.** 3·795	**15.** 3·479
16. 12·25	**17.** 13·42	**18.** 17·89	**19.** 10·95	**20.** 30·98
21. 11·18	**22.** 17·75	**23.** 20·62	**24.** 19·60	**25.** 44·72
26. 54·77	**27.** 70·71	**28.** 34·64	**29.** 42·43	**30.** 57·01

Exercise 2.4g *page 39*

1. 2h 45 min	**2.** 4h 45 min	**3.** 3h 15 min	**4.** 1h 15 min
5. 2h 24 min	**6.** 1h 36 min	**7.** 3h 21 min	**8.** 4h 33 min
9. 45 min	**10.** 51 min	**11.** 48 min	**12.** 36 min

Exercise 2.4h *page 39*

1. 259 km	**2.** 532 km	**3.** 288 km	**4.** 234 km
5. 352 km	**6.** 273 km	**7.** 174 km	**8.** 413 km
9. 184 km	**10.** 135 km	**11.** 117 km	**12.** 110 km

Exercise 2.5a *page 40*

1. £100	**2.** £180	**3.** £200	**4.** £126	**5.** £270
6. £336	**7.** £252	**8.** £297	**9.** £360	**10.** £176
11. £112·50	**12.** £292·80	**13.** £254·40	**14.** £361·20	**15.** £167·40
16. £181·50	**17.** £178·20	**18.** £431·80	**19.** £251·25	**20.** £321·75

Exercise 2.5b *page 40*

1. £360	**2.** £450	**3.** £240	**4.** £320	**5.** £280	**6.** £500
7. £250	**8.** £550	**9.** £420	**10.** £340	**11.** £375	**12.** £286
13. £512	**14.** £336	**15.** £254	**16.** £542	**17.** £192	**18.** £394
19. £285	**20.** £325				

Exercise 2.5c page 40   **1.** 75p   **2.** 80p   **3.** 60p   **4.** 70p   **5.** 90p   **6.** 80p   **7.** 75p   **8.** 60p   **9.** 90p   **10.** 70p   **11.** 60p   **12.** 90p   **13.** 65p   **14.** 80p   **15.** 85p   **16.** 75p   **17.** 80p   **18.** 70p   **19.** 60p   **20.** 90p

Exercise 2.5d page 41   **1.** £1500   **2.** £2250   **3.** £2550   **4.** £2450   **5.** £2250   **6.** £3180   **7.** £2610   **8.** £3190   **9.** £1730   **10.** £2080

Exercise 2.5e page 41   **1.** £945   **2.** £770   **3.** £1085   **4.** £1120   **5.** £847   **6.** £875   **7.** £735   **8.** £1260   **9.** £518   **10.** £980

Exercise 2.5f page 42   **1.** £504, £42   **2.** £1260, £105   **3.** £840, £70   **4.** £1596, £133   **5.** £924, £77   **6.** £364, £7   **7.** £1092, £21   **8.** £1456, £28   **9.** £728, £14   **10.** £1820, £35

Exercise 2.6a page 42   **1.** £48   **2.** £54   **3.** £42   **4.** £49·60   **5.** £59·20   **6.** £63   **7.** £46·20   **8.** £58·80   **9.** £48·30   **10.** £55·44   **11.** £45   **12.** £46·80   **13.** £57·60   **14.** £52·20   **15.** £37·44   **16.** £52·80   **17.** £61·60   **18.** £68·20   **19.** £44·88   **20.** £51·04

Exercise 2.6b page 42   **1.** £1·65, £2·20   **2.** £1·86, £2·48   **3.** £2·55, £3·40   **4.** £1·98, £2·64   **5.** £2·10, £2·80   **6.** £2·04, £2·72   **7.** £1·45, £2·61   **8.** £2·00, £3·60   **9.** £1·80, £3·24   **10.** £1·60, £2·88

Exercise 2.6c page 43   **1.** £62·40   **2.** £61·75   **3.** £60·50   **4.** £65·28   **5.** £73·44   **6.** £55·80   **7.** £56·10   **8.** £62·40   **9.** £65·10   **10.** £65·54

Exercise 2.6d page 43   **1.** £3000   **2.** £3840   **3.** £4260   **4.** £3300   **5.** £3792   **6.** £4128   **7.** £5016   **8.** £3516   **9.** £3846   **10.** £3414   **11.** £3723   **12.** £3303   **13.** £4635   **14.** £3489   **15.** £4473

Exercise 2.6e page 43   **1.** £300   **2.** £375   **3.** £267   **4.** £326   **5.** £319   **6.** £332   **7.** £278   **8.** £284   **9.** £340·25   **10.** £364·25   **11.** £290·75   **12.** £376·75   **13.** £240·50   **14.** £345·50   **15.** £336·50

Exercise 2.6f page 43   **1.** 1st by £4   **2.** 2nd by £8   **3.** 1st by £2   **4.** 2nd by £1   **5.** 1st by £4   **6.** 1st by £5   **7.** 2nd by £12   **8.** 2nd by £3

Exercise 2.7a page 44   **1.** 2263, 674   **2.** 1865, 905   **3.** 455, 1204   **4.** 621, 1542   **5.** 534, 1636   **6.** 627, 516   **7.** 713, 429   **8.** 2321, 762, 548   **9.** 533, 1747, 2455   **10.** 1893, 815, 749, 3086

Exercise 2.7b page 44   **1.** £40·86   **2.** £42·84   **3.** £13·41   **4.** £18·54   **5.** £25·56   **6.** £23·40   **7.** £32·76   **8.** £29·88   **9.** £43·56   **10.** £55·80   **11.** £29·16   **12.** £21·60   **13.** £15·48   **14.** £38·61   **15.** £47·70

Exercise 2.7c page 45   **1.** £45·30   **2.** £12·96   **3.** £11·42   **4.** £39·80   **5.** £37·38, £10·54   **6.** £12·52, £9·00   **7.** £33·20, £34·74   **8.** £30·56, £11·64   **9.** £36·94, £12·96   **10.** £10·98, £9·44   **11.** £16·92, £31·22   **12.** £33·64, £49·92   **13.** £48·38, £15·16, £11·86, £43·32   **14.** £40·68, £14·06, £13·18, £33·20   **15.** £46·18, £14·28, £12·08, £38·26

Exercise 2.8a page 45   **1.** 14.00 h   **2.** 17.00 h   **3.** 18.00 h   **4.** 14.15 h   **5.** 14.40 h   **6.** 17.20 h   **7.** 17.32 h   **8.** 18.30 h   **9.** 18.48 h   **10.** 19.50 h   **11.** 19.05 h   **12.** 16.35 h   **13.** 20.00 h   **14.** 20.30 h   **15.** 21.15 h

16. 22.00 h	**17.** 22.24 h	**18.** 23.45 h	**19.** 23.05 h	**20.** 08.00 h
21. 06.00 h	**22.** 09.15 h	**23.** 08.42 h	**24.** 07.35 h	**25.** 06.08 h
26. 09.03 h	**27.** 10.00 h	**28.** 10.15 h	**29.** 11.55 h	**30.** 11.32 h

Exercise 2.8b page 45

1. 1.00 p.m.	**2.** 4.00 p.m.	**3.** 7.00 p.m.	**4.** 1.30 p.m.	**5.** 1.50 p.m.
6. 7.35 p.m.	**7.** 7.18 p.m.	**8.** 4.25 p.m.	**9.** 4.05 p.m.	**10.** 6.40 p.m.
11. 6.12 p.m.	**12.** 3.55 p.m.	**13.** 2.36 p.m.	**14.** 9.00 p.m.	**15.** 9.45 p.m.
16. 11.00 p.m.	**17.** 11.30 p.m.	**18.** 10.45 p.m.	**19.** 10.06 p.m.	**20.** 9.00 a.m.
21. 7.00 a.m.	**22.** 8.45 a.m.	**23.** 9.53 a.m.	**24.** 8.56 a.m.	**25.** 7.02 a.m.
26. 8.09 a.m.	**27.** 11.00 a.m.	**28.** 11.25 a.m.	**29.** 10.50 a.m.	**30.** 10.36 a.m.

Exercise 2.8c page 46

1. 1 h 45 min	**2.** 1 h 35 min	**3.** 1 h 15 min	**4.** 34 min
5. 2 h 25 min	**6.** 3 h 30 min	**7.** 1 h 50 min	**8.** 1 h 45 min
9. 45 min	**10.** 2 h 45 min		

11. 1 h 12 min, 1 h 45 min, 2 h 0 min, 2 h 40 min
12. 1 h 11 min, 1 h 40 min, 2 h 5 min, 2 h 45 min
13. 2 h 12 min, 2 h 40 min, 3 h 45 min, 6 h 0 min
14. (a) 9.45 p.m., 35 min   (b) 15 min
    (c) 8.15 a.m., 8.45 a.m., 30 min, 4.20 p.m., 4.50 p.m.
    (d) 2.15 p.m., 25 min   (e) 1.05 p.m., 3 h 35 min
15. (a) 1 h 55 min, 3 h 5 min, 2 h 55 min, 5.25 p.m.
    (b) 9.20 a.m., 10.10 a.m., 50 min, 1 h 5 min, 12.10 p.m., 25 min, 1.00 p.m.
    (c) 1 h 5 min, 6.35 p.m.
16. (a) 1.35 p.m., 2.15 p.m., 40 min   (b) 26 min
    (c) 8.28 a.m., 8.50 a.m., 22 min, 6.25 p.m., 55 min
    (d) 5.50 p.m.   (e) 14 min, 3 h 41 min

Exercise 2.9a page 48

1. $392	**2.** $294	**3.** $49	**4.** Fr. 364	**5.** Fr. 2730
6. Fr. 1274	**7.** DM 288	**8.** DM 768	**9.** DM 960	**10.** SF 162
11. SF 378	**12.** Ptas. 2496	**13.** Ptas. 3276	**14.** BF 7500	**15.** BF 4320

Exercise 2.9b page 48

1. £300	**2.** £250	**3.** £75	**4.** £60	**5.** £400
6. £160	**7.** £50	**8.** £350	**9.** £125	**10.** £15
11. £35	**12.** £18	**13.** £24	**14.** £115	**15.** £64

Exercise 2.9c page 48

1. Britain, £25   **2.** France, £40   **3.** Switzerland, 5p
4. Britain, 10p   **5.** Spain, 25p   **6.** Britain, 50p
7. Germany, 25p
8. Spain, £2200; Germany, £2250; Britain, £2400; Switzerland, £2450;
    U.S.A. £2500; Belgium, £2600; France, £2700
9. France, 50p; Belgium, 57p; Britain, 60p; Spain, 63p; Switzerland, 65p
10. Spain, £4.50; U.S.A. £4.75; Switzerland, £4.80; Belgium, £4.90;
    Britain, £5; France, £5.20; Germany, £5.25

PART 3

Exercise 3.1a page 50

1. 37 500 cm^2	**2.** 275 mm^2	**3.** 4·65 cm^2	**4.** 1·25 m^2	**5.** 1·5 cm^2
6. 6750 cm^2	**7.** 1·27 cm^2	**8.** 41 000 cm^2	**9.** 0·0062 m^2	**10.** 0·001 764 m^2

Exercise 3.1b page 50

1. (a) 144 cm², 50 cm   (b) 198 cm², 58 cm   (c) 700 mm², 110 mm
   (d) 960 mm², 128 mm   (e) 300 m², 74 m   (f) 360 m², 78 m
2. (a) 400 cm², 80 cm   (b) 256 cm², 64 cm   (c) 1·96 m², 5·6 m
   (d) 3·61 m², 7·6 m   (e) 324 mm², 72 mm   (f) 625 mm², 100 mm

Exercise 3.1c page 51

1. 100 cm²   2. 10 mm²   3. 1·56 cm²   4. 0·06 m²   5. 350 mm²
6. 276 cm²   7. 350 cm²   8. 304 cm²   9. 300 mm²   10. 300 mm²
11. 9 cm²   12. 1·5 m²   13. 200 cm²   14. 186 cm²   15. 270 mm²
16. 19·2 m²   17. 8·64 m²

Exercise 3.1d page 51

1. 13 cm   2. 17 cm   3. 19 mm   4. 23 mm   5. 15 m
6. 22 cm   7. 14 cm   8. 27 mm   9. 17 m   10. 7m
11. 36 cm   12. 44 cm   13. 60 cm   14. 4·8 m   15. 6·4 m

Exercise 3.1e page 52

1. 24   2. 80   3. 30, £8·40   4. 150, £18   5. 40
6. 48   7. 24   8. 500   9. 1500   10. 2500

Exercise 3.2a page 53

1. 66 cm²   2. 324 cm²   3. 2400 cm²   4. 0·165 m²   5. 240 m²
6. 0·00432 m²   7. 228 mm²   8. 702 mm²   9. 39·2 cm²   10. 18·4 cm²
11. 270 cm²   12. 8 cm²   13. 125 cm²   14. 2·16 m²   15. 0·0336 m²
16. 378 cm²   17. 0·016 m²   18. 44 800 mm²   19. 54 mm²   20. 1788 mm²

Exercise 3.2b page 54

1. 120 cm²   2. 360 cm²   3. 720 cm²   4. 5·6 cm²   5. 0·32 m²
6. 0·076 m²   7. 0·000 54 m²   8. 125 000 mm²   9. 27 000 mm²   10. 1660 mm²

Exercise 3.2c page 54

1. 120 cm²   2. 272 cm²   3. 528 cm²   4. 720 cm²   5. 25 m²
6. 25 m²   7. 18 m²   8. 36 m²   9. 24 400 mm²   10. 20 m²

Exercise 3.2d page 56

1. 840 cm²   2. 2000 cm²   3. 274 cm²   4. 20 400 mm²   5. 90 000 mm²
6. 1620 mm²   7. 270 cm²   8. 24 cm²   9. 7200 mm²

Exercise 3.3a page 57

1. 66 cm   2. 154 cm   3. 220 cm   4. 44 cm   5. 13·2 m
6. 17·6 m   7. 26·4 m   8. 462 mm   9. 330 mm   10. 286 mm
11. 15·7 cm   12. 9·42 cm   13. 34·54 cm   14. 25·12 cm   15. 94·2 cm
16. 1256 mm   17. 1884 mm   18. 2826 mm   19. 4·71 m   20. 7·85 m
21. 314·2 mm   22. 62·84 cm   23. 15·71 m   24. 25·136 cm   25. 31·42 cm
26. 314·2 cm   27. 1256·8 mm   28. 157·1 cm   29. 471·3 mm   30. 942·6 mm

Exercise 3.3b page 58

1. 56 cm, 28 cm   2. 84 cm, 42 cm   3. 112 mm, 56 mm
4. 98 mm, 49 mm   5. 126 mm, 63 mm   6. 154 mm, 77 mm
7. 4·2 m, 2·1 m   8. 1·4 m, 0·7 m   9. 18·2 cm, 9·1 cm
10. 19·6 cm, 9·8 cm

Exercise 3.3c page 58

1. 1386 cm²   2. 154 cm²   3. 3850 mm²   4. 2464 mm²
5. 1·54 m²   6. 3·465 m²   7. $38\frac{1}{2}$ cm²   8. $9\frac{5}{8}$ cm²
9. $\frac{77}{200}$ m²   10. $6\frac{4}{25}$ m²   11. 12·56 cm²   12. 28·26 cm²
13. 78·5 cm²   14. 2826 mm²   15. 1256 mm²   16. 706·5 mm²
17. 1962·5 mm²   18. 7·065 m²   19. 0·785 m²   20. 19·625 m²

Exercise 3.3d page 59

1. 176 cm, 25   2. 44 cm, 75   3. 11 m, 45   4. 4·4 m, 7
5. 30·8 mm, 50   6. 9   7. 12   8. 4200 mm²   9. 231
10. 1·4 cm, 2·8 cm, 8·8 cm, 160   11. 22 cm²
12. 96 800 cm, 0·792 km/h, 554·4 cm²

Exercise 3.4a page 60

1. 330 cm^3 2. 720 cm^3 3. 1200 cm^3
4. 9m^3 5. 28 m^3 6. 36 m^3
7. 729 cm^3 8. 1728 cm^3 9. 216 000 mm^3
10. 512 000 mm^3 11. 72 m^3, 72 000 l   12. 84 m^3, 84 000 l
13. 28 m^3, 28 000 l   14. 27 m^3, 27 000 l   15. 9 m^3, 9000 l
16. 0·024 m^3, 24 l   17. 0·008 m^3, 8 l   18. 12 000 cm^3, 12 l
19. 27 000 cm^3, 27 l   20. 21 000 cm^3, 21 l

Exercise 3.4b page 61

1. 180 cm^3 2. 154 cm^3 3. 540 cm^3 4. 1500 cm^3
5. 75 000 mm^3 6. 18 000 mm^3 7. 12 000 mm^3 8. 5400 mm^3
9. 3 m^3 10. 9 m^3 11. 1848 cm^3 12. 594 cm^3
13. 1100 cm^3 14. 7040 cm^3 15. 8800 mm^3 16. 9240 mm^3
17. 70 400 mm^3 18. 49 500 mm^3 19. 3·08 m^3 20. 1·1 m^3

Exercise 3.4c page 61

1. 36   2. 60   3. 150   4. 250   5. 2160
6. 20 cm   7. 136 m^3 8. 180 l, 12   9. 88 l, 160   10. 7040 m^3, 80

Exercise 3.5a page 62

1. 50 cm^3 2. 189 cm^3 3. 480 cm^3 4. 1200 cm^3 5. 32 000 mm^3
6. 13 500 mm^3 7. 3750 mm^3 8. 0·06 m^3 9. 0·32 m^3 10. 0·36 m^3

Exercise 3.5b page 63

1. 66 cm^3 2. 2970 cm^3 3. 1570 cm^3 4. 9420 cm^3
5. 396 000 mm^3 6. 148 500 mm^3 7. 25 120 mm^3 8. 0·77 m^3
9. 0·154 m^3 10. 0·1256 m^3

Exercise 3.5c page 63

1. 113·04 cm^3 2. 38 808 mm^3 3. 3052 080 mm^3
4. 904·32 cm^3 5. 14·13 cm^3 6. 4·851 cm^3
7. 606$\frac{3}{8}$ cm^3 8. 381·51 cm^3 9. 47·688 75 cm^3
10. 1·766 25 cm^3

Exercise 3.5d page 64

1. 216 m^3 2. 40 800 cm^3 3. 12·6 cm^3, 18·9 g   4. 22·608 l
5. 0·14 l   6. 7065 cm^3 7. 4710 cm^3, 12 cm   8. 80

Exercise 3.6a page 65

1. 3cm   2. 4 cm   3. 5 cm   4. 18 cm   5. 21 cm
6. 5mm   7. 15 mm   8. 30 mm   9. 25 mm   10. 63 mm
11. 3 cm   12. 4 cm   13. 3 cm   14. 6 cm   15. 2 cm

Exercise 3.6b page 65

1. 90°   2. 144°   3. 30°   4. 300°   5. 288°
6. 165°   7. 40°   8. 105°   9. 100°   10. 315°
11. 192°   12. 200°   13. 72°   14. 252°   15. 330°

Exercise 3.6c page 66

1. 120°   2. 54°   3. 300°   4. 60°   5. 108°
6. 90°   7. 45°   8. 45°   9. 225°   10. 144°
11. 55 mm   12. 88 mm   13. 11 mm   14. 77 mm   15. 8·8 cm
16. 4·4 cm   17. 10·56 cm   18. 0·44 m   19. 0·66 m   20. 0·33 m

Exercise 3.6d page 66

1. 8800 km   2. 6600 km   3. 5500 km   4. 7700 km   5. 5280 km
6. 7920 km   7. 5940 km   8. 6160 km   9. 4840 km   10. 8250 km

Exercise 3.6e page 67

1. 2640 km   2. 3300 km   3. 1980 km   4. 3960 km   5. 1980 km
6. 990 km   7. 4950 km   8. 2420 km   9. 1320 km   10. 4400 km

Exercise 4.1a page 68

1. $x + 3$	**2.** $y + 7$	**3.** $t + 5$	**4.** $u + 8$	**5.** $v + 11$
6. $9 + x$	**7.** $4 + y$	**8.** $6 + a$	**9.** $12 + b$	**10.** $1 + c$
11. $x - 4$	**12.** $y - 8$	**13.** $l - 10$	**14.** $m - 2$	**15.** $n - 6$
16. $5 - x$	**17.** $3 - y$	**18.** $11 - p$	**19.** $7 - q$	**20.** $9 - r$

Exercise 4.1b page 68

1. $3x$	**2.** $9y$	**3.** $4z$	**4.** $7a$	**5.** $12b$	**6.** $\frac{x}{5}$	**7.** $\frac{y}{8}$	**8.** $\frac{z}{3}$
9. $\frac{3l}{4}$	**10.** $\frac{5m}{6}$	**11.** $\frac{2n}{5}$	**12.** $\frac{x}{6}$	**13.** $\frac{y}{9}$	**14.** $\frac{z}{12}$	**15.** $\frac{t}{10}$	**16.** $\frac{5}{x}$
17. $\frac{7}{y}$	**18.** $\frac{11}{z}$	**19.** $\frac{8}{u}$	**20.** $\frac{4}{v}$				

Exercise 4.1c page 68

1. $x + b$	**2.** $y + c$	**3.** $m + x$	**4.** $n + y$	**5.** $x - u$
6. $y - v$	**7.** $p - x$	**8.** $q - y$	**9.** ab	**10.** mn
11. uv	**12.** pq	**13.** a^2	**14.** b^2	**15.** t^2
16. $\frac{x}{a}$	**17.** $\frac{y}{b}$	**18.** $\frac{t}{z}$	**19.** $\frac{m}{x}$	**20.** $\frac{n}{y}$

Exercise 4.1d page 68

1. $1000, 5000, 1000z$ **2.** $100, 300, 100s$ **3.** $1000, 8000, 1000a$
4. $100, 600, 100b$ **5.** $60, 240, 60t$ **6.** $10, 90, 10p$
7. $1000, 7000, 1000w$ **8.** $60, 720, 60r$ **9.** $24, 120, 24q$
10. $1000, 11\,000, 1000v$ **11.** $3, \frac{2a}{5}$ **12.** $4, \frac{3l}{4}$ **13.** $9, \frac{3c}{5}$
14. $6, \frac{4d}{5}$ **15.** $3, \frac{2e}{3}$ **16.** $5, \frac{f}{4}$ **17.** $8, \frac{5g}{12}$ **18.** $7, \frac{h}{5}$
19. $12, \frac{5i}{8}$ **20.** $10, \frac{3j}{8}$

Exercise 4.2a page 69

1. $+14$	**2.** $+21$	**3.** $+35$	**4.** $+57$	**5.** $+75$	**6.** $+18$
7. $+22$	**8.** $+28$	**9.** $+33$	**10.** $+36$	**11.** -20	**12.** -24
13. -27	**14.** -45	**15.** -48	**16.** -15	**17.** -25	**18.** -30
19. -35	**20.** -50	**21.** $+4$	**22.** $+18$	**23.** $+16$	**24.** $+15$
25. $+25$	**26.** -15	**27.** -9	**28.** -8	**29.** -21	**30.** -15
31. $+4$	**32.** $+14$	**33.** $+8$	**34.** $+4$	**35.** $+12$	**36.** -7
37. -3	**38.** -4	**39.** -5	**40.** -9		

Exercise 4.2b page 69

1. $+12$	**2.** $+13$	**3.** $+9$	**4.** $+15$	**5.** $+18$	**6.** -15
7. -11	**8.** -5	**9.** -8	**10.** -9	**11.** $+15$	**12.** $+23$
13. $+27$	**14.** $+25$	**15.** $+33$	**16.** $+35$	**17.** $+33$	**18.** $+36$
19. $+32$	**20.** $+45$	**21.** -19	**22.** -21	**23.** -30	**24.** -32
25. -35	**26.** -40	**27.** -31	**28.** -33	**29.** -42	**30.** -44
31. $+5$	**32.** $+12$	**33.** $+15$	**34.** $+14$	**35.** $+16$	**36.** -5
37. -6	**38.** -15	**39.** -8	**40.** -15		

Exercise 4.2c page 70

1. $+24$	**2.** $+60$	**3.** $+80$	**4.** $+42$	**5.** $+72$	**6.** $+9$
7. $+20$	**8.** $+15$	**9.** $+14$	**10.** $+11$	**11.** $+32$	**12.** $+75$
13. $+54$	**14.** $+132$	**15.** $+48$	**16.** $+64$	**17.** $+121$	**18.** $+12$
19. $+5$	**20.** $+15$	**21.** -80	**22.** -75	**23.** -200	**24.** -60
25. -36	**26.** -49	**27.** -144	**28.** -25	**29.** -18	**30.** -10
31. -90	**32.** -200	**33.** -150	**34.** -60	**35.** -120	**36.** -36
37. -81	**38.** -30	**39.** -6	**40.** -21		

Exercise 4.2d page 70

1. $+12$	**2.** $+9$	**3.** $+8$	**4.** $+12$	**5.** $+7$	**6.** $+\frac{1}{5}$
7. $+\frac{1}{9}$	**8.** $+\frac{3}{5}$	**9.** $+1\frac{1}{4}$	**10.** $+1\frac{2}{5}$	**11.** $+16$	**12.** $+13$
13. $+11$	**14.** $+16$	**15.** $+22$	**16.** $+\frac{1}{8}$	**17.** $+\frac{1}{9}$	**18.** $+\frac{1}{5}$

19. $+1\frac{3}{4}$ 20. $+1\frac{3}{5}$ 21. -15 22. -18 23. -19 24. -21

25. -14 26. $-\frac{1}{7}$ 27. $-\frac{1}{12}$ 28. $-\frac{4}{5}$ 29. $-1\frac{1}{2}$ 30. $-1\frac{1}{5}$

31. -22 32. -24 33. -19 34. -13 35. -17 36. $-\frac{1}{4}$

37. $-\frac{1}{10}$ 38. $-\frac{2}{5}$ 39. $-1\frac{1}{4}$ 40. $-1\frac{4}{5}$

Exercise 4.3a page 71

1. 24, 36, 216
2. 25, 75, 500
3. 28, 128, 192
4. 96, 192, 1024
5. 63, 245, 343
6. 33, 54, 135
7. 72, 80, 96
8. 120, 54, 48
9. 180, 648, 729
10. 1000, 700, 6000
11. 48, 80, 192
12. 66, 324, 1080
13. 240, 384, 1536
14. 80, 200, 1250
15. 84, 196, 686
16. 64, 60, 200
17. 150, 108, 108
18. 360, 486, 1458
19. 10 000, 10 000, 10 000
20. 300, 20 000, 40 000

Exercise 4.3b page 71

1. 90, 900, 144
2. 75, 360, 400
3. 144, 896, 100
4. 144, 192, 81
5. 160, 500, 64
6. 126, 162, 900
7. 24, 108, 144
8. 250, 3000, 100
9. 288, 1280, 400
10. 1000, 350, 36
11. 200, 960, 900
12. 18, 72, 144
13. 98, 196, 64
14. 144, 1620, 144
15. 600, 10 000, 400
16. 144, 576, 36
17. 252, 735, 900
18. 420, 720, 400
19. 88, 128, 16
20. 270, 540, 900

Exercise 4.3c page 72

1. 20, 8, 8
2. 45, 350, 25
3. 24, 90, 18
4. 100, 480, 20
5. 50, 104, 16
6. 40, 120, 12
7. 55, 96, 2
8. 125, 350, 5
9. 35, 8, 4
10. 169, 360, 2
11. 25, 28, 2
12. 81, 105, 3
13. 49, 84, 18
14. 64, 168, 128
15. 35, 52, 8
16. 55, 90, 24
17. 1225, 350, 20
18. 91, 114, 27
19. 45, 8, 4
20. 91, 510, 5

Exercise 4.3d page 72

1. $-50, +100, -1000$
2. $-24, +20, -72$
3. $-32, +48, -320$
4. $-66, +144, -432$
5. $-45, +200, -500$
6. $-84, +98, -686$
7. $-90, +99, -162$
8. $-150, +180, -81$
9. $-54, +405, -729$
10. $-72, +640, -1024$
11. $-28, +96, -256$
12. $-40, +100, -375$
13. $-1000, +40 000, -5000$
14. $-50, +60, -96$
15. $-72, +180, -648$
16. $-81, +324, -1458$
17. $-120, +180, -108$
18. $-42, +147, -343$
19. $-88, +320, -512$
20. $-500, +8000, -24 000$

Exercise 4.3e page 73

1. $-120, -60, +100$
2. $-84, -96, +81$
3. $-108, -288, +144$
4. $-88, -80, +400$
5. $-250, -1500, +900$
6. $-72, -324, +16$
7. $+96, -576, +144$
8. $+72, -288, +144$
9. $+96, -384, +100$
10. $+420, -3000, +144$
11. $+144, -324, +36$
12. $+150, -600, +900$
13. $+540, -1800, +144$
14. $+120, -300, +900$
15. $+480, -640, +400$
16. $+144, -1440, +36$
17. $-240, +1600, +400$
18. $-220, +1200, +64$
19. $-135, +810, +900$
20. $-42, +90, +64$

Exercise 4.3f page 73

1. $-32, +20, +20$
2. $-31, +28, +8$
3. $-56, +110, +25$
4. $-33, +60, +18$
5. $+2, -20, +5$
6. $+5, -8, +16$
7. $+2, -12, +2$
8. $-10, +500, +2$
9. $-6, +198, +12$
10. $-11, +168, +4$
11. $-21, +63, -3$
12. $-18, +26, -2$
13. $-25, +30, -24$
14. $+3, -4, -8$
15. $+9, -10, -20$
16. $+5, -32, -128$
17. $-6, +162, -18$
18. $-8, +250, -5$
19. $-5, +64, -4$
20. $-6, +90, -27$

Exercise 4.4a page 74

1. $9a$	2. $15b$	3. $3c$	4. $-4d$	5. $-10l$
6. $-18m$	7. $12n$	8. $8p$	9. q	10. $-2r$
11. $-t$	12. $5u$	13. v	14. 0	15. $4y$

16. $10a + 9b$ 17. $15m + 14n$ 18. $8p + 12q$ 19. $12z + 5z^2$ 20. $15u - 8v$
21. $9x - 10y$ 22. $15b - 8c$ 23. $16a - 4a^2$ 24. $15l + 8m$ 25. $10q + r$
26. $16t - 5u$ 27. $9y - 4z$ 28. $16a + 7b$ 29. $18m + 5n$ 30. $15p + 2q$
31. $14u - 3v$ 32. $11x - 7y$ 33. $16b - c$ 34. $3l + 5m$ 35. $5b + 9b^2$
36. $c - 4c^2$ 37. $11d + d^2$ 38. $4q - 9r$ 39. $7t - 16u$ 40. $y - 10z$

Exercise 4.4b page 75

1. $a^2 + 5ab + b^2$ 2. $m^2 + 6mn + n^2$ 3. $3p^2 + 5pq - 3q^2$
4. $2u^2 + 3uv - 2v^2$ 5. $x^2 - 4xy + y^2$ 6. $5b^2 - 10bc - 5c^2$
7. $2l^2 - 5ml - 2m^2$ 8. $11\,abc$ 9. $5xy^2 + 5x^2y$
10. $5a^2b + 6ab^2$ 11. $2lmn$ 12. $3pq^2 + 4p^2q$
13. $9x^2y - 8xy^2$ 14. pqr 15. $12bc^2 - 5b^2c$
16. $l^2m + 12lm^2$ 17. $-2xyz$ 18. $5qr^2 - 5q^2r$
19. $9u^2v - uv^2$ 20. $-12abc$

Exercise 4.4c page 75

1. $5a + 12$ 2. $7b + 8$ 3. $6c + 15$ 4. $6l - 10$ 5. $7m - 12$
6. $12n + 15$ 7. $14p + 2$ 8. $11q - 12$ 9. $14r - 6$ 10. $22t - 25$
11. $9u + 10$ 12. $16v + 18$ 13. $24x + 10$ 14. $9y + 2$ 15. $14z - 2$
16. $15a - 9$ 17. $15b + 12$ 18. $2c + 9$ 19. $15l + 3$ 20. $3m + 2$

Exercise 4.4d page 75

1. $3p - 6$ 2. $5q - 6$ 3. $2r - 20$ 4. $3t + 8$ 5. $4u + 6$
6. $15 - 2v$ 7. $3x - 3$ 8. $5y - 6$ 9. $3z + 10$ 10. $3a + 8$
11. $3b + 3$ 12. $3c - 3$ 13. $2l - 8$ 14. $2m + 24$ 15. $2n + 4$
16. $3t - 3$ 17. $3u + 3$ 18. $6 - 9v$ 19. $6x + 15$ 20. $20 - 16y$

Exercise 4.4e page 76

1. $14l + 4l^2$ 2. $6m + 6m^2$ 3. $12n^2 - 4n$ 4. $18t - 4t^2$ 5. $3u - 2u^2$
6. $4v - 9v^2$ 7. $22x - 8x^2$ 8. $6y - 20y^2$ 9. $2a + 2a^2$ 10. $5b + 9b^2$

Exercise 4.4f page 76

1. $a^2 + 5a + 6$ 2. $b^2 + 8b + 12$ 3. $c^2 + 12c + 32$
4. $d^2 + 13d + 12$ 5. $x^2 + 7xy + 12y^2$ 6. $l^2 + 3l - 10$
7. $m^2 + 2m - 24$ 8. $n^2 + 7n - 8$ 9. $p^2 + p - 20$
10. $a^2 + 3ab - 28b^2$ 11. $q^2 - 3q - 10$ 12. $r^2 - 4r - 32$
13. $t^2 - 2t - 3$ 14. $u^2 - u - 42$ 15. $m^2 - 2mn - 24n^2$
16. $v^2 - 7v + 10$ 17. $x^2 - 12x + 27$ 18. $x^2 - 11x + 30$
19. $x^2 - 9x + 8$ 20. $p^2 - 6pq + 8q^2$ 21. $10y^2 + 19y + 6$
22. $6z^2 + 11z + 4$ 23. $2a^2 + 11a + 15$ 24. $4m^2 + 11mn + 6n^2$
25. $12p^2 + 29pq + 15q^2$ 26. $12b^2 + 7b - 10$ 27. $10c^2 + 13c - 3$
28. $6d^2 + d - 15$ 29. $5l^2 + 3l - 2$ 30. $2u^2 + 7uv - 15v^2$
31. $36x^2 - 7x - 15$ 32. $8y^2 - 6y - 35$ 33. $35z^2 - z - 12$
34. $4x^2 - 5xy - 6y^2$ 35. $3b^2 - 11bc - 4c^2$ 36. $6a^2 - 17a + 12$
37. $6b^2 - 23b + 7$ 38. $10c^2 - 31c + 15$ 39. $40l^2 - 37lm + 4m^2$
40. $3q^2 - 14qr + 16r^2$

Exercise 4.4g page 76

1. $x^2 - 4$ 2. $y^2 - 16$ 3. $z^2 - 9$
4. $a^2 - 100$ 5. $b^2 - 1$ 6. $16c^2 - 25$
7. $4l^2 - 81$ 8. $49m^2 - 4$ 9. $4n^2 - 1$
10. $16 - 9y^2$ 11. $x^2 + 4x + 4$ 12. $p^2 + 10p + 25$
13. $q^2 + 2q + 1$ 14. $r^2 + 24r + 144$ 15. $u^2 + 20uv + 100v^2$
16. $4t^2 + 20t + 25$ 17. $16a^2 + 8a + 1$ 18. $4b^2 + 12b + 9$
19. $25c^2 + 20cd + 4d^2$ 20. $16x^2 + 24xy + 9y^2$ 21. $x^2 - 2x + 1$
22. $y^2 - 14y + 49$ 23. $z^2 - 6z + 9$ 24. $l^2 - 24l + 144$
25. $m^2 - 10m + 25$ 26. $4p^2 - 20p + 25$ 27. $25q^2 - 10q + 1$

28. $4t^2 - 2t + \frac{1}{4}$ **29.** $16z^2 - 4z + \frac{1}{4}$ **30.** $64a^2 - 8a + \frac{1}{4}$

31. $25b^2 - 30bc + 9c^2$ **32.** $36 - 84y + 49y^2$ **33.** $a^2 + 2 + \frac{1}{a^2}$

34. $x^2 - 2 + \frac{1}{x^2}$ **35.** $x^3 + 5x^2 + 8x + 4$ **36.** $x^3 + x^2 - 4x - 4$

37. $x^3 + 3x^2 + 3x + 1$ **38.** $a^3 - 3a^2 + 3a - 1$ **39.** $x^4 - 1$

40. $a^4 - b^4$

Exercise 4.5a page 77   **1.** 5   **2.** 1   **3.** 0   **4.** -1 **5.** 26   **6.** 11   **7.** 37   **8.** -2
9. -19 **10.** 23   **11.** -1 **12.** 2   **13.** $1\cdot1$ **14.** $4\cdot4$ **15.** $-1\cdot4$

Exercise 4.5b page 77   **1.** 6   **2.** 7   **3.** 10   **4.** 14   **5.** 0   **6.** 1   **7.** -2 **8.** 10
9. 48   **10.** 57   **11.** -3 **12.** 14   **13.** $6\frac{1}{4}$ **14.** $-1\frac{3}{4}$ **15.** $-1\cdot8$

Exercise 4.5c page 77   **1.** 2   **2.** 4   **3.** 5   **4.** -6 **5.** -4 **6.** 2   **7.** 4   **8.** $5\frac{1}{2}$
9. $-3\frac{1}{3}$ **10.** $4\frac{1}{2}$ **11.** 8   **12.** 18   **13.** -6 **14.** 63   **15.** 88   **16.** 10
17. -12 **18.** $-7\frac{1}{2}$ **19.** 15   **20.** -62

Exercise 4.5d page 78   **1.** 2   **2.** 1   **3.** 1   **4.** 4   **5.** 3   **6.** 15   **7.** 2   **8.** 3
9. 16   **10.** 11   **11.** 2   **12.** 5   **13.** 72   **14.** 84   **15.** 6   **16.** -21
17. 8   **18.** 18   **19.** 15   **20.** 1

Exercise 4.5e page 78   **1.** 6   **2.** 8   **3.** 6   **4.** 2   **5.** 3   **6.** -3 **7.** -6 **8.** $10\frac{1}{2}$
9. 8   **10.** -2 **11.** 2   **12.** 2   **13.** 4   **14.** 4   **15.** 4   **16.** 5
17. 6   **18.** -3 **19.** 8   **20.** 2

Exercise 4.5f page 78   **1.** 4   **2.** 3   **3.** 3   **4.** 2   **5.** 4   **6.** 4   **7.** 6   **8.** 44
9. 3   **10.** 3   **11.** 2   **12.** 3   **13.** 2   **14.** 3   **15.** 10   **16.** 3
17. 4   **18.** 6   **19.** 8   **20.** 23   **21.** 2   **22.** 12   **23.** 10   **24.** 10
25. 2   **26.** -4 **27.** 21   **28.** 4   **29.** 7   **30.** 2

Exercise 4.5g page 79   **1.** 3   **2.** 6   **3.** 7   **4.** 6   **5.** 4   **6.** 12   **7.** 12   **8.** 1
9. 3   **10.** 6   **11.** 4   **12.** 3   **13.** 3   **14.** 13   **15.** 5

Exercise 4.5h page 79   **1.** (a)   **2.** (b)   **3.** (a)   **4.** (c)   **5.** (b)   **6.** (c)   **7.** (b)   **8.** (b)
9. (a)   **10.** (b)   **11.** (a)   **12.** (c)

Exercise 4.6a page 80   **1.** 7, 3   **2.** 9, 2   **3.** 12, 3   **4.** 9, 3   **5.** 5, 3   **6.** 7, 4   **7.** 3, 4   **8.** 6, 5
9. 4, 2   **10.** 5, 3   **11.** 3, 4   **12.** 3, 5   **13.** 3, 6   **14.** 2, 9   **15.** 4, 3   **16.** 3, 2
17. 3, 6   **18.** 1, 4   **19.** 2, 4   **20.** 3, 7   **21.** 5, 2   **22.** 5, 3   **23.** 2, 4   **24.** 6, 1
25. 5, 2   **26.** 5, 4   **27.** 7, 3   **28.** 1, 5   **29.** 2, 3   **30.** 6, 1

Exercise 4.6b page 81   **1.** 3, 4   **2.** 1, 2   **3.** 4, 3   **4.** 2, 5   **5.** 3, 2   **6.** 2, 4   **7.** 3, 1   **8.** 5, 2
9. 3, 2   **10.** 2, 3   **11.** 4, 2   **12.** 1, 3   **13.** 2, 1   **14.** 2, 6   **15.** 4, 5   **16.** 2, 1
17. 3, 5   **18.** 3, 4   **19.** 3, 2   **20.** 4, 3   **21.** 1, 3   **22.** 2, 5   **23.** 3, 4   **24.** 2, 3
25. 3, 1   **26.** 4, 2   **27.** 1, 2   **28.** 2, 5   **29.** 3, 2   **30.** 2, 4

Exercise 4.6c page 81   **1.** 3, 2   **2.** 3, 4   **3.** 2, 4   **4.** 1, 2   **5.** 3, 1   **6.** 2, 3   **7.** 2, 2   **8.** 1, 3
9. 2, 1   **10.** 3, 3   **11.** 1, 4   **12.** 4, 2   **13.** 2, 3   **14.** 2, 4   **15.** 3, 5   **16.** 2, 3
17. 3, 5   **18.** 5, 4   **19.** 3, 2   **20.** 5, 1   **21.** 3, 4   **22.** 4, 3   **23.** 4, 1   **24.** 5, 2
25. 5, 2   **26.** 1, 2   **27.** 3, 1   **28.** 2, 2   **29.** 4, 2   **30.** 2, 5

Exercise 4.7a page 82   1. $\dfrac{P}{I}$ 2. $\dfrac{v}{t}$ 3. $\dfrac{v}{l}, \dfrac{v}{f}$ 4. $\dfrac{F}{a}, \dfrac{F}{m}$ 5. $\dfrac{C}{\pi}$

6. $\dfrac{A}{h}, \dfrac{A}{b}$ 7. $\dfrac{a}{t}, \dfrac{a}{f}$ 8. $ST, \dfrac{V}{S}$ 9. $DV, \dfrac{m}{D}$ 10. $vt, \dfrac{s}{v}$

11. $RI, \dfrac{V}{R}$ 12. $ny, \dfrac{x}{n}$

Exercise 4.7b page 82   1. $180° - \hat{X} - \hat{Z}$ 2. $\dfrac{p - m}{2}$ 3. $\dfrac{540° - 2y}{3}$ 4. $\dfrac{P - p}{a}$ 5. $\dfrac{H - h}{b}, \dfrac{H - h}{T}$

6. $\dfrac{s + k}{v}, \dfrac{s + k}{t}$ 7. $\dfrac{l - L}{c}, \dfrac{l - L}{t}$ 8. $\dfrac{v}{m + 1}$ 9. $\dfrac{360}{n + 1}$ 10. $\dfrac{C}{rP}$

11. $\dfrac{P}{Dg}, \dfrac{P}{hg}$ 12. $\dfrac{V}{bh}, \dfrac{V}{lb}$ 13. $\dfrac{RA}{l}, \dfrac{kl}{R}$ 14. $ms\Delta T, \dfrac{U}{ms}$ 15. $\dfrac{RT}{V}, \dfrac{PV}{T}$

Exercise 4.7c page 82   1. $\dfrac{a}{3} - c$ 2. $\dfrac{m}{5} - 2$ 3. $q - \dfrac{p}{4}$ 4. $\dfrac{u}{2} + 5$ 5. $5 - \dfrac{x}{3}$

6. $\dfrac{a}{b} - d$ 7. $\dfrac{A}{180} + 2$ 8. $\dfrac{E}{I} - R$ 9. $\dfrac{2s}{t} - v$ 10. $l - \dfrac{2A}{h}$

11. $\dfrac{L}{4} - b - c$ 12. $\dfrac{l}{n + 4}, \dfrac{l}{m} - 4$ 13. $\dfrac{p}{q} + s, r - \dfrac{p}{q}$ 14. $\dfrac{t}{v - 3}, \dfrac{t}{u} + 3$

15. $\dfrac{x}{5 - z}, 5 - \dfrac{x}{y}$

Exercise 4.7d page 83   1. $\sqrt{A}$ 2. $\sqrt{\dfrac{V}{b}}$ 3. $\sqrt{\dfrac{V}{\pi l}}$ 4. $\sqrt{\dfrac{2s}{a}}$ 5. $\sqrt{\dfrac{3V}{h}}$

6. $\sqrt{\dfrac{3V}{\pi h}}$ 7. $\sqrt[3]{V}$ 8. $\sqrt[3]{\dfrac{3V}{4\pi}}$ 9. $\dfrac{x^2}{b}$ 10. $\dfrac{y^2}{25m}$

11. $z^2 a$ 12. $\dfrac{t^2 l}{9}$ 13. $\dfrac{4\pi^2 l}{T^2}$ 14. $\dfrac{x^3}{b}$ 15. $y^3 a$

Exercise 4.7e page 83   1. 3   2. 4   3. 3   4. 4   5. 3   6. 8   7. 11   8. 30
                             9. 5   10. 0·3, 106   11. 9   12. 5   13. 5   14. 3   15. 3

Exercise 4.8a page 84
1. $3(a + 3b)$ 2. $4(m - 5n)$ 3. $5(3p + q)$
4. $6(4u - v)$ 5. $5(2x + 3y)$ 6. $3(3b - 5c)$
7. $8(3l + 2m)$ 8. $2(4q - 3r)$ 9. $3t(u + v)$
10. $x(y - 9z)$ 11. $4a(b + 3a)$ 12. $8m(2n + m)$
13. $2p(4q - 5p)$ 14. $5u(5v + 3u)$ 15. $x(10y + 9x)$
16. $9b(b - 6a)$ 17. $5q(9q - p)$ 18. $8v(2v + 5u)$
19. $9y(3y - 2x)$ 20. $c(5c - 6b)$ 21. $6t^2(5t + 1)$
22. $7a^2(a + 3)$ 23. $12m^2(2m - 1)$ 24. $9c^2(c - 4)$
25. $p^2 q(16p + 15)$ 26. $5n^2(3n + 4)$ 27. $9u^2 v(5u + 4)$
28. $6x^2 y(4x - 5)$ 29. $6z^2(3z - 2)$ 30. $t^2 u(9t - 16)$

Exercise 4.8b page 84
1. $(u + v)(l + m)$ 2. $(a + b)(q - r)$ 3. $(m + 3)(t + u)$
4. $(5 + n)(y + z)$ 5. $9(b - c)$ 6. $(x - y)(l - m)$
7. $(a - b)(q + r)$ 8. $(p - 2)(t + u)$ 9. $(7 - q)(y - z)$
10. $5(b + c)$ 11. $(3b + 4c)(p + q)$ 12. $(5l + 2m)(u - v)$
13. $(3q + 2)(x - y)$ 14. $(8 + 7r)(a + b)$ 15. $5t(m + n)$
16. $(2y - 7z)(p - q)$ 17. $(b - 3c)(u - v)$ 18. $(5l - 3)(x + y)$
19. $(9 - 4m)(a + b)$ 20. $7u(m - n)$

Exercise 4.8c page 85

1. $(4p + 3q)(r + s)$
2. $(2x - 5y)(z + t)$
3. $(3c + 4d)(3a + 2b)$
4. $5(m + n)(4k + 3l)$
5. $(4r + 3s)(2p + q)$
6. $6(z + t)(x - 3y)$
7. $(c - d)(3a + 7b)$
8. $(2m - 3n)(2k + 3l)$
9. $(r - s)(9p - 5q)$
10. $2(z - t)(5x - 3y)$
11. $(b + d)(5a + 2c)$
12. $(l + n)(12k + 5m)$
13. $(q - s)(2p + 3r)$
14. $(y - t)(8x + 5z)$
15. $(b + d)(4a - 3c)$
16. $3(l + n)(3k - 4m)$
17. $(q - s)(7p - 5r)$
18. $3(y - t)(x - 2z)$
19. $(b + d)(2a + 5c)$
20. $4(l - n)(2k + 3m)$

Exercise 4.8d page 85

1. $(x + 3)(x + 4)$
2. $(x + 2)(x + 6)$
3. $(x + 4)(x + 6)$
4. $(x + 2)(x + 12)$
5. $(a + 2)(a + 9)$
6. $(b + 3)(b + 6)$
7. $(l + 3)(l + 7)$
8. $(m + 1)(m + 21)$
9. $(n + 8)(n + 12)$
10. $(x - 4)(x - 8)$
11. $(x - 16)(x - 2)$
12. $(y - 3)(y - 12)$
13. $(z - 4)(z - 9)$
14. $(p - 4)(p - 12)$
15. $(q - 6)(q - 8)$
16. $(u - 6)(u - 12)$
17. $(v - 8)(v - 9)$
18. $(x - 4)(x + 6)$
19. $(x - 3)(x + 8)$
20. $(y - 4)(y + 7)$
21. $(z - 2)(z + 14)$
22. $(a - 6)(a + 9)$
23. $(b - 2)(b + 6)$
24. $(c - 1)(c + 12)$
25. $(l - 3)(l + 10)$
26. $(m - 2)(m + 15)$
27. $(x - 3)(x + 4)$
28. $(x - 6)(x + 7)$
29. $(y - 5)(y + 6)$
30. $(x - 11)(x + 4)$
31. $(x - 22)(x + 2)$
32. $(y - 16)(y + 2)$
33. $(z - 8)(z + 4)$
34. $(q - 9)(q + 7)$
35. $(r - 21)(r + 3)$
36. $(t - 14)(t + 4)$
37. $(u - 28)(u + 2)$
38. $(x - 10)(x + 9)$
39. $(y - 12)(y + 11)$
40. $(z - 9)(z + 8)$

Exercise 4.8e page 86

1. $(3x + 2)(x + 3)$
2. $(5x + 3)(x + 4)$
3. $(3x + 4)(x + 2)$
4. $(3x + 4)(2x + 3)$
5. $(2a + 3)(2a + 5)$
6. $(3b + 5)(2b + 3)$
7. $(2m + 9)(2m + 1)$
8. $(4n + 3)(2n + 3)$
9. $(5p + 1)(p + 2)$
10. $(5x - 3)(x - 7)$
11. $(4x - 3)(2x - 7)$
12. $(5y - 9)(2y - 1)$
13. $(5z - 1)(z - 9)$
14. $(3q - 2)(q - 9)$
15. $(3r - 2)(2r - 9)$
16. $(2t - 7)(2t - 5)$
17. $(4u - 5)(2u - 7)$
18. $(5x - 3)(x + 2)$
19. $(4x - 3)(x + 2)$
20. $(4y - 3)(y + 3)$
21. $(3z - 1)(2z + 9)$
22. $(4b + 5)(2b - 1)$
23. $(4c - 1)(2c + 5)$
24. $(6l - 5)(l + 2)$
25. $(3m - 2)(3m + 5)$
26. $(5n + 7)(n - 1)$
27. $(5x - 7)(2x + 3)$
28. $(3x - 4)(2x + 3)$
29. $(3y - 2)(y + 1)$
30. $(7x + 1)(x - 2)$
31. $(4x + 1)(3x - 2)$
32. $(4y + 3)(2y - 3)$
33. $(4z + 3)(z - 3)$
34. $(6p + 5)(p - 3)$
35. $(6q + 5)(2q - 3)$
36. $(3u - 7)(2u + 3)$
37. $(4v + 3)(v - 7)$
38. $(4x + 5)(3x - 4)$
39. $(3y - 5)(2y + 3)$
40. $(3z + 8)(z - 3)$

Exercise 4.8f page 86

1. $(x + 3)^2$
2. $(y + 8)^2$
3. $(z + 1)^2$
4. $(a + 10)^2$
5. $(b + 7)^2$
6. $(x - 2)^2$
7. $(y - 5)^2$
8. $(z - 4)^2$
9. $(m - 9)^2$
10. $(n - 12)^2$
11. $(2x + 3)^2$
12. $(3y + 4)^2$
13. $(4z + 1)^2$
14. $(10p + 3)^2$
15. $(5q + 2)^2$
16. $(3x - 1)^2$
17. $(2y - 5)^2$
18. $(4z - 3)^2$
19. $(10u - 1)^2$
20. $(5v - 4)^2$

Exercise 4.8g page 87

1. $(x - 4)(x + 4)$
2. $(y - 5)(y + 5)$
3. $(z - 9)(z + 9)$
4. $(a - 1)(a + 1)$
5. $(b - 6)(b + 6)$
6. $(c - 12)(c + 12)$
7. $(2x - 5)(2x + 5)$
8. $(4y - 3)(4y + 3)$
9. $(3z - 8)(3z + 8)$
10. $(2l - 9)(2l + 9)$
11. $(5m - 4)(5m + 4)$
12. $(2q - 7)(2q + 7)$
13. $(3r - 5)(3r + 5)$
14. $(x - \frac{1}{2})(x + \frac{1}{2})$
15. $(y - \frac{1}{5})(y + \frac{1}{5})$
16. $(z - \frac{1}{4})(z + \frac{1}{4})$
17. $(t - \frac{1}{10})(t + \frac{1}{10})$
18. $(a - \frac{2}{3})(a + \frac{2}{3})$
19. $(b - \frac{4}{5})(b + \frac{4}{5})$
20. $(c - \frac{3}{4})(c + \frac{3}{4})$

Exercise 4.8h page 87

1. 240	**2.** 400	**3.** 80	**4.** 480	**5.** 1620	**6.** 9000
7. 720	**8.** 280	**9.** 1440	**10.** 720	**11.** 72	**12.** 56
13. 14	**14.** 14	**15.** 39	**16.** 22	**17.** 7	**18.** 19
19. 32·5	**20.** 17·5	**21.** 0·4	**22.** 0·8	**23.** 12	**24.** 28
25. 30	**26.** 16	**27.** 7	**28.** 28	**29.** 1·6	**30.** 6·4

Exercise 4.9a page 87

1. $\dfrac{3a}{3b}, \dfrac{au}{bu}, \dfrac{a^2}{ab}, \dfrac{8au}{8bu}, \dfrac{12a^2}{12ab},$

2. $\dfrac{5m}{5n}, \dfrac{mv}{nv}, \dfrac{mn}{n^2}, \dfrac{7mv}{7nv}, \dfrac{3mn}{3n^2}$

3. $\dfrac{8p}{12q}, \dfrac{2px}{3qx}, \dfrac{2p^2}{3pq}, \dfrac{10px}{15qx}, \dfrac{6p^2}{9pq}$

4. $\dfrac{25b}{30c}, \dfrac{5by}{6cy}, \dfrac{5b^2}{6bc}, \dfrac{45by}{54cy}, \dfrac{(5b)^2}{30bc}$

5. $\dfrac{15l}{20m}, \dfrac{3lq}{4mq}, \dfrac{3lm}{4m^2}, \dfrac{9lq}{12mq}, \dfrac{18lm}{24m^2}$

Exercise 4.9b page 88

1. $\dfrac{a}{b}$	**2.** $\dfrac{3m}{4n}$	**3.** $\dfrac{p}{3q}$	**4.** $\dfrac{5u}{6v}$	**5.** $\dfrac{2x}{3y}$	**6.** $\dfrac{b}{4c}$
7. $\dfrac{8l}{15m}$	**8.** $\dfrac{2q}{3r}$	**9.** $\dfrac{9t}{14u}$	**10.** $\dfrac{y}{5z}$	**11.** $\dfrac{5b}{12a}$	**12.** $\dfrac{3n}{5m}$
13. $\dfrac{8}{21q}$	**14.** $\dfrac{1}{6v}$	**15.** $\dfrac{3x^2}{4y^2}$	**16.** $\dfrac{5b^2}{8c^2}$	**17.** $\dfrac{m}{7l}$	**18.** $\dfrac{16r}{25q}$
19. $\dfrac{5u^2}{6t^2}$	**20.** $\dfrac{7z^2}{9y^2}$				

Exercise 4.9c page 88

1. $4b$	**2.** $\dfrac{3n}{m}$	**3.** $\dfrac{qr}{8}$	**4.** $5uv$	**5.** $\dfrac{y}{12x}$	**6.** $\dfrac{5b^2c^2}{6}$
7. $\dfrac{4m^2}{3l}$	**8.** $\dfrac{9t^2u}{32}$	**9.** $4y^2z^2$	**10.** $\dfrac{2b^2}{3a^2}$	**11.** $\dfrac{2n}{5m^2}$	**12.** $\dfrac{5p}{3q}$
13. $\dfrac{4}{3uv}$	**14.** $\dfrac{8xy}{21}$	**15.** $\dfrac{c}{15b}$	**16.** $6c$	**17.** $\dfrac{4}{5n}$	**18.** $\dfrac{3r(p+q)}{8}$
19. $\dfrac{4v^2(t-u)}{3}$	**20.** $\dfrac{8}{x}$				

Exercise 4.9d page 89

1. $\dfrac{ab}{4}$	**2.** $\dfrac{5bc}{2}$	**3.** $\dfrac{mn}{6}$	**4.** $\dfrac{2}{3lm}$	**5.** $\dfrac{4pq^2}{3}$	**6.** $\dfrac{7}{4q^2r}$
7. $\dfrac{uv}{12}$	**8.** $\dfrac{1}{9tu}$	**9.** $\dfrac{3xy^2}{4}$	**10.** $\dfrac{3}{5y^2z}$	**11.** $\dfrac{a^2b}{15}$	**12.** $\dfrac{7}{4bc^2}$
13. $\dfrac{7m^2n^2}{3}$	**14.** $\dfrac{4}{15l^2m^2}$	**15.** $\dfrac{10}{3pq^2}$	**16.** 3	**17.** $1\frac{1}{2}$	**18.** $\dfrac{3(p-q)}{5}$
19. $\dfrac{8(u+v)}{9}$	**20.** $3\frac{3}{4}$				

Exercise 4.9e page 89

1. $\dfrac{3a+b}{12}$	**2.** $\dfrac{5m+3n}{30}$	**3.** $\dfrac{9p+7q}{30}$	**4.** $\dfrac{25u+9v}{60}$
5. $\dfrac{3x+5y}{12}$	**6.** $\dfrac{2b-c}{8}$	**7.** $\dfrac{4l-3m}{36}$	**8.** $\dfrac{8p-3r}{10}$
9. $\dfrac{27t-8u}{30}$	**10.** $\dfrac{7y-3z}{14}$	**11.** $\dfrac{25a-4b}{60}$	**12.** $\dfrac{4q+p}{4pq}$
13. $\dfrac{3v+u}{6uv}$	**14.** $\dfrac{8y+5x}{xy}$	**15.** $\dfrac{14c+9b}{4bc}$	**16.** $\dfrac{9m+10l}{12lm}$
17. $\dfrac{r-48q}{6qr}$	**18.** $\dfrac{20u-11t}{6tu}$	**19.** $\dfrac{15z-16y}{10yz}$	**20.** $\dfrac{7b-8a}{6ab}$
21. $\dfrac{3a}{4}$	**22.** $\dfrac{a}{2}$	**23.** $\dfrac{11}{6a}$	**24.** $\dfrac{4}{3a}$

Exercise 4.9f *page 90*

1. $\dfrac{5x+18}{12}$ 2. $\dfrac{7y+2}{10}$ 3. $\dfrac{7z-1}{12}$ 4. $\dfrac{4t-25}{12}$

5. $\dfrac{7a+b}{24}$ 6. $\dfrac{5m-3n}{12}$ 7. $\dfrac{3p+2q}{6}$ 8. $\dfrac{4u-9v}{6}$

9. $\dfrac{5b-19c}{6}$ 10. $\dfrac{14l-m}{12}$ 11. $\dfrac{3x+8}{10}$ 12. $\dfrac{y-7}{8}$

13. $\dfrac{z+21}{6}$ 14. $\dfrac{t+2}{6}$ 15. $\dfrac{3q+7r}{10}$ 16. $\dfrac{t-7u}{12}$

17. $\dfrac{a+11b}{4}$ 18. $\dfrac{m}{6}$ 19. $\dfrac{9p-2q}{15}$ 20. $\dfrac{3v}{4}$

Exercise 4.10a *page 91*

1. 243 2. 7776 3. 2401 4. 6561 5. 15 625 6. 16 384
7. 1331 8. 19 683 9. 1024 10. 4096 11. 200 12. 256
13. 288 14. 576 15. 432 16. 4^5 17. 5^4 18. 8^6
19. 3^7 20. 12^3 21. 2^9 22. 6^8 23. 7^{10} 24. x^6
25. y^8 26. z^9 27. a^7 28. b^{10} 29. c^5 30. t^{11}

Exercise 4.10b *page 91*

1. x^8 2. y^7 3. z^{10} 4. p^{12} 5. q^9 6. 32
7. 512 8. 729 9. 81 10. 1024 11. $12a^8$ 12. $15b^{10}$
13. $30c^{12}$ 14. $24l^{15}$ 15. $54m^9$ 16. a^7x^5 17. $b^{10}y^6$ 18. c^4z^9
19. m^7u^8 20. n^4v^{10}

Exercise 4.10c *page 91*

1. x^2 2. y^5 3. z^9 4. m 5. 1 6. 16
7. 64 8. 27 9. 125 10. 625 11. $5p^3$ 12. $9q^5$
13. $6r^8$ 14. $\dfrac{u^4}{5}$ 15. $\dfrac{1}{7}$ 16. p^2t^5 17. q^4u^6 18. r^2v^7
19. $\dfrac{a^3}{x^4}$ 20. $\dfrac{b^2}{y}$

Exercise 4.10d *page 92*

1. 64 2. 256 3. 729 4. 625 5. 4096 6. 1024
7. 64 8. 1000 000 9. 512 10. 1 11. $400a^2$ 12. $27b^3$
13. $343c^3$ 14. $25l^4$ 15. $1600m^4$ 16. $8n^6$ 17. $125p^6$ 18. $900q^6$
19. $64r^9$ 20. $216t^9$

Exercise 4.10e *page 92*

1. 9 2. 100 3. 4 4. 2 5. 10 6. $\dfrac{1}{125}$

7. $\dfrac{1}{144}$ 8. $\dfrac{1}{81}$ 9. $\dfrac{1}{15}$ 10. $\dfrac{1}{32}$ 11. $\sqrt{x}$ 12. $\sqrt{ax}$

13. $3\sqrt{x}$ 14. $\sqrt[3]{by}$ 15. $2\sqrt[3]{y}$ 16. $\dfrac{1}{a^3}$ 17. $\dfrac{1}{b}$ 18. $\dfrac{1}{9m^2}$

19. $\dfrac{1}{64n^3}$ 20. $\dfrac{1}{16p^4}$ 21. $5a$ 22. $12b^2$ 23. $7c^4$ 24. $3p$

25. $4q^3$ 26. $\dfrac{1}{9x^8}$ 27. $\dfrac{1}{25y^6}$ 28. $\dfrac{1}{8z^6}$ 29. $\dfrac{1}{125u^{15}}$ 30. $\dfrac{1}{81v^{12}}$

Exercise 4.11a *page 93*

1. $x^2+8x+7=0$ 2. $x^2+5x-14=0$ 3. $y^2-2y+1=0$
4. $z^2-3z-18=0$ 5. $2a^2+5a+2=0$ 6. $3b^2+2b-8=0$
7. $4c^2-8c+3=0$ 8. $5m^2-6m-8=0$ 9. $x^2+2x-15=0$
10. $y^2-7y+12=0$ 11. $2z^2-11z-21=0$ 12. $3t^2+2t-5=0$
13. $a^2-5a+4=0$ 14. $4b^2+15b+9=0$ 15. $c^2-2c-3=0$

Exercise 4.11b page 93

1. $0, 4$ 2. $0, -3$ 3. $0, 5$ 4. $0, -\frac{1}{4}$ 5. $0, -1\frac{1}{2}$ 6. $3, 5$

7. $8, -2$ 8. $1, -5$ 9. $-2, -4$ 10. 6 11. $1\frac{1}{2}, 1\frac{1}{4}$ 12. $1\frac{1}{3}, -\frac{1}{5}$

13. $3\frac{1}{2}, -3$ 14. $5, 1\frac{1}{3}$ 15. $-2\frac{2}{3}, -2\frac{1}{2}$

Exercise 4.11c page 94

1. $0, 5$ 2. $0, 9$ 3. $0, 1$ 4. $0, -7$ 5. $0, -11$ 6. $0, 3$

7. $0, 7$ 8. $0, 12$ 9. $0, -5$ 10. $0, -6$ 11. $0, 2\frac{1}{2}$ 12. $0, 3\frac{1}{2}$

13. $0, -1\frac{2}{3}$ 14. $0, -1\frac{1}{3}$ 15. $0, -1\frac{1}{6}$ 16. $0, 1\frac{1}{4}$ 17. $0, 1\frac{1}{8}$ 18. $0, -1\frac{7}{9}$

19. $0, 2\frac{1}{2}$ 20. $0, 4\frac{1}{2}$

Exercise 4.11d page 94

1. ± 6 2. ± 10 3. ± 7 4. ± 1 5. ± 8 6. ± 11

7. $\pm\frac{1}{2}$ 8. $\pm\frac{1}{5}$ 9. $\pm\frac{1}{3}$ 10. $\pm\frac{1}{10}$ 11. $\pm 2\frac{1}{2}$ 12. $\pm 1\frac{1}{3}$

13. $\pm 1\frac{3}{5}$ 14. $\pm 2\frac{1}{4}$ 15. $\pm 1\frac{5}{7}$ 16. $\pm\frac{3}{8}$ 17. $\pm\frac{2}{9}$ 18. $\pm\frac{1}{6}$

19. $\pm\frac{10}{11}$ 20. $\pm\frac{7}{10}$

Exercise 4.11e page 94

1. $4, 5$ 2. $9, 3$ 3. $5, 1$ 4. 4 5. $-4, -10$ 6. $-9, -1$
7. -9 8. $-8, 6$ 9. $-12, 5$ 10. $-9, 4$ 11. $-8, 7$ 12. $-3, 7$

13. $-6, 12$ 14. $-1, 8$ 15. $-11, 12$ 16. $1\frac{1}{4}, 1\frac{1}{2}$ 17. $2\frac{1}{3}, \frac{1}{4}$ 18. $2\frac{1}{3}, 4$

19. $2\frac{1}{2}$ 20. $-\frac{2}{3}, -1\frac{1}{3}$ 21. $-\frac{2}{5}, -2$ 22. $-\frac{4}{5}$ 23. $\frac{2}{5}, -1\frac{1}{2}$ 24. $\frac{1}{2}, -3\frac{1}{2}$

25. $\frac{1}{3}, -5$ 26. $1\frac{1}{3}, -1\frac{1}{2}$ 27. $1\frac{1}{2}, -1\frac{1}{4}$ 28. $4\frac{1}{2}, -2$ 29. $1\frac{1}{2}, -\frac{5}{6}$ 30. $4, -3\frac{1}{2}$

Exercise 4.11f page 95

1. $-5, 3$ 2. $-2, -8$ 3. $-3, 8$ 4. $2, 1$ 5. $\frac{1}{3}, -2\frac{1}{2}$ 6. $-\frac{2}{3}, -4$

7. $1\frac{1}{4}, -\frac{2}{3}$ 8. $3\frac{1}{2}, 6$ 9. $2, 9$ 10. $-9, 6$ 11. $\frac{4}{5}, -1\frac{1}{2}$ 12. $4\frac{1}{2}, -2$

13. $2, 3$ 14. $-4, 7$ 15. $-1\frac{1}{4}, -2$

Exercise 4.12a page 95

1. 6 2. 3 3. 5 4. 8 5. 63 6. 21 7. 80 8. 48

Exercise 4.12b page 96

1. $15, 16, 17$ 2. $24, 25, 26$ 3. $32, 34, 36$
4. $6, 8, 10$ km   5. $27, 29, 31, 33$ 6. $15, 17, 19$ m
7. $40, 45, 50, 55$ 8. $10, 15, 20, 25$ 9. $24, 27, 30$
10. $6, 9, 12$ m

Exercise 4.12c page 96

1. 1800 cm^2 2. 9 cm   3. 5 cm, $12 \cdot 5 \text{ cm}^2$
4. $75°$ 5. $2, 3, 4$ cm, 24 cm^2 6. $22, 24, 88, 96$ cm
7. 8 cm, $6, 10$ cm; square by 4 cm^2 8. 20 cm, 100 cm; 8000 cm^2
9. 6 cm, 11 cm; 10 cm; rectangle, by 16 cm^2
10. 40 cm, 60 cm; $14\,400 \text{ cm}^2$; 480 cm

PART 5

Exercise 5.1 page 98

1. (a) $3°C, 13°C, 7°C, 3°C, 11°C, 8°C$; (b) 11 a.m. to 12 noon, $5°C$;
   (c) 4 p.m. to 5 p.m., $4°C$
2. (a) 238 V, 247 V, 250 V, 246 V, 241 V, 237 V; (b) 5 p.m. to 6 p.m., 13 V
3. 55 m, 53 m, 30 m, 47 m, 55 m, 31 m, $57\cdot5$ m, $48\cdot5$ m
4. (a) 751 mm, 759 mm, 757 mm, 761 mm, 756 mm, 755 mm; (b) Tues. to
   Wed., 8 mm; (c) Thurs. to Fri., 6 mm

5. (a) 282·5 m, 255 m, 205 m; (b) first; (c) 100 m
6. (a) 410 m, 305 m, 460 m, 265 m, 415 m, 500 m; (b) 1.30 p.m. to 2.00
   p.m., 140 m; (c) 3.30 p.m. to 4.00 p.m., 240 m
7. (a) 6·25 km, 11·75 km, 16·5 km, 4·0 km, 10·5 km, 13·25 km;
   (b) 10.30 a.m. to 11.00 a.m., 1.30 p.m. to 2.00 p.m.;
   (c) 12 noon to 12.30 p.m., 12.30 p.m. to 1.00 p.m.
8. (a) 21 km/l, 20 km/l, 18 km/l, 16 km/l; (b) 54·8 km/h, 74·8 km/h, 82 km/h
9. 4·5, 6·3, 9·5, 10·5, 7·6, 9·3, 10·3, 11·5
10. (a) 14 cm², 30 cm², 53 cm², 77 cm²; (b) 18 cm, 27 cm, 31 cm, 38 cm
11. (a) 30 cm, 17 cm, 16 cm, 13·5 cm; (b) 15 cm, 12 cm, 30 cm
12. 77°C, 87°C, 94°C; (b) 36 s, 174 s

Exercise 5.2d page 104
1. (a) 300 (b) £275   2. (a) 120 (b) 168° (c) 54°
6. (a) 250, 400, 350, 500, 300   7. (a) 100 (b) 80° (c) 60°, 30°

Exercise 5.3a page 107
1. (a) 54 km/h, 126 km/h, 198 km/h; (b) 10 m/s, 25 m/s, 45 m/s
2. (a) 45 km/h, 72 km/h, 99 km/h;   (b) 2 s, 7 s, 10 s
3. (a) 0·4 A, 1·6 A, 2·8 A, 4·4 A; (b) 2 V, 5 V, 8 V, 10·5 V
4. (a) £1·60, £3·60, £4·80;   (b) 30 km, 70 km, 110 km
5. (a) 0·75 cm, 6 cm, 7·5 cm;   (b) 2 cm, 5 cm, 11 cm
6. (a) £12, £28, £64;   (b) £100, £450, £750

Exercise 5.3b page 109
1. 28°, 81 m, 10°, $\frac{2}{3}$ 2. 38 cm, 350 g, 30 cm, $\frac{2}{25}$
3. 830 mm, 68°, 750 mm, 2·5   4. 170 m, 13·5 km, 110 m, $\frac{1}{75}$
5. 1002·7 mm, 30°, 1002 mm, $\frac{1}{100}$ mm

Exercise 5.4a page 110
(3, 2), (1, 4), (0, 1), (2$\frac{1}{2}$, 3$\frac{1}{2}$), (−2, 4), (−3$\frac{1}{2}$ 1$\frac{1}{2}$), (1, −2), (3, 0),
(−3, −4), (−1 −3$\frac{1}{2}$)

Exercise 5.4c page 111
1. The clock shows 3 o'clock.   2. The door of Number 7.
3. Signpost to Hull.   4. Television set.   5. Telephone.
6. Church with steeple.   7. Bird.   8. Weathervane pointing North.
9. Electric kettle.   10. Scissors.

Exercise 5.5a page 114
1. 5, 3   2. 1, 4   3. 4, 2   4. 3, 1   5. 2, 1$\frac{1}{2}$ 6. −7, 2
7. −9, 4   8. −6, 1   9. −2, 2$\frac{1}{2}$ 10. −3, $\frac{1}{2}$

Exercise 5.5b page 115
1. 5, −2   2. 7, −4   3. 3, −5   4. 8, −3   5. 6, −1   6. −2, −1
7. 4, −1$\frac{1}{2}$ 8. 2, −$\frac{1}{2}$ 9. 3, −2$\frac{1}{2}$ 10. −1, −$\frac{1}{2}$

Exercise 5.5c page 115
1. 5$\frac{1}{4}$, ± 1   2. −6$\frac{3}{4}$, ±3   3. −9$\frac{3}{4}$, ±4   4. 2·8, ±$\frac{1}{2}$
5. −4$\frac{3}{4}$, ±2·2   6. −2$\frac{3}{4}$, ±1$\frac{1}{2}$ 7. −1$\frac{1}{4}$, ± 0·8   8. 8$\frac{1}{2}$, ±2·8
9. −3$\frac{1}{2}$, ± 4   10. 8, ±2   11. −3, ±4   12. −2$\frac{1}{4}$, ±3

Exercise 5.5d page 120
1. (a) 1$\frac{1}{4}$ (b) 0·6, 3·4 (c) 1, 3   2. (a) 6$\frac{3}{4}$ (b) 0·2, 4·8   (c) 1, 4
3. (a) 1$\frac{1}{4}$ (b) 0·6, 5·4 (c) 2, 4   4. (a) −3$\frac{3}{4}$ (b) 0·25, 3·75 (c) 0, 4
5. (a) −5$\frac{1}{4}$ (b) −1·7, 4·7 (c) −1, 4 6. (a) −2$\frac{1}{4}$ (b) −3·4, 4·4 (c) −2, 3
7. (a) −1$\frac{3}{4}$ (b) −2·4, 0·4 (c) −3, 1 8. (a) 3$\frac{3}{4}$ (b) −3·2, 2·2 (c) −4, 3

Exercise 5.6a page 122 **1.** 2, 3 **2.** 4, 2 **3.** 3, 1 **4.** 4, 1 **5.** 3, 2 **6.** 2, 9
7. 3, 11 **8.** 2, 7 **9.** 4, 2 **10.** 3, 1

Exercise 5.6b page 123 **1.** (a) 0, 2 (b) −2, 4 **2.** (a) 0, 5 (b) −1, 6 (c) 1, 4 (d) 2, 3
3. (a) 0, 1 (b) −1, 2 **4.** (a) 0, 6 (b) 1, 5
5. (a) ±3 (b) ±2 (c) ±1·4 (d) ±1·75 **6.** (a) −3, 0 (b) −4, 1
7. (a) −2, 0 (b) −3, 1 **8.** (a) −4, 0 (b) −3, −1 **9.** (a) −1, 0 (b) −3, 2

Exercise 5.7a page 124 **1.** 5.30 p.m., 6.15 p.m., 6.30 p.m., 7.00 p.m., 7.15 p.m.
2. 8.15 p.m., 8.30 p.m., 9.15 p.m., 10.15 p.m., 10.30 p.m., 11.00 p.m.
3. 4.30 p.m., 4.45 p.m., 5.15 p.m., 5.30 p.m., 6.15 p.m., 7.00 p.m.
4. 2.30 p.m., 2.45 p.m., 3.30 p.m., 4.00 p.m., 4.15 p.m., 4.30 p.m.
5. 8.00 a.m., 8.30 a.m., 8.45 a.m., 9.15 a.m., 10.00 a.m., 10.45 a.m., 11.15 a.m.
6. 3.30 p.m., 4.15 p.m., 4.30 p.m., 4.45 p.m., 5.15 p.m.
7. 12.45 p.m., 1.15 p.m., 3.00 p.m., 3.30 p.m., 4.00 p.m.
8. 2.15 p.m., 2.45 p.m., 3.30 p.m., 3.45 p.m.

Exercise 5.7b page 126 **1.** 3.15 p.m.; 6 km; 15 min; 45 min; 18 km/h, 12 km/h, 20 km/h
2. 10.10 a.m.; 10.40 a.m.; 14 km; 21 km; 10 min; 1 h 30 min; 84 km/h;
63 km/h, 70 km/h
3. 8.40 a.m.; 9.50 a.m.; 12 km; 30 min; 30 km/h; 36 km/h, 24 km/h
4. 2.30 p.m.; 60 km; 2 h; 50 km/h; 40 km/h, 40 km/h
5. 7.30 p.m.; 2 h 30 min; 140 km; 75 m; 95 km/h; 80 km/h, 100 km/h

Exercise 5.7c page 128 **1.** 2.50 p.m., 5 km **2.** 10.27 a.m., 9 km **3.** 9.30 a.m., 25 km
4. 8.40 a.m, 50 km **5.** 9.30 a.m, 9 km **6.** 3.36 p.m., 30 km
7. 7.54 p.m., 90 km **8.** 4.40 p.m., 16 km **9.** 5.58 p.m., 29 km
10. 7.05 p.m., 91 km **11.** 7.00 p.m., 120 km, 60 km/h, 80 km/h
12. 3.15 p.m., 150 km, 120 km/h, 150 km/h

PART 6

Exercise 6.2a page 132 **1.** 62° **2.** 88° **3.** 57° **4.** 34° **5.** 30°, 150°
6. 36°, 144° **7.** 100°, 60°, 20° **8.** 135°, 30°, 15°
9. 108°, 54°, 18° **10.** 12°, 24°, 144°

Exercise 6.2b page 133 **1.** 78° **2.** 71° **3.** 142° **4.** 105° **5.** 30°, 60°, 120°, 150°
6. 24°, 72°, 96°, 168° **7.** 40°, 80°, 120° **8.** 18°, 54°, 72°, 90°, 126°
9. 20°, 40°, 80°, 100°, 120° **10.** 15°, 30°, 60°, 120°, 135°

Exercise 6.2c page 134 **1.** 40°, 140°, 40° **2.** 85°, 95°, 85° **3.** 100°, 80°, 100°
4. 115°, 65°, 115° **5.** 60°, 120°, 60°, 120° **6.** 30°, 150°, 30°, 150°
7. 45°, 135°, 45°, 135° **8.** 20°, 160°, 20°, 160° **9.** 36°, 144°, 36°, 144°
10. 18°, 162°, 18°, 162°

Exercise 6.2d page 134 **1.** 50°, 18°, $27\frac{1}{2}$°, 10·6°, 84°30′ **2.** 40°, 162°, $77\frac{1}{2}$°, 153·8°, 99°48′

Exercise 6.2e page 135 **1.** 045°, $067\frac{1}{2}$°, $112\frac{1}{2}$°, 225°, 315°, $022\frac{1}{2}$°, $157\frac{1}{2}$°, $202\frac{1}{2}$°, $247\frac{1}{2}$°, $292\frac{1}{2}$°
2. 240° **3.** 255° **4.** 214° **5.** 280° **6.** 305° **7.** 70°
8. 45° **9.** 56° **10.** 130°

Exercise 6.2f page 136 **1.** 195° **2.** 225° **3.** 286° **4.** 107° **5.** 046° **6.** 136°
7. 151° **8.** 148° **9.** 223° **10.** 195°

Answers

Exercise 6.3a page 137
1. $70°, 40°$ 2. $60°, 60°, 120°$ 3. $120°, 120°, 60°$
4. $80°, 40°$ 5. $20°, 30°$ 6. $75°, 15°, 25°, 65°$
7. $40°, 30°, 60°, 50°$ 8. $55°, 45°$ 9. $50°, 75°, 55°, 55°, 130°$
10. $60°, 60°, 40°, 80°, 140°$

Exercise 6.3b page 137
1. $130°$ 2. $50°, 60°$ 3. $140°, 40°$
4. $60°, 120°$ 5. $25°, 100°$ 6. $20°, 130°$
7. $75°, 65°, 40°, 40°$ 8. $30°, 35°, 115°, 115°$ 9. $85°, 75°, 20°, 20°$
10. $95°, 40°, 40°, 140°$

Exercise 6.3c page 139
1. $50°$ 2. $120°$ 3. $115°, 55°$
4. $60°, 95°$ 5. $80°, 130°, 150°$ 6. $110°, 40°, 30°$
7. $110°, 50°, 120°$ 8. $75°, 135°, 120°$ 9. $140°, 75°, 35°$
10. $40°, 70°, 155°, 110°$

Exercise 6.3d page 140
1. $93°, 93°, 93°$ 2. $60°, 60°, 120°, 60°$ 3. $80°$
4. $120°$ 5. $120°, 110°$ 6. $140°$
7. $150°, 120°$ 8. $140°, 130°$ 9. $45°, 60°, 105°$
10. $125°, 55°, 95°$ 11. $50°, 40°, 140°$ 12. $125°, 55°, 35°$
13. $40°, 70°$ 14. $60°, 60°, 30°, 60°$

Exercise 6.4a page 142
1. C   2. F   3. A   4. D   5. C   6. B   7. D   8. E

Exercise 6.4b page 142
1. $60°$ 2. $80°$ 3. $35°$
4. $55°, 55°$ 5. $66°, 66°$ 6. $27°, 27°$
7. $90°, 130°$ 8. $60°, 40°$ 9. $75°, 30°, 105°$
10. $65°, 65°, 115°$ 11. $70°, 70°, 40°$ 12. $55°, 35°$
13. $60°, 30°, 150°$ 14. $65°, 50°, 65°$

Exercise 6.4c page 143
1. $50°$ 2. $65°$ 3. $50°$ 4. $115°$ 5. $25°$ 6. $45°$
7. $60°, 65°, 65°, 55°$ 8. $25°, 15°, 15°, 140°$ 9. $45°, 75°, 75°, 60°$
10. $35°, 25°, 25°, 120°$ 11. $35°, 75°, 75°, 70°$
12. $30°, 55°, 55°, 125°, 90°, 60°, 65°$ 13. $55°, 25°, 55°, 70°, 110°, 45°, 70°$
14. $60°, 80°, 80°, 40°, 60°, 80°, 120°$ 15. $70°, 70°, 50°, 70°, 20°, 50°$
16. $110°, 50°, 50°, 20°, 70°, 110°, 20°, 50°, 130°$

Exercise 6.5a page 145
1. $72°, 108°$ 2. $120°, 60°$ 3. $36°, 144°$
4. $24°, 156°$ 5. $45°, 135°$ 6. $12°, 168°$
7. $15°, 165°$ 8. $10°, 170°$ 9. $8°, 172°$
10. $6°, 174°$ 11. 6   12. 4   13. 9   14. 18
15. 40   16. 16   17. 48   18. 32   19. 25   20. 50

Exercise 6.5b page 145
1. $540°$ 2. $900°$ 3. $1080°$ 4. $1440°$ 5. $2160°$ 6. $3240°$
7. $3960°$ 8. $9000°$ 9. $10\,800°$ 10. $13\,500°$

Exercise 6.5c page 145
1. 6   2. 4   3. 3   4. 9   5. 12   6. 11   7. 17   8. 27
9. 30   10. 60

Exercise 6.6a page 147
1. (c)   2. (b)   3. (b)   4. (c)   5. (a)   6. (b)   7. (a)   8. (b)

Exercise 6.6b page 149
1. SSS   2. SSS   3. no   4. SAS   5. no   6. no   7. SAS   8. AAS
9. no   10. no   11. no   12. RHS   13. no   14. no   15. RHS

Exercise 6.6c page 151
1. (c)   2. (b)   3. (c)   4. (a)   5. (a)   6. (c)

Exercise 6.6d page 153 **1.** (c) **2.** (a) **3.** (b) **4.** (b) **5.** (c) **6.** (a)
7. (a) **8.** (c) **9.** (a) **10.** (c) **11.** (c) **12.** (b)

Exercise 6.6e page 156 **1.** $\frac{4}{5}$ **2.** $\frac{5}{8}$ **3.** $\frac{2}{3}$ **4.** $\frac{5}{6}$ **5.** $\frac{1}{5}$ **6.** $\frac{3}{4}$

Exercise 6.6f page 157 **1.** 33 mm, 24 mm **2.** 15 mm, 40 mm **3.** 32 cm, 16 cm
4. 72 mm, 45 mm, 45 mm **5.** 6 cm, 8·4 cm, 25 mm
6. 54 mm, 20 mm **7.** 27 cm **8.** 36 mm
9. 1·5 m **10.** 37·5 mm

Exercise 6.7a page 158 **1.** 1 **2.** 2 **3.** 1 **4.** 1 **5.** 2 **6.** 1 **7.** 5 **8.** 6
9. 8 **10.** 2 **11.** 1 **12.** 0 **13.** 2 **14.** 1 **15.** 0

Exercise 6.7c page 159 **1.** 4 **2.** 6 **3.** 8 **4.** 5 **5.** 2 **6.** 6 **7.** 4 **8.** 3
9. 2 **10.** 4 **11.** 3 **12.** 4

Exercise 6.8a page 162 **1.** kite, rhombus **2.** isosceles trapezium, parallelogram
3. kite, square, rectangle **4.** kite
5. parallelogram **6.** rectangle
7. rhombus, rectangle, rectangle, rhombus
8. (a) kite, rectangle, parallelogram
    (b) rhombus, rectangle, isosceles trapezium, parallelogram
    (c) parallelogram, kite (d) parallelogram, rhombus

Exercise 6.8b page 163 **1.** 60°, 120° **2.** 50°, 50°, 80° **3.** 70°, 130° **4.** 68°, 113°, 22°
5. 45°, 135°, parallelogram
6. (a) and (b) supplementary (c) and (d) equal
7. 30°, 30° **8.** 110° **9.** 130°
10. 95°, 95°, 30°, 30° **11.** 95°, 40° **12.** 60°, 95°

Exercise 6.9a page 164 **1.** 5 cm **2.** 25 cm **3.** 15 cm **4.** 17 cm
5. 26 cm **6.** 40 mm **7.** 41 mm **8.** 58 mm
9. 61 mm **10.** 150 mm **11.** 1 cm **12.** 3 cm
13. 1 cm **14.** 20 cm **15.** 1·5 cm

Exercise 6.9b page 164 **1.** 4·47 cm **2.** 6·71 cm **3.** 7·07 cm **4.** 3·16 cm
5. 11·2 cm **6.** 18·4 mm **7.** 24·1 mm **8.** 17·9 mm
9. 23·0 mm **10.** 19·2 mm **11.** 1·30 cm **12.** 1·92 cm
13. 0·894 cm **14.** 0·632 cm **15.** 1·92 cm

Exercise 6.9c page 165 **1.** 12 mm **2.** 20 mm **3.** 20 mm **4.** 16 mm
5. 1 cm **6.** 1·2 cm **7.** 1·6 cm **8.** 2 cm
9. 13 mm **10.** 14 mm **11.** 36 mm **12.** 45 mm
13. 2·5 cm **14.** 3 cm **15.** 1·5 cm

Exercise 6.9d page 165 **1.** 6·32 cm **2.** 4·47 cm **3.** 8·94 cm **4.** 11·0 mm
5. 13·4 mm **6.** 1·10 cm **7.** 1·34 cm **8.** 1·55 cm
9. 7·75 cm **10.** 6·32 cm **11.** 12·6 mm **12.** 16·7 mm
13. 8·94 mm **14.** 1·90 cm **15.** 0·775 cm

Exercise 6.9e page 165 **1.** 50 km **2.** 170 km **3.** 250 km **4.** 150 km **5.** 25 km **6.** 130 km
7. 75 km

Exercise 6.9f page 166	**1.** 12 m	**2.** 8 m	**3.** 9 m	**4.** 24 cm, 168 cm²	**5.** 26 m

Exercise 6.9f page 166
1. 12 m   **2.** 8 m   **3.** 9 m   **4.** 24 cm, 168 cm²   **5.** 26 m
6. 29 cm   **7.** 60 m   **8.** 12 cm, 60 cm²   **9.** 18 cm, 162 cm², 2160 cm³

Exercise 6.10a page 167
1. 70°, 55°   **2.** 150°, 75°   **3.** 84°   **4.** 48°, 48°
5. 112°   **6.** 72°, 54°, 18°   **7.** 30°, 70°, 35°   **8.** 10°
9. 140°, 70°, 20°   **10.** 80°, 40°, 40°

Exercise 6.10b page 168
1. 60°, 50°   **2.** 20°, 40°   **3.** 35°, 75°, 75°   **4.** 60°, 40°
5. 50°, 50°, 50°, 50°   **6.** 25°, 25°, 50°, 65°
7. 25°, 55°, 55°, 60°, 40°, 40°   **8.** 60°, 40°, 40°, 30°, 50°, 50°
9. 30°, 60°, 60°, 30°, 30°, 60°

Exercise 6.10c page 168
1. 56°   **2.** 23°   **3.** 38°, 62°   **4.** 55°, 35°, 55°
5. 25°, 25°, 65°   **6.** 35°, 35°   **7.** 60°, 30°, 35°
8. 40°, 40°, 25°, 25°   **9.** 50°, 15°, 75°
10. 30°, 30°, 30°, 30°, 30°, 60°

Exercise 6.10d page 169
1. 75°, 95°   **2.** 100°, 90°   **3.** 105°, 75°, 105°   **4.** 96°, 115°
5. 100°, 80°, 100°   **6.** 95°, 70°   **7.** 102°, 78°, 78°   **8.** 110°, 70°
9. 40°, 80°   **10.** 15°, 75°

Exercise 6.10e page 170
1. 60°, 60°   **2.** 65°, 130°, 25°   **3.** 50°, 85°   **4.** 60°, 35°, 85°
5. 60°, 40°, 40°, 50°   **6.** 100°, 80°, 20°
7. 40°, 50°, 20°, 60°   **8.** 20°, 40°, 60°, 60°
9. 35°, 25°, 25°, 75°   **10.** 65°, 25°, 130°
11. 90°, 30°, 30°, 120°   **12.** 22°, 136°, 68°, 78°, 34°

Exercise 6.11a page 172
1. 34°, 34°, 112°   **2.** 65°, 50°, 25°   **3.** 15°, 75°   **4.** 45°, 45°
5. 40°   **6.** 18°, 72°   **7.** 35°, 70°, 20°
8. 20°, 55°, 55°, 35°   **9.** 65°, 50°, 25°, 40°   **10.** 30°, 60°, 120°

Exercise 6.11b page 173
1. 50°, 50°   **2.** 65°, 65°   **3.** 20°, 80°   **4.** 35°, 35°
5. 36°, 54°   **6.** 50°, 50°, 40°, 40°, 100°   **7.** 65°, 50°
8. 63°, 54°   **9.** 72°, 54°, 108°   **10.** 65°, 25°

Exercise 6.11c page 174
1. 60°, 70°, 50°   **2.** 65°, 60°, 55°   **3.** 65°, 40°, 40°   **4.** 54°, 63°, 63°
5. 75°, 75°, 75°, 30°   **6.** 50°, 80°, 80°, 50°   **7.** 45°, 90°   **8.** 36°, 108°

Exercise 6.11d page 175
1. 55°, 55°, 55°, 55°, 35°, 110°   **2.** 40°, 50°, 50°, 40°, 10°
3. 50°, 25°, 65°, 25°, 65°, 65°   **4.** 60°, 30°, 120°, 35°
5. 70°, 110°, 40°, 70°, 40°, 30°   **6.** 50°, 80°, 30°, 50°, 50°, 80°

Exercise 6.12 page 177
1. square   **2.** square   **3.** equilateral triangle
4. square   **5.** square   **6.** rectangle, equilateral triangle
7. rectangle, isosceles triangle   **8.** regular hexagon
9. rhombus   **10.** kite, equilateral triangle, isosceles triangle

Exercise 7.1a page 178

1. 0·2588	**2.** 0·5000	**3.** 0·9397	**4.** 0·7071	**5.** 0·4226
6. 0·7660	**7.** 0·9511	**8.** 0·9135	**9.** 0·7431	**10.** 0·1392
11. 0·6521	**12.** 0·8171	**13.** 0·2622	**14.** 0·8788	**15.** 0·9568
16. 0·2198	**17.** 0·6143	**18.** 0·1323	**19.** 0·4894	**20.** 0·9999
21. 0·5628	**22.** 0·8859	**23.** 0·7927	**24.** 0·2256	**25.** 0·3196
26. 0·2896	**27.** 0·1343	**28.** 0·9309	**29.** 0·6708	**30.** 0·5507
31. 0·7341	**32.** 0·8354	**33.** 0·9413	**34.** 0·7017	**35.** 0·4360
36. 0·2410	**37.** 0·1834	**38.** 0·4065	**39.** 0·1544	**40.** 0·4318

Exercise 7.1b page 179

1. 80°	**2.** 10°	**3.** 20°	**4.** 35°	**5.** 75°
6. 65°	**7.** 84°	**8.** 56°	**9.** 12°	**10.** 33°
11. 39° 42′	**12.** 75° 12′	**13.** 9° 24′	**14.** 51° 30′	**15.** 40° 18′
16. 81° 48′	**17.** 5° 36′	**18.** 55° 6′	**19.** 66° 54′	**20.** 87° 42′
21. 19° 26′	**22.** 66° 10′	**23.** 55° 17′	**24.** 42° 45′	**25.** 34° 27′
26. 30° 37′	**27.** 39° 51′	**28.** 65° 3′	**29.** 31° 33′	**30.** 66° 40′
31. 39° 9′	**32.** 60° 34′	**33.** 13° 28′	**34.** 26° 40′	**35.** 34° 53′
36. 28° 11′	**37.** 20° 50′	**38.** 38° 31′	**39.** 25° 51′	**40.** 17° 20′

Exercise 7.1c page 179

1. 0·1763	**2.** 0·8391	**3.** 0·4663	**4.** 0·7265	**5.** 0·2493
6. 1·1918	**7.** 2·1445	**8.** 3·7321	**9.** 1·9626	**10.** 6·3138
11. 0·2217	**12.** 0·6669	**13.** 0·4494	**14.** 0·6273	**15.** 0·8662
16. 1·2482	**17.** 1·9292	**18.** 1·1303	**19.** 1·5880	**20.** 2·8083
21. 0·2629	**22.** 0·8219	**23.** 1·0919	**24.** 1·2268	**25.** 0·2908
26. 0·7409	**27.** 0·7006	**28.** 0·6428	**29.** 0·1989	**30.** 1·0996
31. 0·2425	**32.** 1·3671	**33.** 0·4211	**34.** 0·3313	**35.** 0·4820
36. 1·2580	**37.** 0·1847	**38.** 1·1276	**39.** 0·2119	**40.** 0·5445

Exercise 7.1d page 179

1. 35°	**2.** 20°	**3.** 16°	**4.** 8°	**5.** 42°
6. 55°	**7.** 60°	**8.** 70°	**9.** 57°	**10.** 64°
11. 39° 24′	**12.** 15° 48′	**13.** 41° 18′	**14.** 25° 36′	**15.** 5° 6′
16. 52° 42′	**17.** 61° 30′	**18.** 48° 54′	**19.** 71° 12′	**20.** 76° 36′
21. 22° 26′	**22.** 34° 19′	**23.** 50° 31′	**24.** 49° 43′	**25.** 34° 49′
26. 20° 2′	**27.** 8° 1′	**28.** 46° 25′	**29.** 18° 26′	**30.** 13° 39′
31. 16° 15′	**32.** 32° 25′	**33.** 19° 40′	**34.** 38° 45′	**35.** 51° 32′
36. 59° 7′	**37.** 8° 49′	**38.** 50° 7′	**39.** 26° 50′	**40.** 55° 7′

Exercise 7.1e page 180

1. 0·7660	**2.** 0·8660	**3.** 0·3420	**4.** 0·5736	**5.** 0·0872
6. 0·9659	**7.** 0·9135	**8.** 0·3090	**9.** 0·9877	**10.** 0·1045
11. 0·9048	**12.** 0·3338	**13.** 0·8396	**14.** 0·7649	**15.** 0·6388
16. 0·4756	**17.** 0·8771	**18.** 0·9767	**19.** 0·5764	**20.** 0·0837
21. 0·8371	**22.** 0·9421	**23.** 0·5671	**24.** 0·3872	**25.** 0·5294
26. 0·2580	**27.** 0·4904	**28.** 0·8200	**29.** 0·9119	**30.** 0·3803
31. 0·1816	**32.** 0·4329	**33.** 0·8586	**34.** 0·2984	**35.** 0·9144
36. 0·8799	**37.** 0·4433	**38.** 0·0683	**39.** 0·2170	**40.** 0·1604

Exercise 7.1f page 180

1. 50°	**2.** 10°	**3.** 20°	**4.** 35°	**5.** 75°
6. 5°	**7.** 36°	**8.** 56°	**9.** 18°	**10.** 64°
11. 41° 36′	**12.** 19° 6′	**13.** 71° 18′	**14.** 37° 12′	**15.** 26° 24′
16. 64° 30′	**17.** 15° 42′	**18.** 57° 48′	**19.** 32° 54′	**20.** 81° 36′
21. 12° 45′	**22.** 37° 9′	**23.** 57° 26′	**24.** 45° 34′	**25.** 76° 3′
26. 71° 2′	**27.** 32° 57′	**28.** 38° 44′	**29.** 30° 8′	**30.** 39° 45′
31. 78° 50′	**32.** 19° 17′	**33.** 30° 50′	**34.** 73° 51′	**35.** 33° 51′
36. 81° 14′	**37.** 64° 46′	**38.** 85° 4′	**39.** 77° 35′	**40.** 83° 40′

Exercise 7.2a page 181
1. EG, EF, FG 2. RT, RS, ST 3. LN, MN, LM 4. PR, QR, PQ
5. XZ, XY, YZ 6. UW, VW, UV 7. DF, EF, DE 8. SU, ST, TU
9. OQ, PQ, OP 10. AB, AC, BC

Exercise 7.2b page 182

1. $\dfrac{AB}{BC}, \dfrac{BC}{AB}$ 2. $\dfrac{BC}{AB}, \dfrac{CD}{AD}$ 3. $\dfrac{PQ}{PS}, \dfrac{QR}{RS}$ 4. $\dfrac{XZ}{WZ}, \dfrac{YZ}{XZ}$

5. $\dfrac{DH}{DE}, \dfrac{GH}{FG}$ 6. $\dfrac{WY}{VY}, \dfrac{XY}{WY}$ 7. $\dfrac{MP}{LP}, \dfrac{NP}{MP}$ 8. $\dfrac{TU}{ST}, \dfrac{RV}{UV}, \dfrac{RU}{SU}$

9. $\dfrac{AE}{BE}, \dfrac{BF}{CF}, \dfrac{CF}{DF}$ 10. $\dfrac{BD}{AD}$ or $\dfrac{BC}{AB}, \dfrac{BD}{CD},$ or $\dfrac{AB}{BC}$

Exercise 7.2c page 183
1. 3·5010 cm 2. 4·8072 cm 3. 14·4064 cm 4. 12·7375 mm
5. 5·308 m 6. 27·06 mm 7. 5·5026 mm 8. 16·9263 cm
9. 13·3272 cm 10. 2·828 m 11. 3·6944 cm 12. 4·848 cm
13. 4·049 cm 14. 38·4072 mm 15. 24·706 mm 16. 6·738 m
17. 7·216 m 18. 3·057 cm 19. 25·512 m 20. 12·645 m
21. 18·2 m 22. 14·104 m

Exercise 7.2d page 184
1. 21° 48′ 2. 16° 42′ 3. 19° 48′ 4. 41° 11′ 5. 23° 45′
6. 59° 32′ 7. 58° 47′ 8. 34° 13′ 9. 42° 37′ 10. 49° 36′
11. 66° 57′ 12. 53° 16′ 13. 16° 42′ 14. 19° 48′ 15. 21° 48′
16. 16° 42′

Exercise 7.3a page 184

1. $\dfrac{BC}{AC}, \dfrac{AB}{AC}$ 2. $\dfrac{PQ}{QS}, \dfrac{PS}{QS}$ 3. $\dfrac{XY}{WY}, \dfrac{YZ}{WY}$ 4. $\dfrac{MP}{MN}, \dfrac{LP}{LM}$

5. $\dfrac{XY}{WX}, \dfrac{VY}{VW}$ 6. $\dfrac{RT}{RS}, \dfrac{ST}{PS}$ 7. $\dfrac{AB}{AC}, \dfrac{CD}{AC}$ 8. $\dfrac{QR}{PR}, \dfrac{ST}{PS}$

9. $\dfrac{DE}{CE}, \dfrac{BE}{BC}, \dfrac{AB}{BE}$ 10. $\dfrac{LQ}{LP}$ or $\dfrac{LM}{MP}, \dfrac{NQ}{MN}$ or $\dfrac{NP}{MP}$

Exercise 7.3b page 185
1. 5·6568 cm 2. 3·8044 cm 3. 3·6108 cm 4. 2·163 m
5. 8·7327 m 6. 9·192 m 7. 6·81 mm 8. 4·5792 mm
9. 4·33 cm 10. 1·1721 cm 11. 7·364 cm 12. 3·9124 m
13. 5·665 m 14. 5·472 mm 15. 17·05 mm 16. 3·94 m
17. 12·856 m 18. 43·3 m

Exercise 7.3c page 186
1. 30° 2. 13° 18′ 3. 41° 18′ 4. 21° 6′ 5. 11° 32′
6. 48° 35′ 7. 20° 29′ 8. 38° 41′ 9. 54° 6′ 10. 35° 6′
11. 4° 18′ 12. 53° 8′ 13. 64° 9′ 14. 7° 11′ 15. 61° 3′
16. 53° 8′ 17. 48° 35′

Exercise 7.4a page 186

1. $\dfrac{AC}{BC}, \dfrac{AB}{BC}$ 2. $\dfrac{RS}{PR}, \dfrac{QR}{PR}$ 3. $\dfrac{KN}{KM}, \dfrac{KL}{LN}$ 4. $\dfrac{UY}{UV}, \dfrac{XY}{UX}$

5. $\dfrac{AD}{AB}, \dfrac{BD}{BC}$ 6. $\dfrac{RS}{QR}, \dfrac{PT}{QT}$ 7. $\dfrac{UY}{UX}, \dfrac{WY}{VW}$

8. $\dfrac{CD}{AD}$ or $\dfrac{AD}{BD}, \dfrac{BC}{AB}$ or $\dfrac{AB}{BD}$

Exercise 7.4b page 187
1. 2·7615 cm 2. 5·9418 cm 3. 4·8515 cm 4. 3·5163 m
5. 0·3657 m 6. 4·4586 m 7. 3·816 mm 8. 1·5675 mm
9. 21·7512 mm 10. 1·71 cm 11. 1·816 cm 12. 5·456 cm
13. 28·839 mm 14. 8·899 mm 15. 16·038 mm 16. 8·66 m
17. 6·5 cm

Exercise 7.4c page 188
1. 68° 54'   2. 18° 12'   3. 76° 42'   4. 8° 6'   5. 77°
6. 33° 54'   7. 78° 28'   8. 56° 38'   9. 60°   10. 54° 54'
11. 48° 42'   12. 35° 54'   13. 24° 30'   14. 66° 25'   15. 51° 19'

Exercise 7.5a page 188
1. 12 cm   2. 300 mm   3. 400 mm   4. 2 m   5. 40 cm   6. 400 mm
7. 8 m   8. 5 m   9. 40 mm   10. 500 mm   11. 45 cm   12. 4 m
13. 600 mm   14. 300 mm   15. 12 cm

Exercise 7.5b page 189
1. 4·289 m, 4·733 m   2. 4·196 m, 6·527 m   3. 2·747 m
4. 3·536 m, 2·071 m   5. 0·700 m, 4·441 m   6. 6·928 m
7. 12·87 m   8. 9·332 m   9. 2·801 m, 6·062 m
10. 75·32 cm

Exercise 7.5c page 191
1. 45·96 km   2. 45 km, 194·9 km   3. 388·1 km
4. 27·81 km, 85·60 km   5. 36·05 km   6. 29·71 km, 4·176 km
7. 41·04 km   8. 95·38 km, 152·6 km   9. 101·7 km, 228·4 km
10. 36·09 km, 41·51 km   11. 71·68 km, 186·7 km   12. 126·4 km, 167·7 km
13. 76·32 km, 47·69 km

Exercise 7.6a page 192
1. $U\hat{S}Q$, $U\hat{S}V$, $S\hat{U}P$, $T\hat{P}U$ or $W\hat{S}V$ 2. $P\hat{M}K$, $P\hat{M}S$, $M\hat{P}L$, $P\hat{L}K$ or $S\hat{M}N$
3. $G\hat{A}C$, $G\hat{A}F$, $A\hat{G}D$, $B\hat{F}A$ or $C\hat{G}D$ 4. $F\hat{B}D$, $E\hat{F}G$, $F\hat{G}E$, $F\hat{G}E$
5. $Q\hat{M}P$, $P\hat{Q}R$, $Q\hat{R}P$, $Q\hat{R}P$

Exercise 7.6b page 194
1. 5 cm, 21° 48'   2. 7·071 cm, 8·66 cm, 35° 16'
3. 17 cm, 26° 34'   4. 28·28 mm, 35° 16', 45°
5. 22° 37', 13 cm, 15·81 cm, 34° 42'   6. 17° 21', 20 cm, 14° 2'
7. 2 cm, 21° 48', 68° 12'
8. 9·90 cm, 19·80 cm, 17·38 cm, 7 cm, 21° 56', 68° 4'

PART 8

Exercise 8.1b page 196
1. 196   2. 315   3. 82   4. 37   5. 26   6. 14   7. 109   8. 242
9. 186   10. 61   11. (a)   12. (c)   13. (a)   14. (a)   15. (b)   16. (a)
17. (c)   18. (b)   19. (b)   20. (a)

Exercise 8.1c page 197
1. 2303   2. 4211   3. 3220   4. 2040   5. 101110
6. 100111   7. 11001   8. 10011   9. 10332   10. 12233
11. 3000   12. 3205   13. 2010   14. 421   15. 10212
16. 1220   17. 10120   18. 1160   19. 520   20. 323
21. 102   22. 300   23. 159   24. EO   25. 1000
26. 50T

Exercise 8.1d page 197
1. 110001   2. 101010   3. 110100   4. 221   5. 124
6. 131   7. 355   8. 425   9. 330   10. 234
11. 302   12. 76

Exercise 8.2a page 198
1. {1, 3, 5, 7, 9}   2. {2, 4, 6, 8, 10}
3. {5, 10, 15, 20, 25, 30}   4. {1, 4, 9, 16}
5. {Tuesday, Thursday}   6. {A, E, I, O, U}
7. {odd numbers smaller than 8}   8. {multiples of 3 smaller than 16}
9. {days of the week beginning with S}   10. {months of the year beginning with A}
11. {prime numbers smaller than 12}   12. {colours of the spectrum}

Exercise 8.2b page 198
1. { April, June, September, November } 2. { January, June, July }
3. { January, February, May, July} 4. { March, April }
5. {February} 6. { August }
7. { days of the week } 8. { months of the year }
9. { letters of the alphabet } 10. { fruit }
11. { points of the compass } 12. { leap years}

Exercise 8.2c page 198
1, 3 and 5 are empty sets.

Exercise 8.2d page 199
1. (a) {2, 3}   (b) { 1, 2, 3, 5, 7 }
2. (a) {3, 5 }   (b) {1, 2, 3, 5, 7, 8 }
3. (a) {12 }   (b) {4, 6, 8, 12, 16, 18 }
4. (a) { square }   (b) {rectangle, square, rhombus}
5. (a) ϕ (b) { 1, 2, 3, 4, 5, 6 }
6. 5, 3, 1, 7   7. 4, 4, 2, 6   8. 4, 5, 3, 6   9. 5, 4, 3, 6   10. 4, 4, 0, 8
11. (a) {2, 3}   (b) {2, 4}   (c) {2, 3, 4}   (d) {2}   (e) {2}
    (f) {2, 3, 4}
12. (a) {3, 5, 6}   (b) {2, 4, 5}   (c) {2, 3, 4, 5, 6}   (d) {5}   (e) {5}
    (f) {2, 3, 4, 5, 6}
13. (a) {Manchester, Glasgow}   (b) {Cardiff, Edinburgh, Glasgow}
    (c) and (f) {Cardiff, Manchester, Edinburgh, Glasgow}
    (d) and (e) {Glasgow}
14. (a) {4}   (b) {3}   (c) {3, 4}   (d) ϕ (e) ϕ (f) {3, 4}
15. 2, 3, 4, 1, 1, 4   16. 4, 4, 6, 2, 2, 6   17. 1, 2, 3, 0, 0, 3   18. 2, 1, 3, 0, 0, 3
19. (a) {1, 2}   (b) {1, 7}   (c) {1, 3}   (d) {1}   (e) {1, 2, 3, 4, 6, 7}
    (f) {1, 2, 3, 4, 5, 7}   (g) {1, 2, 3, 5, 6, 7}   (h) {1, 2, 3, 4, 5, 6, 7}
    (i) {4, 5, 7, 8}   (j) {3, 5, 6, 8}   (k) {2, 4, 6, 8}
    (l) {2, 3, 4, 5, 6, 7, 8}   (m) {8}
20. (a) {3}   (b) {5}   (c) ϕ (d) ϕ (e) {1, 3, 5, 6}
    (f) {1, 2, 3, 5}   (g) {2, 3, 5, 6}   (h) {1, 2, 3, 5, 6}   (i) {1, 2, 4, 5}
    (j) {2, 4, 6}   (k) {1, 3, 4, 6}   (l) {1, 2, 3, 4, 5, 6}   (m) {4}
21. (a) {1}   (b) {1}   (c) {1}   (d) {1}   (e) {1, 2, 3}   (f) {1, 3, 4}
    (g) {1, 2, 4}   (h) {1, 2, 3, 4}   (i) {3, 4}   (j) {2, 4}
    (k) {2, 3}   (l) {2, 3, 4}   (m) ϕ
22. (a) {3, 4}   (b) {1, 3}   (c) {1, 4}   (d) ϕ (e) ϕ (f) ϕ
23. (a) {ship, aeroplane}   (b) {bus, train, ship}   (c) {bus, train, aeroplane}
    (d) ϕ (e) ϕ (f) ϕ
24. (a) {rugby, cricket, rounders}   (b) {football, rugby}
    (c) {football, rugby, cricket, rounders}   (d) ϕ (e) {rugby}   (f) ϕ

Exercise 8.3d page 201
1. B   2. A$'$ 3. B   4. A or (A $\cup$ B)
5. A$'$ or (A $\cup$ B)$'$ 6. B$'$ 7. A $\cap$ B$'$ 8. (A $\cup$ B $\cup$ C)$'$
9. (A $\cap$ B$'$ $\cap$ C$'$)   10. (A $\cap$ B $\cap$ C$'$)

Exercise 8.3e page 203

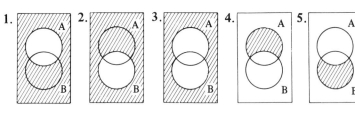

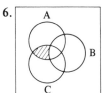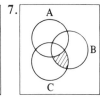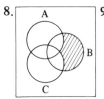

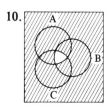

Exercise 8.3f page 203

1. 15, 18, 10, 30, 7   **2.** 50, 8, 18, 100   **3.** 360, 280, 215, 75, 5C
4. 780, 670, 155, 1000
5. yes

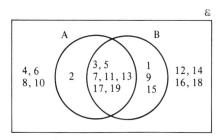

Exercise 8.3g page 204

1. 14, 5, 29, 7   **2.** 270, 150, 990, 10   **3.** 695, 305
4. 425, 75   **5.** 7, 1, 2, 3, 3   **6.** 7
7. 36

Exercise 8.4a page 205

1. >   **2.** <   **3.** =   **4.** <   **5.** =   **6.** >   **7.** >   **8.** =
9. >   **10.** <   **11.** =   **12.** >   **13.** <   **14.** >   **15.** <   **16.** =
17. <   **18.** <   **19.** =   **20.** >

Exercise 8.4b page 205

1. (c)   **2.** (b)   **3.** (b)   **4.** (c)   **5.** (b)   **6.** (a)   **7.** (a)   **8.** (b)
9. (a)   **10.** (b)

Exercise 8.4c page 206

1. {2, 3, 5, 7, 11, 13}   **2.** {1, 4, 9, 16, 25, 36, 49}   **3.** {1, 3, 5, 7}
4. {2, 4, 6, 8}   **5.** {5, 10, 15, 20, 25, 30, 35}   **6.** {6, 12, 18, 24, 30, 36}
7. {1, 2, 3, 4, 5}   **8.** {6, 12, 18}   **9.** {12, 24}
10. {8, 16, 24, 32, 40}   **11.** {13, 14, 15}   **12.** {6, 7}
13. {1, 4, 6, 8, 9}   **14.** {1, 9, 25, 49}   **15.** {4, 16, 64, 100, 144}

Exercise 8.4d page 206   **1.** $x < 5$ **2.** $x \leqslant 6$ **3.** $0 < x \leqslant 4$ **4.** $y \leqslant 6$ **5.** $0 < z < 4$
   6. $0 < t \leqslant 5$ **7.** $m < 9$ **8.** $n > 5$ **9.** $p \geqslant 3$ **10.** $q < 2$
   11. $r \leqslant 3$ **12.** $0 < x \leqslant 2$

Exercise 8.4e page 207   **16.** $x > 5$ **17.** $y \geqslant 2$ **18.** $x - y > 0$ **19.** $x + y \leqslant 4$ **20.** $y - 2x \geqslant 0$

Exercise 8.5a page 208   **1.** yes   **2.** yes   **3.** yes   **4.** yes   **5.** no
   6. no   **7.** yes   **8.** yes   **9.** no   **10.** yes

Exercise 8.5c page 210   **1.** $x \to 4x$ **2.** $x \to \frac{1}{3}x$ **3.** $x \to \frac{3}{2}x$ **4.** $x \to \frac{3}{4}x$ **5.** $x \to x - 2$
   6. $x \to 2x - 3$ **7.** $x \to 3x - 1$ **8.** $x \to \frac{3}{2}x + 1$ **9.** $x \to \frac{1}{4}x + 3$ **10.** $x \to 1/x$
   11. $x \to 6 - x$ **12.** $x \to 4 - \frac{1}{2}x$

Exercise 8.5d page 211   **11.** $x \to x$ **12.** $x \to -x$ **13.** $x \to 4 - x^2$

Exercise 8.7a page 215   **1.** 3×2 **2.** 3×4 **3.** 3×3 **4.** 2×4 **5.** 4×2
   6. 4×3 **7.** 2×5 **8.** 4×4 **9.** 5×2 **10.** 5×3

Exercise 8.7b page 215   **1.** (b)   **2.** (a)   **3.** (a)   **4.** (b)   **5.** (b)   **6.** (c)

Exercise 8.7c page 216

1. $\begin{pmatrix} 11 & 7 \\ 16 & 10 \end{pmatrix}$ **2.** $\begin{pmatrix} 45 & 16 \\ 14 & 4 \end{pmatrix}$ **3.** $\begin{pmatrix} 4 & 10 \\ 7 & 17 \end{pmatrix}$ **4.** $\begin{pmatrix} 4 & 11 & 7 \\ 7 & 18 & 11 \end{pmatrix}$

5. $\begin{pmatrix} 7 & 22 & 26 \\ 2 & 10 & 12 \end{pmatrix}$ **6.** Impossible   **7.** $\begin{pmatrix} 17 & 13 \\ 19 & 17 \\ 14 & 10 \end{pmatrix}$ **8.** $\begin{pmatrix} 4 & 20 \\ 9 & 15 \\ 19 & 5 \end{pmatrix}$

9. Impossible   **10.** $\begin{pmatrix} 5 & 10 & 15 \\ 5 & 5 & 10 \\ 10 & 15 & 25 \end{pmatrix}$ **11.** $\begin{pmatrix} 2 & 1 & 7 \\ 3 & 0 & 6 \\ 1 & 4 & 14 \end{pmatrix}$ **12.** $\begin{pmatrix} 13 & 19 \\ 19 & 22 \end{pmatrix}$

13. $\begin{pmatrix} 4 & 9 \\ 6 & 12 \end{pmatrix}$ **14.** $\begin{pmatrix} 3 & 7 \\ 7 & 3 \\ 9 & 13 \end{pmatrix}$ **15.** Impossible   **16.** $\begin{pmatrix} 1 & 0 \\ 0 & 1 \end{pmatrix}$

17. $\begin{pmatrix} -1 & 0 \\ 0 & -1 \end{pmatrix}$ **18.** $\begin{pmatrix} 1 & 0 \\ 0 & 1 \end{pmatrix}$ **19.** $\begin{pmatrix} -2 & 0 \\ 0 & -2 \end{pmatrix}$ **20.** $\begin{pmatrix} 3 & 0 \\ 0 & 3 \end{pmatrix}$

21. $\begin{pmatrix} 15 & 22 \\ 23 & 34 \end{pmatrix}, \begin{pmatrix} 13 & 16 \\ 29 & 36 \end{pmatrix}$, No.   **22.** $\begin{pmatrix} 1 & 0 \\ 0 & 1 \end{pmatrix}, \begin{pmatrix} 1 & 0 \\ 0 & 1 \end{pmatrix}$, Yes

23. All $\begin{pmatrix} 1 & 0 \\ 0 & 1 \end{pmatrix}$ **24.** $\begin{pmatrix} 5 & 10 \\ 11 & 24 \end{pmatrix}, \begin{pmatrix} 50 & 25 \\ 116 & 59 \end{pmatrix}, \begin{pmatrix} 16 & 9 \\ 17 & 8 \end{pmatrix}, \begin{pmatrix} 50 & 25 \\ 116 & 59 \end{pmatrix}$, Yes

25. $\begin{pmatrix} 15 & 15 \\ 35 & 35 \end{pmatrix}, \begin{pmatrix} 15 & 15 \\ 35 & 35 \end{pmatrix}$, Yes   **26.** 2, 1   **27.** $-5, -10$

28. $\begin{pmatrix} 7 & 10 \\ 15 & 22 \end{pmatrix}, \begin{pmatrix} 9 & 8 \\ 16 & 17 \end{pmatrix}, \begin{pmatrix} 32 & 36 \\ 63 & 77 \end{pmatrix}, \begin{pmatrix} 9 & 8 \\ 19 & 18 \end{pmatrix}, \begin{pmatrix} 7 & 10 \\ 13 & 20 \end{pmatrix}$, Yes

Exercise 8.8a page 218   **1.** $a, -a, b, -b$ **2.** $a, b, -a, -b$ **3.** $a, b, -a, -b$
   4. $a, b, -a, -b$ **5.** $a, -a, b, -b$

Exercise 8.8b page 219   **1.** $a + b, b - 2a, b - a$ **2.** $-a - b, 3a - b, 2a - b$
   3. $a - b, 2a + b, a + b$ **4.** $a + b, 3a - b, 2a - b, b - 2a$
   5. $2a + b, b - 3a, b - a$ **6.** $-2a - 2b, a - 2b, -a - 2b, a + 2b$
   7. $-a - b, b - 2a, b - a, a - b$ **8.** $b - 2a, b - a, 2b, 2b - 3a$

Exercise 8.8c page 220

1. $-z, -y, 0$　　2. $0, q, -q$　　3. $-u, w, x$　　4. $-t, p, s$　　5. $-b, e, 0$
6. $-u, w, 0$　　7. $m, -p, l$　　8. $-c, f, a$　　9. $-u, y, -v$　　10. $-r, -p, m$

Exercise 8.8d page 221

1. $a, b, -b, -a, a + b, b - a$　　　　　2. $b, a, -a, a - b, b + a$
3. $b, a, -b, b - a, -a - b$　　　　　　4. $b, c, b, c - b, a - b$
5. $b, -a, -b, a, a - b, a + c$　　　　　6. $2a, 2b, a, a + b, a + b, b - a, b - a$

Exercise 8.8e page 221

1. 10　　2. 13　　3. 25　　4. 17　　5. 29　　6. 41　　7. 34　　8. 58
9. 26　　10. 50

Exercise 8.8f page 222

1. $\begin{pmatrix} 3 \\ 2 \end{pmatrix}$　2. $\begin{pmatrix} 4 \\ 1 \end{pmatrix}$　3. $\begin{pmatrix} 1 \\ -3 \end{pmatrix}$　4. $\begin{pmatrix} -2 \\ 2 \end{pmatrix}$　5. $\begin{pmatrix} -4 \\ -3 \end{pmatrix}$　6. $\begin{pmatrix} 2 \\ 1 \end{pmatrix}$　7. $\begin{pmatrix} 5 \\ 1 \end{pmatrix}$　8. $\begin{pmatrix} -2 \\ 4 \end{pmatrix}$

9. $\begin{pmatrix} -2 \\ -4 \end{pmatrix}$　10. $\begin{pmatrix} -5 \\ 1 \end{pmatrix}$　11. $\begin{pmatrix} 7 \\ -7 \end{pmatrix}$　12. $\begin{pmatrix} 2 \\ 3 \end{pmatrix}$　13. $\begin{pmatrix} 1 \\ 4 \end{pmatrix}$　14. $\begin{pmatrix} -10 \\ 7 \end{pmatrix}$　15. $\begin{pmatrix} -5 \\ -5 \end{pmatrix}$　16. $\begin{pmatrix} 5 \\ 7 \end{pmatrix}$

17. $\begin{pmatrix} 6 \\ 2 \end{pmatrix}$　18. $\begin{pmatrix} 3 \\ 3 \end{pmatrix}$　19. $\begin{pmatrix} 1 \\ -6 \end{pmatrix}$　20. $\begin{pmatrix} -5 \\ -2 \end{pmatrix}$

Exercise 8.8g page 222

1. $\begin{pmatrix} 6 \\ 8 \end{pmatrix}$　2. $\begin{pmatrix} 5 \\ 2 \end{pmatrix}$　3. $\begin{pmatrix} 3 \\ 5 \end{pmatrix}$　4. $\begin{pmatrix} 2 \\ 5 \end{pmatrix}$　5. $\begin{pmatrix} 7 \\ -2 \end{pmatrix}$　6. $\begin{pmatrix} -1 \\ 6 \end{pmatrix}$

Exercise 8.9a page 223

1. 87, 13　　　　　2. 41, 9　　　　　3. 88, 92　　　　　4. 35, 70
5. 80, 40　　　　　6. 4, 720　　　　　7. 125·6, 62·8　　　8. 4, 40, 20
9. 0·2, 1·2, 120　　10. 50, 100, 10

Exercise 8.9b page 225

1.

A	B
1	1
2	8
3	27
4	64
5	125
6	216

2.

A	B	C
3	1	180
4	2	360
5	3	540
6	4	720
7	5	900
8	6	1080
9	7	1260
10	8	1440

3.

A	B	C
1	1	0·5
2	4	2
3	9	4·5
4	16	8
5	25	12·5

4.

A	B	C
1	1	2
2	4	8
3	9	18
4	16	32
5	25	50

5.

A	B	C	D
1	1	10	5
2	4	40	20
3	9	90	45

6.

A	B	C	D
20	400	0·8	57·8
30	900	1·8	58·8
40	1600	3·2	60·2
50	2500	5·0	62·0
60	3600	7·2	64·2
70	4900	9·8	66·8
80	6400	12·8	69·8
90	8100	16·2	73·2
100	10 000	20·0	77·0

7.

B	C	D	E
1	0·15	1·15	115
2	0·30	1·30	130
3	0·45	1·45	145
4	0·60	1·60	160
5	0·75	1·75	175

8.

A	B	C
100	0·1	100·1
200	0·2	100·2
300	0·3	100·3
400	0·4	100·4
500	0·5	100·5
600	0·6	100·6
700	0·7	100·7
800	0·8	100·8
900	0·9	100·9
1000	1·0	101·0
1100	1·1	101·1
1200	1·2	101·2
1300	1·3	101·3
1400	1·4	101·4
1500	1·5	101·5

9.

A	B	C
1	0·1	0·9
2	0·2	0·8
3	0·3	0·7
4	0·4	0·6
5	0·5	0·5
6	0·6	0·4
7	0·7	0·3
8	0·8	0·2
9	0·9	0·1

10.

A	B	C
1	2	180
2	3	120
3	4	90
4	5	72
5	6	60

Exercise 8.9c page 228

1. (b) → (c) → (g) → (a) → (f) → (e) → (d) **2.** (b) → (f) → (d) → (g) → (c) → (e) → (a)
3. (a) → (d) → (c) → (b) **4.** (a) → (d) → (g) → (f) → (b) → (e) → (c)
5. (d) → (b) → (c) → (a) → (f) → (e) **6.** (d) → (a) → (c) → (b)
7. (f) → (b) → (a) → (d) → (c) → (e) **8.** (b) → (d) → (e) → (a) → (c)
9. (c) → (a) → (e) → (b) → (d) **10.** (b) → (d) → (a) → (c)

Exercise 8.11a page 233

1. 13, 6, 9   **2.** 8, 4, 6   **3.** 10, 4, 8   **4.** 6, 3, 5   **5.** 12, 7, 7
6. 9, 4, 7   **7.** 10, 5, 7   **8.** 10, 4, 8

Exercise 8.11b page 235

7. 15   **8.** 5   **9.** 13   **10.** 18   **11.** 20   **12.** 13   **13.** 24   **14.** 20
15. 28   **16.** 24

Exercise 8.11c page 236

1. yes, two odd junctions   **2.** yes, all junctions even
3. no, four odd junctions   **4.** yes, two odd junctions
5. yes, two odd junctions   **6.** no, four odd junctions
7. yes, all junctions even   **8.** yes, all junctions even

Exercise 8.11d page 237

1. $\begin{pmatrix} 0 & 1 & 2 & 0 \\ 1 & 0 & 0 & 2 \\ 2 & 0 & 0 & 1 \\ 0 & 2 & 1 & 0 \end{pmatrix}; \begin{pmatrix} 5 & 0 & 0 & 4 \\ 0 & 5 & 4 & 0 \\ 0 & 4 & 5 & 0 \\ 4 & 0 & 0 & 5 \end{pmatrix}$

2. $\begin{pmatrix} 0 & 2 & 2 & 0 \\ 2 & 0 & 0 & 1 \\ 2 & 0 & 0 & 2 \\ 0 & 1 & 2 & 0 \end{pmatrix}; \begin{pmatrix} 8 & 0 & 0 & 6 \\ 0 & 5 & 6 & 0 \\ 0 & 6 & 8 & 0 \\ 6 & 0 & 0 & 5 \end{pmatrix}$

3. $\begin{pmatrix} 0 & 3 & 1 & 0 \\ 3 & 0 & 0 & 1 \\ 1 & 0 & 0 & 1 \\ 0 & 1 & 1 & 0 \end{pmatrix}; \begin{pmatrix} 10 & 0 & 0 & 4 \\ 0 & 10 & 4 & 0 \\ 0 & 4 & 2 & 0 \\ 4 & 0 & 0 & 2 \end{pmatrix}$

4. $\begin{pmatrix} 0 & 1 & 3 & 0 \\ 1 & 0 & 0 & 2 \\ 3 & 0 & 0 & 1 \\ 0 & 2 & 1 & 0 \end{pmatrix}; \begin{pmatrix} 10 & 0 & 0 & 5 \\ 0 & 5 & 5 & 0 \\ 0 & 5 & 10 & 0 \\ 5 & 0 & 0 & 5 \end{pmatrix}$

5. $\begin{pmatrix} 0 & 3 & 1 & 0 \\ 3 & 0 & 0 & 2 \\ 1 & 0 & 0 & 1 \\ 0 & 2 & 1 & 0 \end{pmatrix}; \begin{pmatrix} 10 & 0 & 0 & 7 \\ 0 & 13 & 5 & 0 \\ 0 & 5 & 2 & 0 \\ 7 & 0 & 0 & 5 \end{pmatrix}$

6. $\begin{pmatrix} 0 & 2 & 2 & 0 \\ 2 & 0 & 1 & 1 \\ 2 & 1 & 0 & 1 \\ 0 & 1 & 1 & 0 \end{pmatrix}; \begin{pmatrix} 8 & 2 & 2 & 4 \\ 2 & 6 & 5 & 1 \\ 2 & 5 & 6 & 1 \\ 4 & 1 & 1 & 2 \end{pmatrix}$

7.
$$\begin{pmatrix} 0 & 2 & 0 & 0 & 2 \\ 2 & 0 & 1 & 1 & 0 \\ 0 & 1 & 0 & 1 & 1 \\ 0 & 1 & 1 & 0 & 1 \\ 2 & 0 & 1 & 1 & 0 \end{pmatrix} ; \begin{pmatrix} 8 & 0 & 4 & 4 & 0 \\ 0 & 6 & 1 & 1 & 6 \\ 4 & 1 & 3 & 2 & 1 \\ 4 & 1 & 2 & 3 & 1 \\ 0 & 6 & 1 & 1 & 6 \end{pmatrix}$$

8.
$$\begin{pmatrix} 0 & 1 & 0 & 1 & 0 \\ 1 & 0 & 1 & 1 & 1 \\ 0 & 1 & 0 & 0 & 1 \\ 1 & 1 & 0 & 0 & 2 \\ 0 & 1 & 1 & 2 & 0 \end{pmatrix} ; \begin{pmatrix} 2 & 1 & 1 & 1 & 3 \\ 1 & 4 & 1 & 3 & 3 \\ 1 & 1 & 2 & 3 & 1 \\ 1 & 3 & 3 & 6 & 1 \\ 3 & 3 & 1 & 1 & 6 \end{pmatrix}$$

Exercise 8.11e page 237   **1.** 3 routes   **2.** 4 routes   **3.** 2 routes   **4.** 3 routes   **5.** 3 routes
6. 2 routes   **7.** 2 routes   **8.** 4 routes   **9.** 3 routes   **10.** 4 routes

PART 9

Exercise 9.2a page 244   **1.** $3°C$ **2.** 2   **3.** none   **4.** 6   **5.** none   **6.** 28
7. 8   **8.** 1H, 2T   **9.** 14   **10.** 27

Exercise 9.2b page 245   **1.** 32   **2.** 73   **3.** $22°C$ **4.** 80 p   **5.** 23·4 s   **6.** 38 min
7. 29 p   **8.** 39 kg.   **9.** 66 cm   **10.** 8

Exercise 9.2c page 246   **1.** 14   **2.** £57·47   **3.** 46 min   **4.** 2h 49 min   **5.** 13 724   **6.** 84 p
7. 47 s   **8.** 7·5   **9.** 164 cm   **10.** 181 cm

Exercise 9.2d page 247   **1.** 25 p   **2.** 14   **3.** 30   **4.** 5   **5.** 4   **6.** $20·4°C$
7. 1·76   **8.** 0·32   **9.** 0·56   **10.** 761 mm Hg

Exercise 9.2e page 248   **1.** D   **2.** I   **3.** A   **4.** J   **5.** H   **6.** C   **7.** G   **8.** B
9. E   **10.** F

Exercise 9.2f page 248   **1.** mode   **2.** mean   **3.** mode   **4.** mode   **5.** median
6. median   **7.** mode   **8.** mean   **9.** mode   **10.** median
11. mode   **12.** mode   **13.** median   **14.** mode   **15.** mode

Exercise 9.4a page 253   **1.** $\frac{2}{3}, \frac{1}{3}$ **2.** $\frac{1}{2}, \frac{1}{3}$ **3.** $\frac{3}{4}, \frac{1}{4}$ **4.** $\frac{1}{13}, \frac{3}{13}, \frac{9}{13}$ **5.** $\frac{5}{8}, \frac{3}{8}$

6. $\frac{5}{9}, \frac{4}{9}$ **7.** $\frac{3}{5}, \frac{4}{15}, \frac{2}{15}$ **8.** $\frac{3}{4}, \frac{1}{2}$ **9.** $\frac{1}{3}, \frac{5}{18}, \frac{2}{9}, \frac{1}{6}$ **10.** $\frac{1}{4}, \frac{1}{4}, \frac{1}{2}$

Exercise 9.4b page 254   **1.** $\frac{12}{13}, \frac{9}{13}, \frac{5}{13}, 1$ **2.** $\frac{5}{6}, \frac{1}{2}, \frac{2}{3}, \frac{1}{2}$ **3.** $\frac{11}{15}, \frac{3}{5}, \frac{2}{3}$
4. $\frac{2}{11}, \frac{7}{11}, \frac{6}{11}, \frac{9}{11}$ **5.** $\frac{1}{2}, \frac{3}{5}, \frac{9}{10}$

Exercise 9.4c page 255   **1.** $\frac{9}{25}, \frac{4}{25}, \frac{12}{25}, \frac{1}{3}, \frac{2}{15}, \frac{8}{15}$ **2.** $\frac{1}{64}, \frac{1}{4}, \frac{1}{4}, \frac{1}{16}, \frac{1}{28}, \frac{1}{32}, \frac{1}{28}$ **3.** $\frac{1}{18}, \frac{1}{36}, \frac{5}{18}, \frac{1}{4}, \frac{1}{6}, \frac{25}{36}$
4. $\frac{1}{4}, \frac{4}{9}, \frac{1}{9}, \frac{4}{9}, \frac{2}{5}, \frac{8}{15}$ **5.** $\frac{1}{16}, \frac{1}{16}, \frac{1}{4}, \frac{9}{16}, \frac{1}{4}, \frac{1}{16}$ **6.** $\frac{4}{25}, \frac{9}{100}, \frac{9}{100}, \frac{3}{25}, \frac{2}{15}$

Exercise 9.4d page 257   **6.** $\frac{3}{7}, \frac{2}{21}, \frac{5}{21}$ **7.** $\frac{2}{5}, \frac{3}{25}, \frac{6}{25}, \frac{6}{25}, \frac{12}{25}$ **8.** $\frac{7}{40}, \frac{3}{10}, \frac{21}{40}$
9. $\frac{5}{28}, \frac{9}{28}, \frac{1}{4}, \frac{1}{2}$ **10.** $\frac{8}{15}, \frac{1}{15}, \frac{2}{5}$ **11.** $\frac{1}{2}, \frac{1}{14}, \frac{3}{14}, \frac{1}{2}$
12. $\frac{5}{9}, \frac{1}{18}, \frac{7}{18}, \frac{4}{9}$ **13.** $\frac{5}{12}, \frac{1}{12}, \frac{1}{2}, \frac{11}{12}$ **14.** $\frac{1}{28}, \frac{27}{28}, \frac{5}{14}$
15. $\frac{3}{5}, \frac{2}{5}, \frac{43}{45}$